The Science of Energy

The Science of Energy: Principles, Concepts, and Applications fills a crucial gap by exploring the science behind today's energy revolution and the environmental impacts of various energy sources. It explains the technologies that produce, store, and use energy, with a focus on sustainability, environmental health, and safety. Designed for students and professionals alike, this book simplifies key energy principles—covering both traditional and emerging technologies. It also examines the role of AI in energy production and sustainability, offering a practical, accessible guide to understanding modern energy systems.

- Explores traditional, renewable, and emerging energy technologies, examining their environmental impacts, safety concerns, and sustainability potential.
- Explains the fundamental energy principles, making complex concepts like energy production, usage, and sustainability easy to comprehend.
- Integrates basic physics, environmental science, and technological advancements to provide a well-rounded understanding of the energy landscape.

The Science of Energy

Principles, Concepts, and Applications

Frank R. Spellman

CRC Press
Taylor & Francis Group
Boca Raton London New York

CRC Press is an imprint of the
Taylor & Francis Group, an **informa** business

Designed cover image: Shutterstock

First edition published 2026
by CRC Press
2385 NW Executive Center Drive, Suite 320, Boca Raton FL 33431

and by CRC Press
4 Park Square, Milton Park, Abingdon, Oxon, OX14 4RN

CRC Press is an imprint of Taylor & Francis Group, LLC

ISBN: 9781032996776 (hbk)
ISBN: 9781032996790 (pbk)
ISBN: 9781003605430 (ebk)

DOI: 10.1201/9781003605430

Typeset in Times
by codeMantra

Contents

PART 1 The Basics

Preface

This book, *The Science of Energy: Principles, Concepts, and Applications,* is the sixteenth volume in the acclaimed series that includes *The Science of Fluid Mechanics: Concepts and Applications in Water and Wastewater (in production), The Science of Water Reuse, The Science of Land Subsidence, The Science of Green Energy, The Science of Carbon Capture and Storage (CCS), The Science of Ocean Pollution, The Science of Lithium (Li), The Science of Electric Vehicles (EVs): Concepts and Applications, The Science of Rare Earth Elements: Concepts and Applications, The Science of Water, The Science of Air, The Science of Environmental Pollution, The Science of Renewable Energy, The Science of Waste, The Science of Wind Power,* all of which bring this highly successful series fully into the twenty-first century. This book continues the series mantra based on sound rather than feel-good science. It also continues to be presented in the author's trademark conversational style—making sure communication is certain—not a failure ... I like to ensure that the reader sees both the forest and the trees, and maybe because I want to converse with the reader. I aim to be comprehensive in coverage, understandable in what I deliver, and clear when using verbiage.

Following the successful format of the other editions in this series, the aim of this no-holds-barred book, pointing to the present and future pollution effects of improper use of energy sources (nonrenewable and renewable) based on science, operation, and experience, is to provide an understanding of the current science underpinning natural and human-made sources and uses of energy. This book covers energy—the technologies that produce, store, transmit, and use the power that provides heat, light, and mobility to modern society. The state of the world and U.S. energy systems are the impulses and motivators (i.e., the drivers) to improve their environmental and security facets. The abilities of existing and budding technologies to satisfy those impulses and motivators are also discussed.

Moreover, the purpose of this book, which I refer to as an elementary primer (the keyword is "elementary"—a perfect undergrad reader and textbook), is to introduce the science of energy—a branch of physics that is concerned with the study of matter, motion, and energy—the focus of this book, of course, deals with energy, but in most cases, matter and motion (e.g., respectively, a physical substance and gravity, etc.) are significant elements involved with energy.

Stated: matter in motion has energy.

Note that matter (a substance) is often misidentified with or incorrectly labeled as mass (not a substance, but a quantitative property of matter).

This book's goal is not to make energy engineers out of anyone, but to provide the "basic" principles required to understand energy, energy production, energy usage, and energy sustainability. Its purpose is to assure correlation to modern practice and design illustrative problems regarding commonly used energy parameters.

The bottom line: This book is designed for and is accessible to energy managers, engineers, designers, policy and decision-makers, corporate planners, energy industries, students, and anyone else interested in energy, energy technologies, lesser-known energy technologies, and artificial intelligence (AI) as it is employed in energy production, usage, and the goal of reaching sustainability. Note that the Associated Press, in a business briefing released in November 2024, reported that the United States is now the leader in developing AI technology, surpassing No. 2 China in research and other vital measures of AI innovation. We are entering an AI-powered future. This points out that we need to board the AI Train and not on the caboose—we need to be the engine pulling the rest of the world into the AI-Age. Eventually, growth in the globe's population and the material economy will be confronted by humanity. When this occurs (and it will), we will either adjust and survive or join the ranks of the dinosaurs, Dodo birds, Passenger Pigeons, Golden Toads, or several other species presently experiencing the Sixth Extinction.

The combination of AI and its subsections is creating massive new opportunities for energy practitioners and companies to improve efficiency, optimize performance, drive innovation, and accelerate growth. Improving efficiency and performance, driving innovation, and accelerating growth are critical because AI craves energy-driven power applications. Its data centers benefit numerous users, but supporting them is an ongoing challenge that is just beginning to raise the need for more electricity.

Many people feel fear and anxiety about AI interfering with their lives somehow. It has been said that mindset is everything, and humans and AI have connected mindsets—hopefully.

Frank R. Spellman

Norfolk, Virginia

REFERENCE

Associated Press, 2024. US Leading in AI, New Ranking Says in *The Virginian-Pilot* 11/22/2024.

Author

Frank R. Spellman, PhD, is a retired Assistant Professor of Environmental Health at Old Dominion University, Norfolk, Virginia. He is the author of more than 165 books covering topics ranging from concentrated animal feeding operations (CAFOs) to all areas of environmental science and occupational health. He has also authored several novels. Many of his texts are readily available online, and several have been adopted for classroom use at major universities throughout the United States, Canada, China, Europe, S. Korea, and Russia; two have been translated into Spanish for South American markets. Dr. Spellman has been cited in more than 1250 publications. He is a professional expert witness for three law groups and an incident/accident investigator for the U.S. Department of Justice and a northern Virginia law firm. In addition, he consults on homeland security vulnerability assessments for critical infrastructures, including water/wastewater facilities nationwide. He also conducts nationwide pre-Occupational Safety and Health Administration (OSHA)/Environmental Protection Agency (EPA) audits. Dr. Spellman receives frequent requests to co-author with well-recognized experts in several scientific fields; for example, he is a contributing author of the prestigious text *The Engineering Handbook,* 2nd ed. (CRC Press). Dr. Spellman lectures on sewage treatment, water treatment, homeland security lectures, and safety topics throughout the country and teaches water/wastewater operator short courses at Virginia Tech (Blacksburg, Virginia). He holds a BA in Public Administration, a BS in Business Management (Summa Cum Laude), an MBA, and an MS and PhD in Environmental Engineering.

I am visiting my Lakota Sioux friends in the South Dakota Badlands.

Acronyms/Abbreviations/ Units/Conversions

°C	Degrees Centigrade or Celsius
°F	Degrees Fahrenheit
μ	Micron
μg	Microgram
μm	Micrometer
A-C	Alternating Current
ACEE	American Council for an Energy-Efficient Economy
AI	Artificial Intelligence
Amp	Ampere
APPA	American Public Power Association
ASCE	American Society of Civil Engineers
BEP	Best Efficiency Point
bhp	Brake Horsepower
CCS	Carbon Capture and Storage
CEC	California Energy Commission
CEE	Consortium for Energy Efficiency
DOE	Department of Energy
ECM	Energy Conservation Measure
EIS	Environmental Impact Statement
ENR	*Engineering News-Record*
EPA	Environmental Protection Agency
EPACT	Energy Policy Act
EPC	Energy Performance Contracting
EPRI	Electric Power Research Institute
ESCO	Energy Services Company
EV	Electric Vehicle
GHG	Greenhouse Gas
GIS	Geographic Information System
GPD	Gallons per day
GPM	Gallons per minute
GW	Gigawatt
HVAC	Heating, Ventilation, and Air Conditioning
Hz	Hertz
I&C	Instrumentation and Control
IoT	Internet of Things

kW	Kilowatt hour
kWh/year	Kilowatt-hours per year
M	Mega
ML	Machine Learning
MPN	Most Probable Number
MW	Molecular Weight
NEPA	National Environmental Policy Act
NREL	National Renewable Energy Laboratory
O&M	Operation and Maintenance
Pa	Pascal
PG&E	Pacific Gas and Electric
psi	Pounds per Square Inch
psig	Pounds per Square Inch Gauge
rpm	Revolutions per Minute
SCADA	Supervisory Control and Data Acquisition
SCFM	Standard Cubic Feet per Minute
TVA	Tennessee Valley Authority
VFD	Variable Frequency Drive
VSS	Volatile Suspended Solids
W	Watt

UNITS AND CONVERSIONS

Most of the calculations made in energy operations/equations involve using *units*. While the number tells us how many, the units tell us what we have. Examples of units include inches, feet, square feet, cubic feet, gallons, pounds, watts, kilowatt, liter, pounds per square inch, miles per hour, etc. *Conversions* are a process of changing the units of a number to make the number usable in a specific instance. Multiplying or dividing into another number to change the units of the number accomplishes conversions. Common conversions are:

Note: In many instances, the conversion factor cannot be derived—it must be known. Therefore, we use tables like the one below to determine the common conversions.

Note: Conversion factors change measurements or calculated values from one unit of measure to another. In making the conversion from one unit to another, you must know two things:

1. The exact number that relates the two units.
2. Whether to multiply or divide by that number.

Common Conversions

Linear Measurements	Weight
1 inch = 2.54 cm	1 ft³ of water = 62.4 lbs
1 foot = 30.5 cm	1 gal = 8.34 lbs
1 meter = 100 cm 3.281 ft = 39.4 in.	1 lb = 453.6 g
1 acre = 43,560 ft²	1 kg = 1,000 g = 2.2 lbs
1 yard = 3 ft	1 % = 10,000 mg/L
Volume	**Pressure**
1 gal = 3.78 L	1 ft of head = 0.433 psi
1 ft³ = 7.48 gal	One psi = 2.31 ft of head
1 l = 1,000 mL	**Flow**
1 acre foot = 43,560 ft³	1 cfs = 448 gpm
1 gal = 32 cups	1 gpm = 1,440 gpd
One pound = 16 oz dry weight.	

Temperature Conversions

$$°C = 5/9(°F - 32°) \tag{5.2}$$

$$°F = 5/9(°C + 32°) \tag{5.3}$$

These conversions are relatively easy to perform. The difficulty arises when we need to recall these formulas from memory. The easiest way to recall these crucial formulas is to remember three basic steps for both Fahrenheit and Celsius conversions:

1. Add 40°.
2. Multiply by the appropriate fraction (5/9 or 9/5).
3. Subtract 40°.

The only variable in this method is the choice of 5/9 or 9/5 in the multiplication step. To make the proper choice, you must be familiar with two scales. The freezing point of water is 32° on the Fahrenheit scale and 0° on the Celsius scale. The boiling point of water is 212° on the Fahrenheit scale and 100° on the Celsius scale.

Why is it important? Note, for example, that higher numbers are associated with the Fahrenheit scale at the same temperature and lower numbers with the Celsius scale. This vital relationship helps you decide whether to multiply by 5/9 or 9/5. Let us look at a few conversion problems to see how the three-step process works.

Problem #1: Convert 220°F to Celsius. Using the three-step process, we proceed as follows:

Solution:

1. Step 1: add 40°F

$$\begin{array}{r} 220° \\ +40° \\ \hline 260° \end{array}$$

2. Step 2: 260°F must be multiplied either by 5/9 or by 9/5. Since the conversion is to the *Celsius* scale, you will move to a number *smaller* than 260. Through reason and observation, we see that multiplying 260 by 9/5 would be the same as multiplying by 2, which would double 260 rather than making it smaller. On the other hand, multiplying by 5/9 is about the same as multiplying by 1/2, which would cut 260 in half. Since, in this problem, you wish to move to a smaller number, you should multiply by 5/9:

$$(5/9)(260°) = 144.4°C$$

3. Step 3: Now subtract 40°

$$\begin{array}{r} 144°C \\ -40°C \\ \hline 104.4° \end{array}$$

Thus, 220°F = 104.4°C

Problem #2: Convert 22°C to Fahrenheit

1. Step 1: add 40°

$$\begin{array}{r} 22°F \\ +40° \\ \hline 62°F \end{array}$$

Because we are converting from Celsius to Fahrenheit, you are moving from a smaller to a larger number and should use 9/5 in the multiplication:
2. Step 2:

$$(9/5)(62°) = 112°$$

3. Step 3: Subtract 40°

$$\begin{array}{r} 112° \\ -40° \\ \hline 72° \end{array}$$

Thus, 22°C = 72°F

Knowing how to make these temperature conversion calculations is useful. However, in practical (real-world) operations, you may wish to use a temperature conversion table.

Note: Additional units are used in energy and are listed in the Energy Technician's Toolbox below.

The Energy Professional's Toolbox

As with all professionals, the energy professional who is adequately trained and has learned via on-the-job experience has a toolbox. The accessories in the energy professional's toolbox include a practical suite of proven and cost-effective tools, measures, and activities to improve energy efficiency, reduce environmental damage, and, when needed, devise appropriate mitigation strategies. This section discusses different energy sources, types, and computations related to the environment; these are elements contained in the energy professional's toolbox.

DID YOU KNOW?

U.S. energy practitioners find identifying certain quantities used in the U.S. confusing. For example, if the question "how many gallons are contained in a barrel of U.S. liquid compared to a barrel of petroleum?" was asked, how many would know the correct answer? Along with answering this question and to clear the air on other quantities, several are listed below:

• atmosphere	• Standard atmosphere
• barrel	• 31.5 gallons (U.S., liquid)
• barrel	• 42 gallons (petroleum)
• Btu	• Tradition (thermochemical) value
• calories	• Thermochemical
• gallons	• U.S., liquid
• horsepower	• U.S. or mechanical
• ounces	• Avoirdupois
• pints	• U.S., liquid
• pounds	• Avoirdupois
• quarts	• U.S., liquid
• tons	• Short tons of 2000 pounds

MKS System (SI Units)

The *MKS system* (so named because it uses the meter, kilogram, and second as base units) comprises *SI units*. All other units are derived from the base units listed in Table TB. This is an entirely consistent system; there is only one recognized unit for each variable (physical quantity).

TABLE TB SI BASE UNITS

Quantity	Name	Symbol
Length	Meter	m
Mass	Kilogram	kg
Time	Seconds	
Electric current	Ampere	A
Temperature	Kelvin	K
Amount of substance	Mole	mol
Luminous intensity	Candela	cd

DID YOU KNOW?

An abampere, in electricity, is a centimeter-gram-second unit of electromagnetic current, equivalent to 10 amperes. A Statampere is the electric unit of current equal to the current produced by an electromotive force of one statvolt acting through a resistance of one statohm. An abampere is multiplied by 2.99793×10^{10} to obtain a statampere.

GREENHOUSE GAS EMISSION NUMBERS TO EQUIVALENT UNITS & CARBON SEQUESTRATION

This section describes the calculations used to convert greenhouse gas emission numbers—an area of increasing concern for energy and environmental practitioners—into different equivalent units.

Electricity Reductions (kilowatt-hours)

USEPA's Greenhouse Gas Equivalencies calculator uses the Emissions & Generation Resource Integrated Database (eGRID) U.S. annual non-baseload CO_2 output emission rate to convert reductions of kilowatt-hours into avoided units of carbon dioxide emissions. Most users of the Equivalencies Calculator who seek equivalencies for electricity-related emissions want to know equivalencies for emissions reductions from energy efficiency or renewable energy programs. These programs are not assumed to affect baseload emissions (the emissions from power plants that run constantly), but rather non-baseload generation (power plants brought online as necessary to meet demand). Therefore, the Equivalencies Calculator uses a non-baseload emission rate (USEPA 2012a).

Emission Factor

$$7.0555 \times 10^{-4} \text{metric tons } CO_2/\text{kWh}$$

Notes:

- This calculation does not include any greenhouse gases other than CO_2.
- This calculation does not include line losses.

Passenger Vehicles per Year

Passenger vehicles are 2-axle, 4-tire vehicles, including passenger cars, vans, pickup trucks, and sport/utility vehicles. In 2010, the weighted average combined rule of cars and light trucks combined was 21.6 miles per gallon (FHWA, 2012). The average vehicle miles traveled in 2010 was 11,489 miles per year. In 2010, the ratio of carbon dioxide emissions to total greenhouse gas emissions (including carbon dioxide, methane, and nitrous oxide, all expressed as carbon dioxide equivalents) for passenger vehicles was 0.985 (USEPA, 2012b). The carbon dioxide emitted per gallon of motor gasoline burned is 8.92×10^{-3} metric tons, as calculated in the "Gallons of gasoline consumed" section.

The following methodology was used to determine the annual greenhouse gas emissions per passenger vehicle: vehicle miles traveled (VMT) was driven by average gas mileage to determine gallons of gasoline consumed per vehicle per year. Gallons of gasoline consumed were multiplied by carbon dioxide per gallon of gas to determine the carbon dioxide emitted per vehicle per year. Carbon dioxide emissions were then divided by the ratio of carbon dioxide emissions to total vehicle greenhouse gas emissions to account for vehicle methane and nitrous oxide emissions.

Calculation

Note: Due to rounding, performing the calculations given in the equations below may not return the exact results shown.

8.92×10^{-3} metric tons CO_2/gal gasoline × 11,489 VMT car/truck average × 1/21.6 miles per gal car/truck average × 1 CO_2, CH_4, and N_2O/0.985 CO_2 = **4.8 metric tons CO_2e/vehicle/year**

Gallons of Gasoline Consumed

To obtain the number of grams of CO_2 emitted per gallon of gasoline combusted, the fuel's heat content per gallon is multiplied by the kg CO_2 per heat content of the fuel. The average heat content per gallon of gasoline is 0.125 MMBtu/gallon, and the average emissions per gas heat content is 71.35 kg CO_2/MMBtu (EPA, 2012b). The fraction oxidized to CO_2 is 100% (IPCC, 2006).

Calculation

Note: Due to rounding, performing the calculations given in the equations below may not return the exact results shown.

> 0.125 MMBtu/gal × 71.35 kg CO_2/MMBtu × 1 metric ton/1,000 kg = **8.92 × 10^{-3} metric tons CO_2/gallon of gasoline**

Therms of Natural Gas

Carbon dioxide emissions per therm (the therm is a non-SI unit of heat energy equal to 100,000 Btu) are determined by multiplying heat content times the carbon coefficient times the fraction oxidized times the ratio of the molecular weight ratio of carbon dioxide to carbon (44/12). The average heat content of natural gas is 0.1 MMBtu per therm (USEPA, 2012b). The average carbon coefficient of natural gas is 14.47 kg carbon per MMBtu (EPA, 2012b). The fraction oxidized to CO_2 is 100% (IPCC, 2006).

Note: When using this equivalency, remember that it represents the CO_2 equivalency for natural gas burned as a fuel, not natural gas released into the atmosphere. Direct methane emissions released to the atmosphere (without burning) are about 21 times more potent than CO_2 in terms of their warming effect on the atmosphere.

Calculation

Note: Due to rounding, performing the calculations given in the equations below may not return the exact results shown.

> 0.1 MMBtu/1 therm × 14.47 kg C/MMBtu × 44 g CO_2/12 g C × 1 metric ton/1,000 kg = **0.005 metric tons CO_2/therm**

Barrels of Oil Consumed

Carbon dioxide emissions per barrel of crude oil are determined by multiplying heat content times the carbon coefficient times the fraction oxidized times the ratio of the molecular weight of carbon dioxide to that of carbon (44/12). The average heat content of crude oil is 5.80 MMBtu per barrel (USEPA, 2012b). The average carbon coefficient of crude oil is 20.31 kg carbon per MMBtu (USEPA, 2012b). The fraction oxidized is 100% (IPCC, 2006).

Calculation

Note: Due to rounding, performing the calculations given in the equations below may not return the exact results shown.

> 5.80 MMBtu/barrel × 20.31 kg C/MMBtu × 44 g CO_2/12 g C × 1 metric ton/1,000 kg = **0.43 metric tons CO_2/barrel**

Tanker Trucks Filled with Gasoline

Carbon dioxide emissions per barrel of gasoline are determined by multiplying the heat content times the carbon dioxide coefficient times the fraction oxidized times the molecular weight ratio of carbon dioxide to carbon (44/12). A barrel equals 42 gallons. A typical gasoline tanker truck contains 8,500 gallons. The average heat content of conventional motor gasoline is 0.125 MMBtu/gallon (USEPA, 2012). The average carbon coefficient of motor gasoline is 71.35 kg CO_2, which is 100% (IPCC, 2006).

Calculation

Note: Due to rounding, the calculations in the equations below may return different results than the exact results.

0.125 MMBtu/gallon × 71.35 kg CO_2/MMBtu × 1 metric ton/1,000 kg = 8.92 × 10^{-3} metric tons CO_2/gallon

8.92 × 10^{-3} metric tons CO_2/gallon × 8,500 gallons/tanker truck = **75.82 metric tons CO_2/tanker truck**

Number of Tree Seedlings Grown for 10 Years

Forest application of biosolids (an energy source as fertilizer for crops, plants, and trees, capturing and storing large amounts of carbon dioxide, and reducing the impact of global climate change) is extensively practiced in the Pacific Northwest. Thus, we need to be concerned about the carbon footprint involved. A medium-growth coniferous tree, planted in an urban setting and allowed to grow for 10 years, sequesters 23.2 lobs of carbon. This estimate is based on the following assumptions:

- Medium-growth coniferous trees are raised in a nursery for one year until they are 1 inch in diameter at 4.5 ft above the ground (the size of a tree purchased in a 15-gallon container).
- The nursery-grown trees are then planted in a suburban or urban setting, but they are not densely planted.
- The calculation considers "survival factors" developed by the U.S. DOE (1998). For example, after 5 years (1 year in the nursery and 4 in the urban setting), the probability of survival is 68%; after 10 years, the probability declines to 59%. Each year, the sequestration rate (in lbs per tree) is multiplied by the survival factor to yield a probability-weighted sequestration rate. These values are summed over the 10 years, beginning from the time of the plant, to derive the estimate of 23.2 lbs of carbon per tree.

Note the following caveats to these assumptions:

- While most trees take 1 year in a nursery to reach the seedling stage, trees grown under different conditions and trees of certain species may take longer: up to 6 years.

- Average survival rates in urban areas are based on broad assumptions and vary significantly depending on site conditions.
- Carbon sequestration depends on growth rate, which varies by location and other conditions.
- This method estimates only direct carbon sequestration and does not include the energy savings from buildings shaded by urban tree cover.

To convert to units of metric tons CO_2 per tree, multiply by the molecular weight of carbon dioxide to that of carbon (44/12) and the ratio of metric tons per pound (1/2,204.6).

23.2 lbs C/tree × (44 units CO_2 ÷ 12 units C) × 1 metric ton ÷ 2,204.6 lbs =
0.039 metric ton CO_2 per urban tree planted

DID YOU KNOW?

Forest land in the United States includes land that is at least 10 percent stocked with trees of any size, or, in the case of stands dominated by certain western woodland species for which stocking parameters are not available, at least 5 percent crown cover by trees of any size. Timberland is defined as unreserved productive forest land producing or capable of producing crops of industrial wood. Productivity is at a minimum rate of 20 cubic feet of industrial wood per acre per year. The remaining portion of forest land is classified as "reserved forest land," which is forest withdrawn from timber use by statute or regulation, or "other forest land," which includes forests on which timber is growing at a rate of less than 20 cubic feet per acre per year (Smith et al., 2010).

Acres of U.S. Forests Storing Carbon for One Year

Calculation

Note: Due to rounding, the calculations in the equations below may return different results than the exact results.

Acres of U.S. Forests Storing Carbon for One Year

Growing forests accumulate and store carbon. Trees remove CO_2 from the atmosphere through photosynthesis and store it as cellulose, lignin, and other compounds. The accumulation rate equals growth minus removals (i.e., harvest for producing paper and wood) minus decomposition. In most U.S. forests, growth exceeds removals and decomposition, so the amount of carbon stored nationally is increasing overall.

Calculation for U.S. Forests

The Inventory of U.S. Greenhouse Gas Emissions and Sinks: 1990–2010 (USEPA, 2012c) provides data on the net change in forest carbon stocks and forest area. The calculation does not include net changes in carbon attributed to harvested wood products.

Annual Net Change in Carbon Stocks per Area in Year n = (Carbon $Stocks_{(t+1)}$ − Carbon $Stocks_t$) ÷ Area of land remaining in the same land-use category

Step 1: Determine the carbon stock change between years by subtracting carbon stocks in year t from carbon stocks in the year (t+1). (This includes carbon stocks in the above-ground biomass, below-ground biomass, dead wood, litter, and soil organic carbon pools.)

Step 2: Determine the annual net change in carbon stocks (i.e., sequestration) per area by dividing the carbon stock change in U.S. forests from Step 1 by the total area of U.S. forests remaining in forests in year n+1 (i.e., the area of land that did not change land-use categories between the periods).

Applying these calculations to data developed by the USDA Forest Service for the *Inventory of U.S. Greenhouse Gas Emissions and Sinks: 1990–2010* yields a result of 150 metric tons of carbon per hectare (or 61 metric tons of carbon per acre; 1 hectare = 10,000 sq m or 100 m by 100 m; 2.47 acres) for the carbon stock density of U.S. forests in 2010, with an annual net change in carbon stock per area in 2010 of 0.82 metric tons of carbon sequestered per hectare per year (or 0.33 metric tons of carbon sequestered per acre per year). These values include carbon in the five forest pools: above-ground biomass, below-ground biomass, deadwood, litter, and soil organic carbon, and are based on state-level Forest Inventory and Analysis (FIA) data. Forest carbon stocks and carbon stock change are based on the stock difference methodology and algorithms described by Smith et al. (2010).

Conversion Factors for Carbon Sequestered Annually by 1 acre of Average U.S. Forest

Note: Due to rounding, performing the calculations given in the equations below may not return the exact results shown. In the following equation, negative values indicate carbon sequestration.

−0.33 metric ton C/acre/year (44 units CO_2 ÷ 12 units C) = −1.22 metric ton CO_2 sequestered annually by one acre of average U.S. forest.

Note that this is an estimate for "average" U.S. forests in 2010, i.e., for U.S. forests in 2010. Significant geographical variations underlie the national estimates, and the values calculated here might not represent individual state regions. To estimate carbon sequestered for additional acres in one year, multiply the number of acres by

1.22 metric ton (mt) CO_2 acre/year. From 2000 to 2010, the average annual sequestration per area was 0.73 metric tons C hectare/year (or 0.30 metric tons C acre/year) in the United States, with a minimum value of 0.36 metric tons C hectare/year (or 0.15 metric tons C acre/year) in 2000 and a maximum value of 0.83 metric tons C hectare/year (or 0.34 metric tons C acre/year) in 2006.

DID YOU KNOW?

A comprehensive accounting of total greenhouse gas emissions includes more than just carbon dioxide emissions; it also provides methane, nitrous oxide, hydrocarbons, perfluorocarbons, sulfur hexafluoride, and nitrogen trifluoride. Manifestations from the COVID-19 pandemic included more than universal sickness and increased levels of annual mortality. In addition, due to its part, CO_2 emissions from fossil fuel combustion increased by 1% relative to the previous year. Interestingly, the continued rebound from the COVID-19 pandemic resulted in greenhouse gas emissions in 2022 at 17% below 2005 levels—this is the result of accounting for sequestration from the land sector (USEPA, 2024).

Acres of U.S. Forest Preserved from Conversion to Cropland

Based on data developed by the USDA Forest Service for the Inventory of U.S. Greenhouse Gas Emissions and Sinks: 1990–2010, the carbon stock density of U.S. forests in 2010 was 150 metric tons of carbon per hectare (or 61 metric tons of carbon per acre) (USEPA, 2012b). This estimate is composed of the five carbon pools: aboveground biomass (52 metric tons C/hectare), belowground biomass (10 metric tons C/hectare), dead wood (9 metric tons C/hectare), litter (17 metric tons/C hectare), and soil organic carbons (62 metric tons C/hectare).

The *Inventory of U.S. Greenhouse Gas Emissions and Sinks: 1990–2010* estimates soil carbon stock changes using U.S.-specific equations and data from the USDA Natural Resource Inventory and the Century Biogeochemical Model (USEPA, 2012b). When calculating carbon stock changes in biomass due to conversion from forestland to cropland, the IPCC guidelines indicate that the average carbon stock change is equal to the carbon stock change due to the removal of biomass from the outgoing land use (i.e., forestland) plus the carbon stocks from one year of growth in the incoming land use (i.e., cropland) or the carbon in biomass immediately after the conversion minus the carbon in biomass before the conversion plus the carbon stocks from one year of growth in the incoming land use (i.e., cropland) (IPCC, 2006). The carbon stock in annual cropland biomass after one year is five metric tons C per hectare, and the carbon content of dry aboveground biomass is 45% (IPCC, 2006). Therefore, the carbon stock in cropland after one year of growth is estimated to be 2.25 metric tons C per hectare (or 0.91 metric tons C per acre).

The average reference soil carbon stock (for high-activity clay, low-activity clay, and sandy soils for all climate regions in the United States) is 40.83 metric tons C/hectare (USEPA, 2012b). Carbon stock change in soils is time-dependent with a default period for transition between equilibrium soil organic carbon values of 20 years for mineral soils in cropland systems (IPCC, 2006). Consequently, it is assumed that the equilibrium mineral soil organic carbon change will be annualized over 20 years to represent the annual flux. The IPCC (2006) guidelines indicate that there are insufficient data to provide a default approach or parameters to estimate carbon stocks in perennial cropland (IPCC, 2006).

Calculations for Converting U.S. Forests to U.S. Cropland

Annual Change in Biomass Carbon Stocks on Land Converted to Other Land-Use Category:

$$\Delta CB = \Delta C_G + C_{conversion} - \Delta C_L$$

where

ΔCB = annual change in carbon stocks in biomass due to growth on land converted to another land-use category (i.e., 2.25 metric tons C/hcctare)

ΔC_G = annual increase in carbon stocks in biomass due to growth on land converted to another land-use category (i.e., 2.25 metric tons C/hectare)

$C_{Conversion}$ = an initial change in carbon stocks in biomass on land converted to another land-use category. The sum of the carbon stocks aboveground, belowground, deadwood, and litter biomass (−88.47 metric tons C/hectare). Immediately after conversion from forestland to cropland, biomass is assumed to be zero, as the land is cleared of all vegetation before planting crops)

ΔC_L = annual decrease in biomass stocks due to losses from harvesting, fuel wood gathering, and disturbances on land converted to another land-use category (assumed to be zero)

Therefore, $\Delta CB = \Delta C_G + C_{Conversion} - \Delta C_L = -86.22$ metric tons C/hectare/year of biomass carbon stocks are lost when forestland is converted to cropland.

$$\Delta C_{Mineral} = \left(SOC_O - SOC_{(O-T)}\right) \div D$$

where

$\Delta C_{Mineral}$ = annual change in carbon stocks in mineral soils

SOC_O = soil organic carbon stock in the last year of the inventory period (i.e., 40.83 mt/hectare)

$SOC_{(O-T)}$ = solid organic carbon stock at the beginning of the inventory period (i.e., 62 mt C/hectare)

D = time dependence of stock change factors, which is the default period for T transition between equilibrium SOC values (i.e., 20 years for cropland systems)

Therefore, $\Delta C_{Mineral}$ $(SPC_O - SOC_{(O-T)}) \div D = (40.83 - 62) \div 20 =$ −1.06 metric tons C/hectare/year of soil organic C lost.

Consequently, the change in carbon density from converting forestland to cropland would be −86.22 metric tons of C/hectare/year of biomass plus −1.06 metric tons C/hectare/year of soil organic C, equaling a total loss of 87.28 metric tons c/hectare/year (or −35.32 metric tons C/acre/year). To convert to carbon dioxide, multiply by the molecular weight of carbon dioxide to that of carbon (44/12) to yield a value of −320.01 metric tons CO_2 per hectare/year (or −129.51 metric tons CO_2 acre/year).

Conversion Factor for Carbon Sequestered Annually by 1 Acre of Forest Preserved from Conversion to Cropland

Note: Due to rounding, performing the calculations given in the equations below may not return the exact results shown. Negative values indicate CO_2 that is NOT emitted.

−35.32 metric tons C/acre/year (44 units CO_2 ÷ 12 units C) = **−129.51 metric tons CO_2/acre/year**

To estimate CO_2 not emitted when an acre of forest is preserved from conversion to cropland, multiply the number of acres not converted by −129.51 mt CO_2/area/year. Note that this calculation method assumes that all forest biomass is oxidized during clearing (i.e., burned biomass remains as charcoal or ash). Also note that this estimate only includes mineral soil carbon stocks, as most forests in the contiguous United States are growing on mineral soils. In the case of mineral soil forests, soil carbon stocks could be replenished or even increased, depending on the starting stocks, how the agricultural lands are managed, and the timeframe over which lands are managed.

Propane Cylinders Used for Home Barbecues

Propane is 81.7% carbon. The fraction oxidized is 100% (USEPA, 2012c; IPCC, 2006). Carbon dioxide emissions per pound of propane were determined by multiplying the weight of propane in a cylinder by the carbon content percentage times the fraction oxidized times the ratio of the molecular weight of carbon dioxide to that of carbon (44/12). Propane cylinders vary in size; for this equivalency calculation, a typical cylinder for home use was assumed to contain 18 pounds of propane.

Calculation

Note: Due to rounding, the calculations in the equations below do not return the exact results.

18 pounds propane/1 cylinder × 0.817-pound C/pound propane × 0.4536 kg/pound × 44 kg CO_2/12 kg C × 1 metric ton/1,000 kg = **0.024 metric tons CO_2/cylinder**

Railcars of Coal Burned

The average heat content of coal in 2009 was 27.56 MMBtu per metric ton. The average carbon coefficient of coal in 2009 was 25.34 kg of carbon per MMBtu (USEPA, 2011). The fraction oxidized is 100% (IPCC, 2006). Carbon dioxide emissions per ton of coal were determined by multiplying heat content times the carbon coefficient times the fraction oxidized times the ratio of the molecular weight of carbon dioxide to that of carbon (44/12). The amount of coal in an average railcar was assumed to be 100.19 short tons, or 90.89 metric tons (Hancock, 2001).

Calculation

Note: Due to rounding, performing the calculations given in the equations below may not return the exact results shown.

> 27.56 MMBtu/metric ton coal × 25.34 kg C/MMBtu × 44 g CO_2/12g C × 90.89 metric tons coal/railcar × 1 metric ton/1,000 kg = **232.74 metric tons CO_2/railcar**

Tons of Waste Recycled Instead of Landfilled

To develop the conversion factor for recycling rather than landfilling waste, emission factors from EPA's WAste Reduction Model (WARM) were used (USEPA, 2012e). These emission factors were developed following a life-cycle assessment methodology using estimation techniques developed for national inventories of greenhouse gas emissions. According to WARM, the net emission reduction from recycling mixed recyclables (e.g., paper, metals, and plastics), compared with a baseline in which the materials are landfilled, is 0.73 metric tons of carbon equivalent per short ton. This factor was then converted to metric tons of carbon dioxide equivalent by multiplying by 44/12, the molecular weight ratio of carbon dioxide to carbon.

Calculation

Note: Due to rounding, the calculation given in the equation below may not return the exact results shown.

> 0.73 metric tons of carbon equivalent/ton × 44 g CO_2/12 g C = **2.67 metric tons CO_2 equivalent/ton of waste recycled instead of Landfilled**

Coal-Fired Power Plant Emissions for One Year

In 2009, 457 power plants used coal to generate at least 95% of their electricity (USEPA, 2012a). These plants emitted 1,614,625,638.1 metric tons of CO_2 in 2009. Carbon dioxide emissions per power plant were calculated by dividing the total emissions from power plants whose primary fuel source was coal by the number of power plants.

Calculation

Note: Due to rounding, performing the calculations given in the equations below may not return the exact results shown.

1,614,625,638.1 metric tons of CO_2 × 1/457 power plants = **3,533,098 metric tons CO_2/ power plant**

REFERENCES

EIA, 2012. 2009. *Residential Energy Consumption Survey, Table CE2.6, Household Fuel Consumption in the U.S., Totals and Averages, 2009, Physical Units, Preliminary.*

FHWA, 2012. *Highway Statistics 2010. Office of Highway Policy Information*, Federal Highway Administration. Table VM-1.

Hancock, K. and Sreekanth, A., 2001. *Conversion of Weight of Freight to Number of Railcars. Transportation Research Board*, Paper 01-2056, 2001.

IPCC, 2006. *2006 IPCC Guidelines for National Greenhouse Gas Inventories.* Geneva, Switzerland: Intergovernmental Panel on Climate Change.

Rishel, J.B., 2001. Wire-to-water efficiency of pumping systems. *ASHRAE Journal*, April.

Smith, J., Heath, L., & Nichols, M., 2010. *U.S. Forest Carbon Calculation Tool User's Guide: Forestland Carbon Stocks and Net Annual Stock Change.* General Technical Report NRS-13 revised. U.S. Department of Agriculture Forest Service, Northern Research Station.

USEPA, 1998. *Method for Calculating Carbon Sequestration by Trees in Urban and Suburban Settings, Voluntary Reporting of Greenhouse Gases.* Washington, DC: U.S. Department of Energy, Energy Information Administration.

USEPA, 2011. *Inventory of U.S. Greenhouse gas Emissions and Sinks: Fast Facts 1990–2009. Conversion Factors to Energy Units (Heat Equivalents) Heat Contents and Carbon Coefficients of Various Fuel Types.* Washington, DC: U.S. Environmental Protection Agency, USEPA $430-F-11-007.

USEPA, 2012a. *eGrid2012 Version 1.0, US. Annual Non-Baseload CO_2 Output Emission Rate, Year 2009, Year Data.* Washington, DC: U.S. Environmental Protection Agency.

USEPA, 2012b. *Inventory of U.S. Greenhouse Gas Emissions and Sinks: 1990–2010*, Chapter 3 (Energy), Tables 3-12, 3-13, and 3–14. Washington, DC: U.S. Environmental Protection Agency, The US.EPA #430-R-12-001 (PDF) (481 pp., 16.6 MB).

USEPA, 2012c. *Inventory of U.S. Greenhouse gas emissions and Sinks: 1990–2010, Annex 2 (Methodology for estimating CO_2 emissions from fossil fuel combustion), P.A.-85.* Washington, DC: U.S Environmental Protection Agency, The US.EPA #430-R-12-001 (PDF) (389 pp, 10.6 MB).

USEPA, 2012d. *Inventory of U.S Greenhouse Gas Emissions and Sinks:* 1990–2010. *Annex 2 (Methodology for Estimating CO_2 Emissions from Fossil Fuel Combustion), P.A.-68.* Washington, DC: U.S Environmental Protection Agency, U.SEEPA #430-R-12-001.

USEPA, 2012e. *WAste Reduction Model (WARM).* Washington, DC: U.S. Environmental Protection Agency.

USEPA, 2024. About the Inventory of Greenhouse Gas Emissions and Sinks. Accessed 10/09/24 @ https://epa.gov/ghgemissions /inventory-us-greenhouse-gas-emissions-and-sinks-1990-2200>.

Prologue

DURING A RECENT RABBIT AND GRASSHOPPER CONVERSATION

Grasshopper told his friend, Rabbit: "To fix human energy, population, unemployment, and many of their economic problems, what them human-kind need right now, my friend, is innovation, innovation, innovation…."

After deliberate and well-practiced thumping of his foot, Rabbit replied to his friend, Grasshopper: "No, my long-legged friend…to fix all of humankind's economic problems, what they need first is discovery, discovery, discovery and then invention, invention, invention … followed up by innovation, innovation, innovation…hmmmm, human leadership, brain power, common sense, (hmmm, common sense is not that common with them folks) anyway accountability would also help."

Grasshopper: hmmmmmmmmmm, buzzzzzzzz, etc., "Well, sometimes they ain't too smart … them humans.… All they need to do is ask us. We know how to economize … how to be efficient and produce usable energy and use it—well, at least for our use—and how to do the rest."

Rabbit replied: "Right on, Grasshopper!"

–A friend of all grasshoppers and rabbits, 2024

Part 1

The Basics

1 Setting the Stage

1.1 ENERGY CURRENCY

The term "energy currency" was coined by Henry Ford in 1921. Ford's energy currency proposal was about forming the basis of a new monetary system that was the forerunner, the antecedent, the pioneer, and the predecessor to current Bitcoin. To say Henry Ford was a far thinker means underplaying his importance as a visionary. His goal was to use energy currency to replace gold. Ford thought that creating energy currency and using it to replace gold would work to break banking elites' grip on global wealth and put an end to wars. He envisioned energy currency becoming the world's most extraordinary power plant and creating a new currency system.

A few of my associates have told me that Ford's energy currency concept is "currently" a work in progress. However, as far-thinking as Ford's economics might be or is ongoing, a work in progress, prevalent, prevalent, prevailing, popular, and circulating, is one thing, an important thing. But this book is about a different type of energy currency.

Key Point: Power is the rate at which work is done. Note that work is only an active measure of energy and not a form of energy itself; this is a crucial point to remember.

Okay, let us begin with "current"—this term can have several meanings and uses in the present lexicon. We can say that the term itself, current, originates in Latin, currere—"run"—or from the French, "corant"—"running"—from courre—"run."

These origins of the term "current" or currency are fitting. In this book, when stating in a compound manner as energy currency, we are not talking about Ford's definition of energy currency; instead, we are talking about energy flow. We need an energy source to have a flow of energy—gas, oil, electricity (electrons), etc. Energy flows are provided by gas and oil naturally. However, note that electricity flow results from a secondary energy source (a prime mover such as a diesel engine or kinetic energy provided by flowing water, solar energy, for example); that is, a primary energy source, such as a hydro or motor-driven generator, must produce electrical flow (electricity). Electricity must be made or produced. Electricity is transformative and is the future of energy production, regardless of the primary mover (windmills, turbines, water wheels, and internal combustion engines).

In this book, we examine the concept of energy, which is the most far-reaching concept in science (physics). It would be extremely difficult or impossible to find anyone who does not have an awareness of energy and its meaning. We must pay for energy to get things done (to accomplish work). This work usually remains in the background, but even so, we recognize what a gallon of diesel or gasoline is, what one Btu of heating gas is, what each calorie of food value is, and what each kilowatt-hour of electricity represents.

DOI: 10.1201/9781003605430-2

The bottom line is that energy is the universal current in countless currencies, and physical processes represent a conversion from one currency to another.

Note: It is essential to begin at the beginning before entering the modern energy era, and the next section accomplishes this.

1.2 IN THE DISTANT PAST

By today's calendar, it is June 26, 15,543 BC. One of the inhabitants was a male cave dweller, a troglodyte, as shown in Figure 1.1.

The place is a large natural cave set deep under a solid outcropping that formed a significant mountain meadow before the last glacial ice sheet gouged, gorged, ground, and pulverized it down to its present size and shape.

FIGURE 1.1 Cave Dweller. (Artwork by Kat Welch and F.R. Spellman.)

The colossal sheet of ice is in retreat. At its full width, depth, and length, it extended several hundred miles beyond the cave site to a V-shaped valley, where glacial melt fed a youthful, raging river that ran through the valley's bottomland.

A small, steady stream of melt-water courses in a straight line past one side of the cave, down toward that valley, where it will join and feed that river.

On the other side is a sloping field of young grass, brushes, and flowers everywhere. Up close, we see the stark remnants of the terminal moraine that formed this abrupt slope with its fresh cover of grass and blossoms. A closer look reveals a dark heap at the base of the hill—a pile of trash: skin, ligament, bone, decaying corpses, burnt remnants of past hunts and feasts. We know only too well the refuse, filth, and discards that people leave behind as their foulest signature; somebody, many bodies, lives close by. Where? Of course—in the cave.

Let's look inside.

We wander up to the massive hole in the rock that forms the mouth of the cave. We tread carefully—we do not want to disturb (startle, frighten—anger) the occupants. Remember, we are talking about cave dwellers here—no language in common. There is no culture in common. Could we have anything in common with such … primitive people?

Well, we do need to keep warm and be able to cook our food; remember that cave dwellers and cavewomen had these needs, too.

Ok, back to our ancient friends in the cave.

But all is quiet—and with no overt threat present, our curiosity overcomes our caution, and we walk into the opening chamber.

And something reaches out and grabs us—not caveman, cavewoman, Saber Tooth Tiger, Cave Bear, but a stench—a horrible stench too horrible to describe. With our fingers clamped tightly to our noses, crushing to the Cartlidge, we move on, too interested, despite the reek, despite nose pain, to retreat.

We can see well in this chamber because daylight pours through the entrance. We take a few steps and look around, religiously not breathing through our aching noses.

The walls of the cave are covered with black soot. A pit near the cave wall to the right, under attack by millions of flies and other insects, provides much of the stench—the latrine. Like the dump outside, a heap of detritus offers the rest of the reek.

It dawns on us that this cave is abandoned. We do not doubt why. Humankind's most significant byproduct has taken over—the cave is a garbage dump.

Back in other chambers, deeper in the cave, there are cave paintings and remnants of the ceremony; however, the cave dweller shown in Figure 1.2 is no longer present. But we do not have the tools to explore the remnants with us today, so we retreated, grateful for the fresh air.

Outside, a few hundred feet from the smelly cave and within sight of the garbage heap, we contemplate what we have seen. Seventeen thousand years from now, archaeologists will find this cave and explore it thoroughly, learning information that will give us insights into the world of the people whose former home we have just visited. However, the remains the archaeologists will find will be altered by 17,000 years of history. The picture they see will be incomplete, scattered by the natural interferences life causes, giving mystery to our ancestors' short and brutal lives.

FIGURE 1.2 Cave Dweller keeps warm from the fire's thermal energy. (Artwork by Kat Welsh and Frank R. Spellman.)

But here, right now, we see similarities. We foul our environment in the same ways and more. But the cave dwellers had a massive advantage over modern humans. When his living quarters became too foul for comfort, he could pick up whatever he considered of value he and his tribe could carry and move on. A fresh site was always just around the next bend in the river. The pollution he created was ultimately (eventually) naturally biodegradable—in a few years, this cave could house humans again.

Although we have similarities with those far-off ancestors, one stark difference is plain: Modern man cannot destroy and pollute his environment with impunity. We can no longer pick up the stakes and move on. What we do to our environment has ramifications on a scale that we cannot ignore—or avoid.

1.3 FAST FORWARD TO THE PRESENT

In the introductory paragraph, we ask the question, "Could we have anything in common with such … primitive people?" That is, with our ancient cave-dwelling ancestors. Of course, we do have commonalities; the roots of our culture are deep in this ancient past. However, from the description of the cave's condition, similarities should be acutely apparent to all of us: We all need a roof over our heads. We all need a fire (a type of energy) to keep warm to cook food. We all pollute by accident or otherwise in the area(s) where we visit, work, and reside. Yes, we are all still members of the same throw-away society. When we no longer need an item, we dispose of it—often placing it in the trash or on the ground. The difference in methods is a matter of scale and perspective.

What pollutants would that cave contain? Let us discuss the cave for a moment, including its various contaminants.

I stated previously that pollution is an issue of judgment—what is pollution to one person may not be pollution to another. For example, when two people visit a particular stream and both see fishkill floating down the stream belly-up, one person may only see dead fish. That might be as far as their thought process and reactions go. Viewing the same scene, the same dead fish, the other person may see something entirely different. They might see dead fish (of course), but what they comprehend is the pollution, the affected biota (the dead fish)—or the pollution/pollutants/contaminants they believe caused the fishkill.

Indeed, pollution is a matter of judgment—with each person having a different judgment or perception. One thing is sure, however, as the scene in the cave demonstrates, we all share some judgments about pollution. When a substance, a material, a rotting organic mess, or a pollutant reaches out and grabs our attention, few of us would have difficulty agreeing that it is (whatever it is) offensive to us. This is the real essence of pollution and pollutants—when the substance, the material, or the rotting organic mess offends us, it is pollution waste material, a pollutant that we can all agree upon.

Let us take a closer look at what we saw in the cave.

Beyond the fact that the cave was an indoor air quality specialist's worst nightmare (e.g., soot and air pollution from too many cave fires—again, an energy source), the cave was highly offensive to most of us for other reasons. Why? First, the stench. Human beings will voluntarily put up with many unpleasant situations—disgusting, overwhelming, and noxious odors are not one of them. Have you ever visited a garbage dump or followed a garbage truck and had the wind turn so that it pulled the stench of rotting garbage toward you … over you … encapsulating you, making you want to vomit … making you want to get away from the stench? If you have, then you know … and understand the point. Simply put, most of us do not tolerate those odors that offend us—that attack our olfactory sense (sense of smell)—disgusting.

Beyond being disgusting primarily because of its odor, there was more, wasn't there? Yes, there was more—the in-cave restroom and a reserved section of the interior reserved for deposits, for example. The question you might be asking yourself, beyond the horrendous odor, is why would primitive humans foul their living space by defecating in the same place they lived? Good question—and the answer is quite simple … more a matter of self-preservation than anything else. Picture this: You wake up in the middle of the night and have the urgent, pressing need to relieve yourself. You have two simple choices: you can do your business right there in the relative warmth and safety of the cave, or you can step outside into the cold, cruel environment beyond the cave's confines. Outside? Into the world where Smilodon and their mates roam at will? If you are not familiar with Smilodon, let us fill you in. Smilodon (Saber Tooth Tiger) was present when primitive man occupied this cave. The Saber Tooth Tiger was one of the most ferocious, stabbing, meat-eating animals of this period or any other time. Imagine an attack by those two powerful and efficient elongated scimitar-like canine teeth attached to that muscle-bound mouth, open like a Venus fly trap. Smilodon stabbed, tore, and ripped the flesh to disable its victims, then consumed them, body, blood, and entrails. And Smilodon was not the only risk out there—tenacious, ferocious cats, bears, and for-real monsters (e.g., Panthera spelaea, the cave lion bigger than today's lion) roamed the land then. No, sir. I will take the warmth and safety of the cave and the accompanying malodorous ambiance anytime. Wouldn't you?

The presence of "night soil" might make more sense to you now, but you might ask, "Okay, I understand the indoor restroom, but what about all the garbage—the skins, ligaments, bones, decaying corpses, burnt remnants of past hunts and feasts? Why would our primitive ancestors put up living with such items—items that were simply nothing more than disgusting, filthy waste, trash ... messy, filthy trash?"

Good question. We have no easy answer for this one; it is only speculation. But again, remember that human energy is nothing new, and many would say that the accompanying pollution from its use is a judgment. The primitive man did not consider such items as trash, pollutants, and pollution. Hunter societies use all the parts of their kill more than we imagine. And our reaction to the cave is relative to our own experience. For example, farmers become accustomed to manure smells that offend city dwellers. Modern standards of cleanliness are just that—modern and indoor plumbing and water supply are the chief reasons for our standards of cleanliness. Also, remember, when the cave became so offensive that primitive humans could no longer manage it, they picked up their possessions and moved on to another cave ... "a fresh site was always just around the next bend in the river." We do not know the answers for sure. However, we do know that when archaeologists excavate such sites, they do find remnants of these items.

Again, we go back to the earlier statement: "Could we have anything in common with such ... primitive people?" For our uses, the commonality resides in our need to keep warm and cook food and our tendency to throw things away. When we do not want it anymore, we discard it. When it offends us, we dispose of it. Throughout history, a common practice was, "I don't want it any longer. Take it down to the river and dump it." When mile-long caravans of wagon trains crossed the tremendous American prairies, discarding anything and everything that over-burdened the wagon was a common practice for survival—homemaking materials, furnishings that made the trip more complex—some of them necessary to homesteading, but too heavy to carry. Lifelong treasures once so vital that they could not be left behind now became burdens (and waste) to the American pioneering waggoners ... things to be disposed of to make their lives easier. This pioneering throwaway tendency has been called the Frontier Mentality. We can throw anything away here in the Frontier (the prairie) ... there is so much space, this piece of trash placed in the vast, endless prairie will not hurt a thing ... it is just like urinating in the ocean ... will not hurt a thing! This ignores the accumulation effect, eventually catching up with all that space.

Isn't this what we are all about—utilizing energy and creating waste—the throwaway society? We do whatever is necessary to make our lives better ... more tenable, more comfortable. The problem is that our world is not an endless prairie nor an endless supply of specific energy sources. We can no longer throw away, discard, dump, dispose, hide, or ignore the trash and pollutants we produce. Instead, we must recognize pollutants for what they are—byproducts of civilization (no matter what the timeframe) that must be disposed of, recycled—or not produced in the first place. But if pollution is a judgment, how do we come to a consensus on what needs our attention and what does not? And in using energy, we need to ensure that sources are sustainable.

1.4 THE BOTTOM LINE

Contrary to what some might think, this book is not about cave dwellers, Saber Tooth Tigers, or Cave Bears and the filth and trash associated with their activities and the era they represent. The fact is that humans have always polluted their environments in one way or another—this horrendous and unfeeling behavior is discussed in detail in other works. Instead, this opening chapter sets the stage for what is to follow regarding energy and a few side effects of energy usage.

This book begins with a stitch of time and an odd place foreign to us today. However, it is interesting that plant biomass (firewood) was the primary energy source throughout human history. In ancient times and locations (and a few 3rd World locations to this day), biomass was burned for warmth and light, to cook food, and to feed the animals people used for transportation and plowing. It was in the early 1800s when nonrenewable energy began replacing most renewable energy in the United States. By the early 1900s, fossil fuels were the primary source of energy. In the United States, we use a mix of energy sources. Now, we are working to replace nonrenewable energy sources and associate pollution with renewable energy sources.

Artificial Intelligence (AI) is helping to ensure that whatever energy source we use is more efficient (which reduces pollution and cleans up the caves and modern abodes as well), creating a win-win situation for life on Earth.

2 Forms of Energy[1]

2.1 INTRODUCTION

Whenever I taught environmental engineering, environmental health, and even during some nursing courses when my lecture series focused on energy, in one manner or form or another, I always began my lectures by asking the students in mass but one at a time what their definition of energy is. You might think defining energy is easy; it is not. One of the classic replies I most heard (and appreciated) was that "energy is that stuff we have to pay for to do things or get things done." Again, this is a standard reply to my question, mainly because anyone who must commute from place to place, from home to school, might use some form of transportation to get to and from—transportation requires energy in the form of fuel and, therefore, is currently expensive (i.e., relative to historical times of lower costs). The point is that this discussion and many of the answers I receive make it clear to the students that energy is a universal currency (recall Henry Ford's definition of energy currency).

It is common and expected to consider power and energy to be exact synonyms. They are closely related, but not the same type of physical quantity. In physics and this book, *energy* (unit joules = watt-seconds or joule = Newton-meter, common symbol W) is defined as the amount of work that can be performed by force, whereas *power* (unit watt = joules/second, common symbol P) is defined as the rate at which work is performed. It is interesting to note that I have heard a few people refer to energy as work. *Work* is the amount of "energy" transferred into or out of a specific mechanical system through mechanical forces.

Because this book focuses on energy, it notes that defining energy can be accomplished by providing a technical definition or characterization. Because this book aims to reach technical readers and a wide range of general readers, definitions provided herein and hereafter are best described as technical-nontechnical-based. Energy can be defined in several ways. In the broad sense, energy means the capacity of something, a person, an animal, or a physical system (machine) to do work and produce change. Energy is the amount of force or power that, when applied, can move one object from one position to another. It can also describe someone doing energetic things such as running, talking, and acting lively and vigorously. Science uses it to tell how much potential a physical system must change. It is also used in economics (recall Henry Ford's energy currency definition) to describe the part of the market where energy is harnessed and sold to consumers. In this section, we define energy in technical-nontechnical-based terms as something that can do work or the capacity of a system to do work.

2.2 FORMS OF ENERGY

There are two primary forms of energy: kinetic and potential energy. *Kinetic energy* is energy at work or in motion; that is, moving energy—a car in motion or a rotating shaft has kinetic energy. In billiards, a player gives the cue kinetic energy when she

DOI: 10.1201/9781003605430-3

strikes the ball with the cue. As the ball rolls, it exerts kinetic energy. When the ball meets another ball, it transmits its kinetic energy, accelerating the next ball. *Potential energy* is stored energy, like the energy stored in a coiled or stretched spring or an object stationed above a table. A roller coaster has the most potential energy when stopped at the top of a long drop. Another potential energy source is when a carbonated soda can remain unopened. The can is pressurized with gas that is not in motion but has potential energy. The gas is released once the can is opened, and the potential energy is converted to kinetic energy.

2.2.1 Potential Energy

Potential energy is stored energy and the energy of position with categories that include (EIA, 2023):

- **Chemical energy** is stored in the bonds of atoms and molecules. Batteries, biomass, petroleum, natural gas, and coal are examples of chemical energy. For instance, chemical energy is converted to thermal energy when people burn wood in a fireplace or gasoline in a car's engine.
- **Mechanical energy** is energy stored in objects by tension. Compressed springs and stretched rubber bands are examples of stored mechanical energy.
- **Nuclear energy** is energy stored in the nucleus of an atom—the energy that holds the nucleus together. Large amounts of energy can be released when the nuclei are combined or split apart. Presently, there is a growing movement to reactivate or to build nuclear power plants because of the increasing need for more transformative energy (electricity) to power artificial intelligence (AI) data centers, which use huge amounts of electricity—they also use increasing amounts of water for cooling purposes.
- **Gravitation energy** is energy stored in an object's height. The higher and heavier the object, the more gravitational energy it stores. When a person rides a bicycle down a steep hill and picks up speed, the gravitation energy is converted to motion energy. Hydropower is another example of gravitation, where gravity forces water down through a hydroelectric turbine to produce electricity.

2.2.2 Kinetic Energy

Kinetic energy is the motion of waves, electrons, atoms, molecules, substances, and objects.

- **Radiant energy** is electromagnetic energy that travels in transverse waves. Radiant energy includes visible light, X-rays, gamma rays, and radio waves. Light is one type of radiant energy. Sunshine is radiant energy, which provides the fuel and warmth that make Earth possible.
- **Thermal energy**, or heat, is the energy that comes from the movement of atoms and molecules in a substance. Heat increases when these particles move faster. Geothermal energy is the thermal energy in the earth.

- **Motion energy** is energy stored in moving objects. The faster an object moves, the more energy is stored.
- **Sound** is energy moving through substances in a longitudinal (compression/rarefaction) wave.
- **Electrical energy** is delivered by tiny, charged particles called electrons that typically move through a conductor; electricity is the flow of electrons.

According to the Conservation Law of Energy, energy cannot be made or destroyed but can be made to change forms (transformation). Moreover, energy stays the same when energy changes from one form to another. Let us consider an example of the Conservation Law of Energy: The energy of something is measured to start with; the energy changes from potential (stored) energy to kinetic (moving) and back again; at the end, the energy is measured again. The energy measured at the start is the same as that measured at the end; it will always be the same. One caveat to this explanation is that we now know that matter can be made into energy through processes modified or amplified to become the Law of Conservation of Matter and Energy.

2.3 TYPES OF ENERGY

The primary sources of green energy are wind, solar, and hydroelectric power (including pumped hydropower storage systems and tidal energy, which uses ocean energy from the ties in the sea). Both solar and wind power can be produced on a small scale for people's private use or business, or they can be designed and constructed on a larger, industrial scale. Note that various devices can convert energy from one form into another. Power cannot be converted or transformed. Energy generated can be stored, whereas power cannot. There are many types of energy. For example:

- Kinetic (motion) energy
- Water energy
- Potential (at rest) energy
- Elastic energy
- Nuclear energy
- Chemical energy
- Sound energy
- Internal energy
- Heat/Thermal energy
- Light (radiant) energy
- Electric energy

Energy sources can also be categorized as renewable or nonrenewable. EIA (2009) points out that when we use electricity in our homes, the electrical power is generated by burning coal, a nuclear reaction, or a hydroelectric plant at a dam. Therefore, coal, nuclear, and hydropower are called energy sources. When we fill up a gas tank, the source might be petroleum or ethanol made by growing and processing corn.

As mentioned, energy sources are divided into two groups—*renewable* (an energy source that can be quickly replenished—including Green Energy) and *nonrenewable* (an energy source that we are using up and cannot recreate; petroleum, for example, formed millions of years ago from the remains of ancient sea plants and animals). In the United States, most of our energy comes from nonrenewable sources. Coal, petroleum, natural gas, propane, and uranium are nonrenewable energy sources. They are used to make electricity, to heat our homes, to move our cars, and to manufacture all kinds of products. Green (Renewable) and nonrenewable energy sources can produce secondary energy sources, including electricity and hydrogen.

2.4 EIGHT ENERGY LITERACY PRINCIPLES

Before the introduction of AI and its expanding usage, it was common for energy practitioners to be trained and aware of seven energy literacy principles as follows (EERE, 2013):

1. Energy is a physical property that follows precise laws; it is a quantity transferred from system to system. The energy of a system or object that results in the temperature is called thermal energy. Energy is neither created nor destroyed. The energy available to do practical work decreases as it is transferred from system to system. Energy comes in different forms and can be divided into categories of energy, including light energy, elastic energy, chemical energy, and more. Many various or other units are used to quantify energy. Again, power is not energy; instead, it is a measure of energy transfer rates (one joule of energy is transformed in one second and is called a Watt, where 1 J/s = 1 /W).
2. Physical properties on Earth result from energy flow through the Earth system. Earth is constantly changing as energy flows through the system. The physical processes on Earth are driven by sunlight, gravitational potential, decay of radioactive isotopes, and rotation of the Earth.
3. Biological Processes depend on energy flowing through the Earth's system. Energy flows through all food webs, beginning with the Sun, the primary energy source for organisms and the ecosystems in which they live. Food is a biofuel organism used to acquire energy for internal living processes. Energy availability to do practical work decreases as it is transferred from organism to organism. Energy flows through food webs in one direction, from producers to consumers and decomposers. Ecosystems are affected by changes in the availability of energy and matter. Remember that humans are part of Earth's ecosystems and influence energy flow through these systems.
4. Various energy sources can power human activities, and often, this energy must be transferred from source to destination. Humans transform energy from the environment into forms needed for human endeavors. The primary energy sources in the environment include nonrenewable fuels like oil, coal, natural gas, and uranium and renewables like biomass. Renewable energy sources besides biomass include wind, water, sunlight, and geothermal energy. Human use of energy is subject to the availability of energy resources constrained by the distribution of natural resources, availability

of affordable technologies (i.e., are recharge stations readily available and accessible?), socioeconomic policies, and socioeconomic status (i.e., can you purchase a costly electric vehicle?). The amount and kind of energy available is essentially available. Fossil fuels and biofuels are organic matter that contains energy captured from sunlight. Humans transport this energy from place to place. Electrical energy can be generated from various energy resources and transformed into almost any other form of energy. Note that electricity is not a primary energy source but an energy carrier. Electricity is generated in multiple ways, including electromagnetism, electrochemical, piezoelectric, and thermoelectric. Energy is intentionally stored in several different ways. Examples include batteries, water reservoirs, compressed air, hydrogen, and thermal storage.

5. Energy decisions are influenced by economic, political, environmental, and social factors. Many levels are involved when it comes to making energy use decisions. Each level of decision has some common and some unique aspects. Beyond the individual level, decision-making often involves a formally established process. Energy infrastructure is said to have inertia. The point is that today's energy infrastructure has the inherent property that makes it oppose and force that would cause a change in direction.

   Why?

   The large amounts of money, time, and technology invested in energy infrastructure make changing it difficult—not impossible. And needed—think the U.S. electrical grid with its potential for extreme weather events, aging, underinvestment, increased blackouts from grid failure, and cyber-warfare. Another issue is the risk of electromatic pulses (EMP), a brief burst of electromagnetic energy. Energy decisions are based on economics and politics and are influenced by environmental factors. Monetary costs of energy affect energy decision-making at all levels. Politics plays a role in energy decision-making, including governmental structure and power balances, actions taken by politicians, and partisan-based or self-serving actions taken by individuals and groups. Environmental factors come into play in energy decision-making because all energy decisions have positive or negative ecological consequences.
6. The amount of energy human society uses depends on many factors. First, energy conservation comes to mind because it has two distinct meanings. One meaning is the physical law of energy conservation—whereby the total amount of energy in the universe is constant. On the other hand, the second meaning of conserving energy is commonly used to mean the decreased use of energy resources. This means reducing wasteful energy use and using energy for a given purpose more efficiently—basically, it means reducing energy use altogether. Human use of energy is increasing. Increases in population, industrialization, and socioeconomic development drive the need for more energy. One of the problems is supply—Earth has limited energy resources, increasing the stresses on its availability. Technological innovation (e.g., artificial intelligence and its applications data centers) also drives increased energy use. This increased energy use continues, is ongoing, and is growing exponentially.

7. Energy choices affect individuals' and societies' quality of life and economic security. Some populations are more vulnerable to the impacts of energy choices than others. Note that energy decisions have financial, social, and environmental consequences. Poor, marginalized, or underdeveloped populations can most benefit from positive outcomes and are the most susceptible to negative consequences.

   Note: With the introduction of AI, the eighth energy literacy principle has been developed, in this author's opinion, to obtain the best results from energy usage.
8. AI applications in energy production, distribution, and usage.

2.5 KEY ENERGY DEFINITIONS

Energy efficiency—provides the same or better services using less energy.
Energy currency—forms of energy that are easy to use, transport, or store (e.g., electricity, electron flow in a conductor).
Primary energy—coal, gas, oil.
Energy conversion—power station then grid.

2.6 U.S. PRIMARY ENERGY CONSUMPTION BY SOURCE, 2022

As shown in Figure 2.1, energy consumption in 2022 was primarily by nonrenewable petroleum, natural gas, coal, and nuclear, with renewables, biomass, hydropower, wind, solar, and geothermal contributing most of the rest. Figure 2.1 does not include net electricity imports and coal coke, which accounted for less than 1% of U.S. energy consumption in 2022.

DID YOU KNOW?

A list of U.S. primary energy consumption by energy source in r 1.055 × 101[joule (a quad is a unit of energy equal to 10 Btu, or 1.055 × 10 J (1.053 exajoules in SI units)] in 2023 shows the following (EIA, 2023):

- Petroleum 38%
- Natural gas 36%
- Nuclear 9%
- Coal 9%
- Renewable energy 9%

2.7 ENERGY STORAGE

2.7.1 Thermal Store

Those next to a rock-lined roaring campfire are aware of or become accidentally introduced to the heat retained in the rocks. The heat in the stones is mechanical thermal heat, known as *thermal storage*; more specifically, it is known as sensible

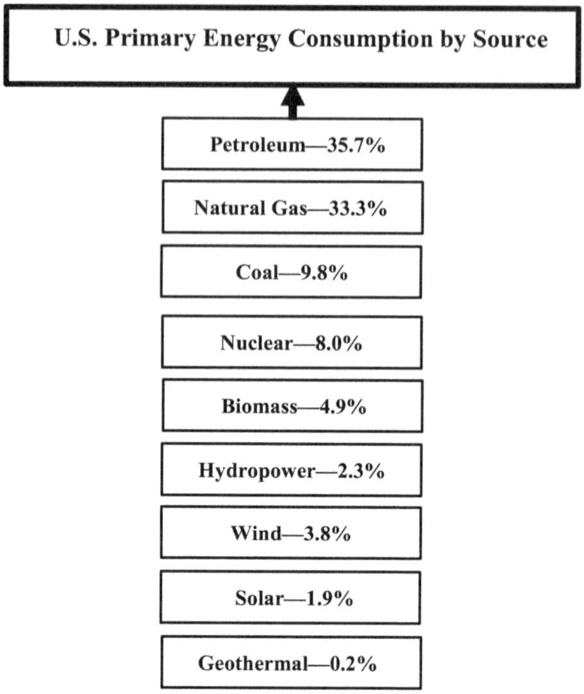

FIGURE 2.1 U.S. primary energy consumption by source, 2022. (Adaptation from U.S. Energy Information Administration (EIA), *Monthly Energy Review*, Table 1.3 and 10.2, April 2023.)

heat storage, which also applies to heated concrete, sand, or liquid media such as molten salt. Energy stored using phase change materials (e.g., metals, organics, and salts) is *latent heat storage.* The energy stored in chemical bonds is known as *thermochemical storage.*

2.7.2 Batteries: Electrochemical Storage Devices

If someone asked what the standard means of storing energy is, those who are unaware would scratch their heads, throw their hands in the air, and eventually say gasoline, diesel, kerosene, white gas, coal, and nuclear. However, others would answer that the battery is the primary storehouse of energy. Storage batteries indeed store energy, contrary to those who think batteries create energy, which they do not—store, yes, but battery devices create no energy. Electrochemical batteries store energy in a form that can be easily converted into electrical energy.

Electricity—What is it? Employees in industrial complexes have little difficulty in recognizing electrical equipment. Electrical equipment is everywhere and is easy to spot. For example, many typical industrial sites are outfitted with electrical equipment to:

- Generate electricity (a generator or emergency generator)
- Store electricity (batteries)
- Change electricity from one form to another (transformers)
- Transport or transmit and distribute electricity throughout the plant site (wiring distribution systems)
- Measure electricity (meters)
- Convert electricity into other forms of energy (rotating shafts—mechanical energy, heat energy, light energy, chemical energy, or radio energy)
- Protect other electrical equipment (fuses, circuit breakers, or relays)
- Operate and control other electrical equipment (motor controllers)
- Convert some condition or occurrence into an electric signal (sensors)
- Convert some measured variable to a representative electrical signal (transducers or transmitters).

Recognizing electrical equipment is easy because we use so much of it. Suppose we ask typical workers where such equipment is in their plant site. They know because they operate these devices or their ancillaries. If we ask these same operators, they could tell us what a particular electrical device does. If we were to ask if their plant electrical equipment was essential to plant operations, the chorus would resound, "Absolutely."

However, another question does not always result in such a resounding note of assurance. If we asked these same workers to explain in fundamental terms how electricity works to make their worksite equipment operate, the answers we would receive would be varied, messy, disjointed—and not all that accurate. Even on a more fundamental level, how many workers would be able to answer the question: What is electricity?

Very few. Why do so many workers in many industries know so little about electricity? Part of the answer is that many industrial workers are expected to know so much (and they do), but are given minimal opportunity to be trained appropriately. We all know that experience is excellent training. For example, let us look at what an operator assigned to change the bearings on a 5-hp three-phase motor would need to know to accomplish this task. (Remember, it is not uncommon for workers in many fields to maintain and operate industrial equipment). The operator would have to know:

1. How to de-energize the equipment (i.e., proper lockout/tagout procedures).
2. After the equipment has been de-energized, how to properly disassemble the motor coupling from the device it operates (e.g., a motor coupling from a pump shaft) and the proper tools to use.
3. When the equipment has been uncoupled, how to properly disassemble the motor end-bells (preferably without damaging the rotor shaft).
4. When the equipment has been disassembled, how to recognize if the bearings are really in need of replacement (though once removed from the end bells, the bearings are typically replaced).

Questions the worker would need to know how to find answers for include the following:

1. If the bearings need replacement, how are they to be removed without causing damage to the rotor shaft?
2. Once removed, what bearings should be used to replace the old bearings?
3. How will the proper bearings be installed correctly when identified and obtained?
4. How can the motor be reassembled properly when the bearings are appropriately replaced?
5. When the motor is correctly put back together, how is it properly aligned to the pump and reconnected?
6. What is the test procedure to ensure the motor has been appropriately restored to full operational status?

Every one of the steps and questions on the above procedures is essential, and obviously, errors at any point in the procedure could cause damage (more damage than occurred in the first place).

Note to Reader: Let us examine batteries and their function using a simple electrical circuit for illustrative purposes, battery terminology, and essential operation. This seems an appropriate digression (although crucial), especially considering the current move toward electric vehicles (EVs).

Important Note: In studies of electricity and electronics, many circuits composed of specially designed resistive components are analyzed. These components are called **resistors**. The resistive element will consist of a physical resistor for the duration of the remaining analysis. However, the resistive component could be any one of several electrical devices.

Note: Battery-supplied direct current (D-C) electricity (batteries do not store A-C) is widely used in countless applications. Applications include providing electrical energy in workplace vehicles and emergency diesel generators, material handling equipment (forklifts), portable electric/electronic equipment, backup emergency power for light-packs, hazard warning signal lights and flashlights, and standby power supplies or uninterruptible power supplies (UPS) for computer systems. In some instances, they are used as the only power source, while in others (as mentioned above), they are used as a secondary or standby power supply.

2.7.2.1 Battery Terminology

- A voltaic cell is a combination of materials used to convert chemical energy into electrical energy.
- A battery is a group of two or more connected voltaic cells.
- An electrode is a metallic compound, or metal, with an abundance of electrons (negative electrode) or positive charges (positive electrode).
- An electrolyte is a solution capable of conducting electric current.
- Specific gravity is the ratio comparing any liquid's weight to an equal volume of water.
- An ampere-hour is defined as a current of 1 ampere flowing for 1 hour.

2.7.3 The Voltaic Cell

The simplest cell (a device that transforms chemical energy into electrical energy) is a voltaic (or galvanic) cell (see Figure 2.2). It consists of a piece of carbon (C) and a piece of zinc (Zn) suspended in a jar containing a solution of water (H_2O) and sulfuric acid (H_2SO_4).

Note: A simple cell consists of two strips, or electrodes, placed in a container that holds the electrolyte. A battery is formed when two or more cells are connected.

The electrodes are the conductors by which the current leaves or returns to the electrolyte. The simple cell described above contains carbon and zinc strips in the electrolyte. Zinc includes an abundance of negatively charged atoms, while carbon has an abundance of positively charged atoms. When the plates of these materials are immersed in an electrolyte, chemical action between the two begins. In the dry cell (see Figure 2.3), the electrodes are the carbon rod in the center and the zinc container in which the cell is assembled.

The electrolyte is the solution that acts upon the electrodes placed in it. It may be salt, an acid, or an alkaline solution. The electrolyte is liquid in the simple voltaic cell and the automobile storage battery, while in the dry cell (see Figure 2.3), it is a moist paste.

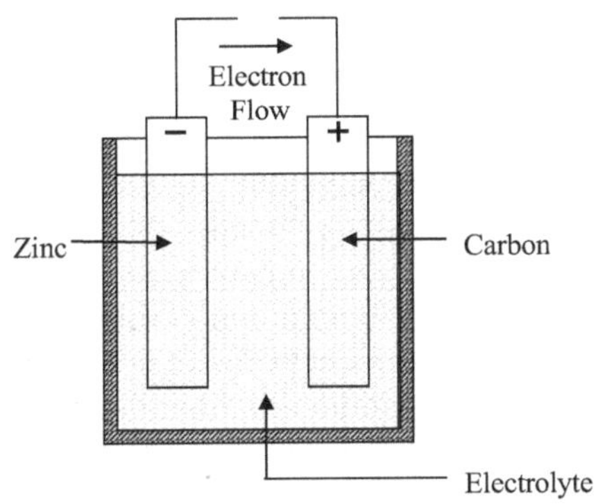

FIGURE 2.2 Simple voltaic cell.

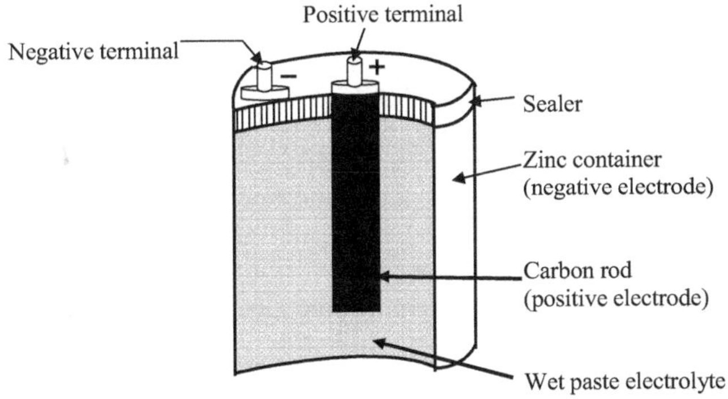

FIGURE 2.3 Dry cell (cross-sectional view).

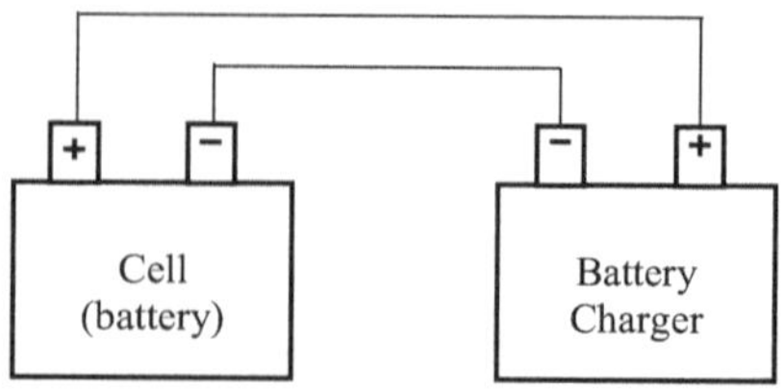

FIGURE 2.4 Hookup for charging a secondary cell with a battery charger.

2.7.4 Primary and Secondary Cells

Primary cells usually cannot be recharged or returned to good condition after their voltage drops too low. Dry cells in flashlights and transistor radios are examples of primary cells. Some primary cells have been developed to a state where they can be recharged. A *secondary cell* is one in which the electrodes and the electrolyte are altered by the chemical action that occurs when the cell delivers current. These cells are rechargeable. The chemicals that provide electric energy are restored to their original condition during recharging. Recharging is accomplished by forcing an electric current through them in the opposite direction of discharge. Connecting, as shown in Figure 2.4, recharges a cell. Some battery chargers have a voltmeter and an ammeter that indicate the charging voltage and current. The automobile storage battery is the most common example of a secondary cell.

Batteries consist of two or more cells placed in a standard container. Depending on the battery's voltage and current requirements, the cells are connected in series, parallel, or some combination of series and parallel.

The chemical reaction within a battery provides voltage. This occurs when a conductor is connected externally to the electrodes of a cell, causing electrons to flow under the influence of a difference in potential across the electrodes from the zinc (negative plate) through the external conductor to the carbon (positive), returning within the solution to the zinc. After a short period, the zinc will begin to waste away because of the acid.

The voltage across the electrodes depends upon the materials from which the electrodes are made and the composition of the solution. The difference in potential between the carbon and zinc electrodes in a dilute solution of sulfuric acid and water is about 1.5 V.

The current that a primary cell may deliver depends upon the resistance of the entire circuit, including that of the cell itself. The internal resistance of the primary cell relies on the size of the electrodes, the distance between them in the solution, and the resistance of the solution. The larger the electrodes and the closer they are in solution (without touching), the lower the internal resistance of the primary cell and the more current it can supply to the load.

Note: When current flows through a cell, the zinc gradually dissolves in the solution, neutralizing the acid.

In many operations, battery-powered devices may require more electrical energy than one cell can provide. Various devices may require either a higher voltage or more current, and in some cases, both. Under such conditions, combining or

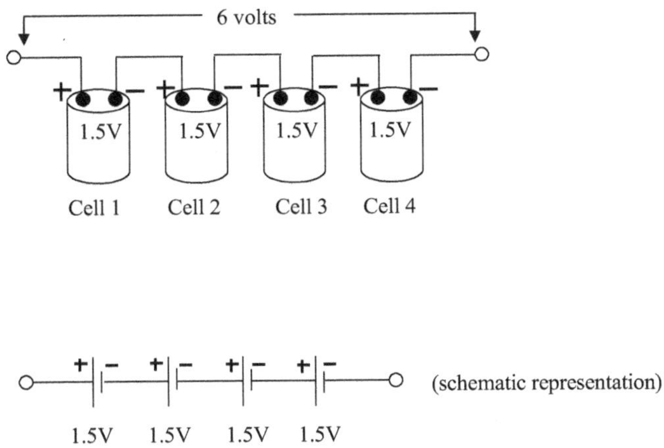

FIGURE 2.5 Cells in series.

interconnecting enough cells to meet the higher requirements is necessary. Cells connected in series provide a higher voltage, while cells connected in parallel provide a higher current capacity. A combination series-parallel network of cells must be interconnected to provide adequate power when both voltage and current requirements are more significant than the capacity of one cell.

When cells are connected in series, the total voltage across the battery of cells is equal to the sum of the voltage of each of the individual cells. In Figure 2.5, the four 1.5-V cells in series provide a total battery voltage of 6 V. When cells are placed in series, the positive terminal of one cell is connected to the negative terminal of the other. The positive electrode of the first cell and the negative electrode of the last cell then serve as the power takeoff terminals of the battery. The current flowing through such a battery of series cells is the same as from one cell because the same current flows through all the series cells.

A battery has cells connected in parallel to obtain a more significant current, as shown in Figure 2.6. In this parallel connection, all the positive electrodes are connected to one line, and all negative electrodes are connected to the other. Any point on the positive side can serve as the battery's positive terminal, and any point on the negative side can be the negative terminal. The total voltage output of a battery of three parallel cells is the same as that for a single cell (Figure 2.6), but the available current is three times that of one cell; the current capacity has been increased. Identical cells supply equal parts of the current to the load in parallel. For example, three parallel cells produce a load current of 210 milliamperes (mA), and each cell contributes 70 mA.

Figure 2.7 depicts a schematic of a series-parallel battery network supplying power to a load requiring both a voltage and current greater than one cell can provide. To provide the necessary increased voltage, groups of three 1.5-V cells are connected in series. Four series groups are connected in parallel to provide the required increased amperage.

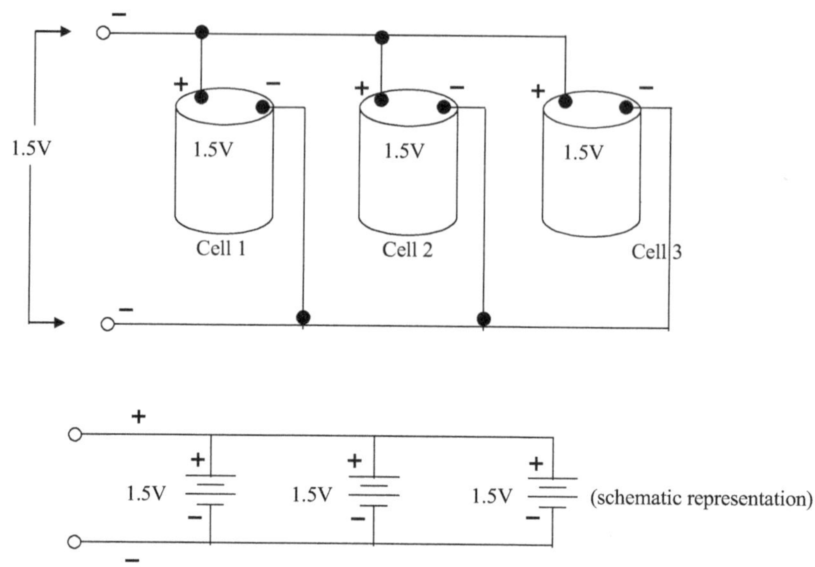

FIGURE 2.6 Cells in parallel.

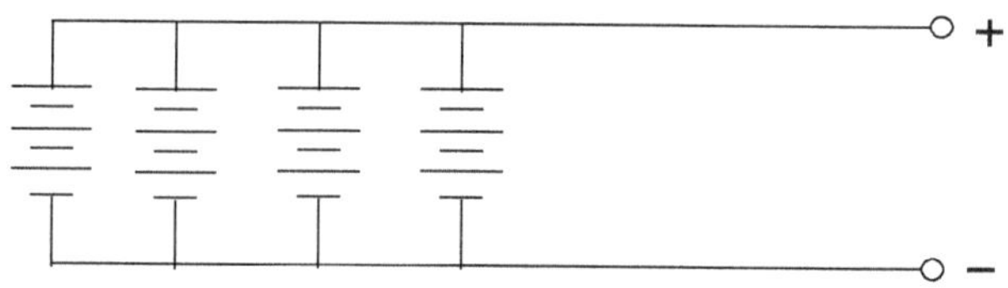

FIGURE 2.7 Series-parallel connected cells.

Several different types of batteries have been developed in the past 25 years. This text briefly discusses five types: the dry cell, lead-acid battery, alkaline cell, nickel-cadmium battery, and mercury cell.

The *dry cell*, or carbon-zinc cell, is named after its electrolyte, which is not liquid (however, it is a moist paste). The dry cell battery is one of the oldest and most widely used commercial types of dry cell. The carbon, in the form of a rod placed in the cell's center, is the positive terminal. The cell case is made of zinc, the negative terminal (see Figure 2.3). Between the carbon electrode and the zinc case is the electrolyte of a moist chemical paste-like mixture. The cell is sealed to prevent the liquid in the paste from evaporating. The voltage of a cell of this type is about 1.5 V.

The *lead-acid battery* is a secondary cell, commonly termed a *storage battery*, that stores chemical energy until it is released as electrical energy.

Note: The lead-acid battery differs from the primary cell type battery because it may be recharged, whereas most primary cells are not recharged. In addition, the term "storage battery" is deceiving because this battery does not store electrical energy but is a source of chemical energy that produces electrical energy.

As the name implies, the lead-acid battery consists of several lead-acid cells immersed in a dilute sulfuric acid solution. Each cell has two groups of lead plates: one set is the positive terminal and the other is the negative terminal. Active materials within the battery (lead plates and sulfuric acid electrolyte) react chemically to produce a flow of direct current whenever current-consuming devices are connected to the battery terminal posts. This current is made by a chemical reaction between the active material of the plates (electrodes) and the electrolyte (sulfuric acid). This type of cell produces slightly more than 2 V. Most automobile batteries contain six cells connected in series so that the output voltage from the battery is slightly more than 12 V. In addition to being rechargeable, the main advantage of the lead-acid storage battery over the dry cell battery is that the storage battery can supply current for much longer than the average dry cell.

Safety Note: Whenever a lead-acid storage battery charges, the chemical action produces dangerous hydrogen gas; thus, the charging operation should only occur in a well-ventilated area.

An *alkaline cell* is a secondary cell named after its alkaline electrolyte, potassium hydroxide. Another type of battery, sometimes called an *alkaline battery*, has a negative zinc electrode and a positive manganese dioxide electrode. It generates 1.5 V.

The *nickel-cadmium cell*, or Ni-Cad cell, is the only dry battery with a proper storage battery with a reversible chemical reaction, allowing recharging many times. In the secondary nickel-cadmium dry cell, the electrolyte is potassium hydroxide, the negative electrode is nickel hydroxide, and the positive electrode is cadmium oxide. The operating voltage is 1.25 V. Because of its rugged characteristics (it stands up well to shock, vibration, and temperature changes) and availability in various shapes and sizes, it is ideally suited for powering portable communication equipment.

The *mercury cell* was developed because of space exploration activities, the development of small transceivers, and miniaturized equipment where a power source of miniaturized size was needed. In addition to reduced size, the mercury cell has a good shelf life and is very rugged; it also produces a constant output voltage under different load conditions. Two types of mercury cells are available. One is a flat cell shaped like a button, while the other is a cylindrical cell that looks like a standard flashlight cell. The advantage of the button-type cell is that several can be stacked inside one container to form a battery. A mercury cell produces 1.35 V.

2.7.5 Battery-Supplied Electricity

Battery-supplied direct current electricity has many applications and is widely used in household, commercial, and industrial operations. Applications include providing electrical energy in industrial vehicles and emergency diesel generators, material handling equipment (forklifts), portable electric/electronic equipment, backup emergency power for light-packs, hazard warning signal lights and flashlights, and standby power supplies or UPS for computer systems. In some instances, they are used as the only power source, while in others (as mentioned above), they are used as a secondary or standby power supply. In renewable energy applications, batteries are used to store electrical energy. A battery pack stores electricity produced by a solar electric system. Today, they are commonly used in specialized applications such as plug-in hybrid electric vehicle energy storage systems. Batteries are often used in wind-hybrid systems to store excess wind energy and provide supplementary energy when the wind cannot generate sufficient power to meet the electric load. Batteries are also used in roadside solar-charging systems; that is, solar energy is used to charge a battery pack that supplies electrical power to some electrical signaling device, navigation buoys, or other remote application (e.g., emergency telephone) where it would be impractical and costly to run miles of electrical supply cable.

In the past 30 years, several different types of batteries have been developed. This text briefly discusses the dry cell battery and those currently used to store electrical energy in a reversible chemical reaction applied to renewable energy production. Renewable energy sources (solar, wind, or hydro) produce energy, and the battery stores it when there is or is no renewable energy production. The types of batteries used in this application and discussed here include lead-acid, alkaline, nickel-cadmium, mercury, nickel-metal hydride, lithium-ion, and lithium-ion polymer batteries. Remember that a battery does not create energy; instead, it stores energy. For most renewable energy applications, the preferred battery type is deep-cycle. A deep-cycle battery is designed to deliver a constant voltage as the battery discharges. A car starting battery, in contrast, is designed to provide sporadic current spikes. Battery-driven vehicles, such as forklifts, golf carts, and floor sweepers, commonly use deep-cycle batteries. Deep cycle batteries can be charged with a lower current than regular batteries.

- **Dry cell—**The dry cell, or carbon-zinc cell, is so called because its electrolyte is not liquid (however, it is a moist paste). The dry cell battery is one of the oldest and most widely used commercial types of dry cell. The carbon, in the form of a rod placed in the cell's center, is the positive terminal. The cell case is made of zinc, the negative terminal (see Figure 2.3). Between the carbon electrode and the zinc case is the electrolyte of a moist chemical paste-like mixture. The cell is sealed to prevent the liquid in the paste from evaporating. The voltage of a cell of this type is about 1.5 V.
- **Lead-acid battery—**A *lead-acid battery* is a secondary cell—commonly termed storage or rechargeable—that stores chemical energy until it is released as electrical energy. The lead-acid battery differs from the primary cell type battery mainly because it may be recharged, whereas most primary cells are generally not recharged. As the name implies, the lead-acid battery consists of several lead-acid cells immersed in a dilute sulfuric acid solution. Each cell has two groups of lead plates: one set is the positive terminal and the other is the negative terminal. Active materials within the battery (lead plates and sulfuric acid electrolyte) react chemically to produce a flow of direct current whenever current-consuming devices are connected to the battery terminal posts. This current is made by a chemical reaction between the active material of the plates (electrodes) and the electrolyte (sulfuric acid). This type of cell produces slightly more than 2 V. Most automobile batteries contain six cells connected in series so that the output voltage from the battery is slightly more than 12 V. Besides being rechargeable, the main advantage of the lead-acid storage battery over the

dry cell battery is that the storage battery can supply current for a much longer time than the average dry cell. Lead-acid batteries can be designed to be high power and are inexpensive, safe, and reliable. Recycling infrastructure is in place for them. However, low specific energy, poor cold temperature performance, and short calendar and cycle life are still impediments to their use. Advanced high-power, deep-cycle lead-acid batteries are being developed for Hybrid Electrical Vehicle (HEV) applications. However, lead-acid batteries are used for residential solar electric systems because of their low maintenance requirements and cost.

Safety Note: Whenever a lead-acid storage battery charges, the chemical action produces dangerous hydrogen gas; thus, the charging operation should only occur in a well-ventilated area.

- **Alkaline cell**—The *alkaline cell* is a secondary cell named after its alkaline electrolyte, potassium hydroxide. Another type of battery, sometimes called the "alkaline battery," has a negative electrode of zinc and a positive electrode of manganese dioxide. It generates 1.5 V.
- **Nickel-cadmium cell**—The *nickel-cadmium cell*, or Ni-Cad cell, is the only dry battery that is an actual storage battery with a reversible chemical reaction, allowing recharging many times. In the secondary nickel-cadmium dry cell, the electrolyte is potassium hydroxide, the negative electrode is nickel hydroxide, and the positive electrode is cadmium oxide. The operating voltage is 1.25 V. Because of its rugged characteristics (it stands up well to shock, vibration, and temperature changes) and availability in various shapes and sizes, it is ideally suited for powering portable communication equipment. NiCad batteries are costly. Moreover, although nickel-cadmium batteries, used in many electronic consumer products, have higher specific energy and better life cycle than lead-acid batteries, they have low efficiency (65%–80%), do not deliver sufficient power, and are not considered for HEV applications. Cadmium is a toxic heavy metal that is very expensive to dispose of, reducing its desirability for hybrid applications (NREL, 2009).
- **Mercury cell**—The *mercury cell* was developed due to space exploration activities, the development of small transceivers, and miniaturized equipment where a power source of miniaturized size was needed. In addition to reduced size, the mercury cell has a good shelf life and is very rugged; it also produces a constant output voltage under different load conditions. There are two different types of mercury cells. One is a flat cell shaped like a button, while the other is a cylindrical cell that looks like a standard flashlight cell. The advantage of the button-type cell is that several can be stacked inside one container to form a battery. A cell produces 1.35 V.
- **Nickel-metal hydride**—Nickel-metal hydride batteries, used routinely in computer and medical equipment, offer reasonable specific energy and specific power capabilities. Their components are recyclable, but a recycling structure is not yet in place. Nickel-metal hydride batteries have a much longer life cycle than lead acid batteries and are safe and abuse-tolerant.

These batteries have been used successfully in producing electric vehicles and recently in the low-volume production of HEVs. The main challenges with nickel-metal hydride batteries are their high cost, high rate of self-discharge, very high gas consumption, and heat generation at high temperatures. There is a need to control hydrogen losses and their low cell efficiency (maybe as low as 50%, typically 60%–65%) (NREL, 2009).

- **Lithium-ion batteries**—Lithium-ion batteries are rapidly entering the laptop and cell phone markets because of their high specific energy. They also have high specific power, high energy efficiency, high-temperature performance, and low self-discharge. Components of lithium-ion batteries could also be recycled. These characteristics make lithium-ion batteries suitable for HEV applications. However, further development is needed to make them commercially viable for HEVs, including improved calendar and cycle life, higher cell and battery safety, abuse tolerance, and acceptable cost (NREL, 2009). A few important properties of lithium are given in Table 2.1.
- **Lithium-ion polymer batteries**—Lithium-ion polymer batteries with high specific energy (i.e., high energy per unit mass), initially developed for cell phone applications, also have the potential to provide high particular power for HEV applications. The other key characteristics of the lithium polymer

TABLE 2.1
Properties of Lithium

Property	Value
Atomic mass	6.941 g
Density	0.535 g/cm^3
Atomic number	3
Atomic symbol	Li
Appearance	Silvery white
Atomic radiance	152 pm
Melting point	180.5°C (356.9°F)
Boiling point	1,342°C (2,448°F)
Critical point	3,220K, 67 MPa
Heat of fusion	3.00 kJ/mol
Heat of vaporization	136 kJ/mol
Molar heat capacity	24.860 J/(mol·K}
Oxidation state	+1
Electronegativity[a]	0.98 (Pauling Scale)
Electronic configuration	[He]2s^1

[a] Electronegativity, symbolized by Y, is the tendency for an atom of a given chemical element to attract shared electrons while forming a chemical bond. As shown in Table 2.1, Lithium's electronegativity is 0.98 on the Pauling Scale. This value, 0.98, is based on calculation because electronegativity can't be measured, it must be calculated.

are safety and good cycle life. The battery could be commercially viable if the cost is lowered and higher specific power batteries are developed (NREL, 2009).

DID YOU KNOW?

Lithium (Li) is very reactive and may catch fire if it cannot sustain normal room temperature. Kerosine oil ignites a fire but is a haven for storing reactive chemicals such as Li.

The chemical reaction within a battery provides voltage. This occurs when a conductor is connected externally to the electrodes of a cell, causing electrons to flow under the influence of a difference in potential across the electrodes from the zinc (pessimistic) through the external conductor to the carbon (positive), returning within the solution to the zinc. After a short period, the zinc will begin to waste away because of the acid.

The voltage across the electrodes depends upon the materials from which the electrodes are made and the composition of the solution. The difference in potential between the carbon and zinc electrodes in a dilute solution of sulfuric acid and water is about 1.5 V.

The current that a primary cell may deliver depends upon the resistance of the entire circuit, including that of the cell itself. The internal resistance of the primary cell relies on the size of the electrodes, the distance between them in the solution, and the resistance of the solution. The larger the electrodes and the closer they are in solution (without touching), the lower the internal resistance of the primary cell and the more current it can supply to the load.

Note: When current flows through a cell, the zinc gradually dissolves in the solution, and the acid is neutralized.

In many operations, battery-powered devices may require more electrical energy than one cell can provide. Various devices may require either a higher voltage or more current, and in some cases, both. Under such conditions, combining or interconnecting enough cells to meet the higher requirements is necessary. Cells connected in series provide a higher voltage, while cells connected in parallel provide a higher current capacity. A combination series-parallel network of cells must be interconnected to provide adequate power when both voltage and current requirements are more significant than the capacity of one cell.

When cells are connected in **series**, the total voltage across the battery of cells is equal to the sum of the voltage of each of the individual cells. Four 1.5-V cells in series provide a total battery voltage of 6 V. When cells are placed in series, the positive terminal of one cell is connected to the negative terminal of the other. The positive electrode of the first cell and the negative electrode of the last cell then serve as the power takeoff terminals of the battery. The current flowing through such a battery of series cells is the same as that from one cell because the same current flows through all the series cells.

A battery connects cells in parallel to obtain a more significant current. In this parallel connection, all the positive electrodes are connected to one line, and all negative electrodes are connected to the other. Any point on the positive side can serve as the battery's positive terminal, and any point on the negative side can be the negative terminal.

The total voltage output of a battery of three parallel cells is the same as that for a single cell, but the available current is three times that of one cell; the current capacity has been increased.

Identical cells supply equal parts of the current to the load in parallel. For example, three different parallel cells produce a current load of 210 mA, and each cell contributes 70 mA.

A **series-parallel** battery network supplies power to a load requiring both voltage and current greater than one cell can provide. For example, groups of three 1.5-V cells are connected in series to provide the necessary increased voltage, and four series groups are connected in parallel to provide the required increased amperage.

Note that batteries are generally classified by their various characteristics. Parameters such as internal resistance, specific gravity, capacity, shelf life, "c" rate, MPV, gravimetric energy density, volumetric energy density, constant-voltage charges, constant-current charge, and specific power are used to describe and classify batteries by operation and type.

Regarding **internal resistance**, it is essential to remember that a battery is a DC voltage generator. As such, the battery has internal resistance or equivalent series resistance (ESR). In a chemical cell, the resistance of the electrolyte between the electrodes is responsible for most of the cell's internal resistance. Because any current in the battery must flow through internal resistance, this resistance is in series with the generated voltage. With no current, the voltage drop across the resistance is zero, so the full generated voltage develops across the output terminals. This is the open-circuit voltage or no-load voltage. If a load resistance is connected across the battery, the load resistance is in series with internal resistance. When the current flows in this circuit, the internal voltage drop decreases the battery's terminal voltage.

The ratio of a specific volume of liquid to the weight of the same volume of water is called the **specific gravity** of the liquid. Pure sulfuric acid has a specific gravity of 1.835 since it weighs 1.835 times as much water per unit volume. The specific gravity of a mixture of sulfuric acid and water varies with the strength of the solution from 1.000 to 1.830.

The specific gravity of the electrolyte solution in a lead-acid cell ranges from 1.210 to 1.300 for new, fully charged batteries. The higher the specific gravity, the less internal resistance the cell has and the higher the possible load current. As the cell discharges, the water formed dilutes the acid, and the specific gravity gradually decreases to about 1.150, at which time the cell is considered fully discharged.

The specific gravity of the electrolyte is measured with a **hydrometer**, which has a compressible rubber bulb at the top, a glass barrel, and a rubber hose at the bottom of the barrel. In taking readings with a hydrometer, the decimal point is usually omitted. For example, a specific gravity 1.260 is read simply as "twelve-sixty." A hydrometer reading of 1,210–1,300 indicates full charge; about 1,250 is half-charge; and 1,150–1,200 is complete discharge.

The **capacity** of a battery is measured in ampere-hours (Ah). The ampere-hour capacity is equal to the product of the current in amperes and the time in hours during which the battery supplies this current and is defined as the amount of current that a battery can deliver for 1 hour before the battery voltage reaches the end-of-life point. In other words, 1 Ah is the equivalent of drawing 1 amp steadily for 1 hour or 2 amps for half an hour. A typical 12-V system may have 800 Ah of battery capacity. The battery can draw 100 amps for 8 hours if fully discharged and starting from a fully charged state. This equals 1,200 watts for 8 hours (power in watts = amps × volts). The ampere-hour capacity varies inversely with the discharge current. The size of a cell is determined generally by its ampere-hour capacity.

The capacity of a storage battery determines how long it will operate at a given discharge rate and depends upon many factors, the most important of these are as follows:

- The area of the plates in contact with the electrolyte.
- The quantity and specific gravity of the electrolyte.
- The type of separators.
- The general condition of the battery (degree of sulfating, plates bucked, separators warped, sediment in the bottom of cells, etc.).
- The final limiting voltage.

A cell's shelf life is when it can be stored without losing more than approximately 10% of its original capacity. The loss of capacity of a stored cell is due primarily to the drying out of its electrolyte in a wet cell and to chemical actions that change the materials within the cell. A cell's shelf life can be extended in a cool, dry place.

The **"c" rate** is a current numerically equal to the Ah-rating of the cell. Charge and discharge currents are typically expressed in fractions of multiples of the c rate.

The **MPV (mid-point voltage)** is the cell's nominal voltage, measured when the battery has discharged 50% of its energy.

The **gravimetric energy density** of a battery measures how much energy it contains compared to its weight.

The **volumetric energy density** of a battery measures how much energy it contains compared to its volume.

A **constant-voltage charger** is a circuit that recharges a battery by sourcing only enough current to force the battery voltage to a fixed value.

A **constant-current charger** is a circuit that charges a battery by supplying a fixed current, regulating battery voltage.

In batteries, **specific power** usually refers to the power-to-weight ratio, measured in kilowatts per kilogram (generally, kW/kg).

2.7.6 Capacitors, Ultracapacitors, Supercapacitors

Electrical capacitors used for commercial applications are divided into two major groups—fixed and variable—and are named according to their dielectric. The most common are air, mica, paper, ceramic, and electrolytic capacitors. Table 2.2 compares these types.

TABLE 2.2
Comparison of Capacitor Types

Dielectric	Construction	Capacitance Range
Air	Meshed plates	10–400 pF
Mica	Stacked plates	10–5,000 pF
Paper	Rolled foil	0.001–1 μF
Ceramic	Tubular	0.5–1,600 pF
	Disk	0.002-0.1 μF
Electrolytic	Aluminum	5–1,000 μF
	Tantalum	0.01–300 μF

The fixed capacitor has a set capacitance value determined by its construction. The construction of the variable capacitors allows a range of capacitances. Within this range, the desired value of capacitance is obtained by some mechanical means, such as by turning a shaft (as in turning a radio tuner knob, for example) or adjusting a screw to adjust the distance between the plates.

The electrolytic capacitor consists of two metal plates separated by an electrolyte. The electrolyte, either paste or liquid, is in contact with the negative terminal, forming the negative electrode. The dielectric is a thin film of oxide deposited on the positive electrode, an aluminum sheet. Electrolytic capacitors are polarity sensitive (i.e., they must be connected in a circuit according to their polarity markings) and are used where a large amount of capacitance is required.

NREL (2009) points out that, like batteries, ultracapacitors, also known as supercapacitors, pseudocapacitors, electric double-layer capacitors, or electrochemical double-layer capacitors (EDLCs), are energy storage devices. They use electrolytes and configure various-sized cells into modules to meet various applications' power, energy, and voltage requirements. However, storage devices differ from batteries because ultracapacitors store energy electrostatically, whereas batteries store energy chemically. Ultracapacitors provide quick bursts of energy and improve capacitance by about two to three orders of magnitude (compared to an average capacitor) with a lower working voltage. Moreover, they can withstand hundreds of thousands of charge/discharge cycles without degrading. As an alternative energy source, ultracapacitors have proven reliable energy storage components for various electronic and portable devices such as AM/FM radios, flashlights, cell phones, and emergency kits. As ultracapacitor technology matures, it is being developed to function as batteries; for example, the vehicle industry is deploying ultracapacitors to replace chemical batteries. In the automotive industry, they have been integrated into EVs and HEVs to help alleviate stress and extend the life of the batteries.

An ultracapacitor polarizes an electrolytic solution to store energy electrostatically. Though it is an electrochemical device, no chemical reactions participate in its energy storage mechanism. This highly reversible mechanism allows the ultracapacitor to be charged and discharged hundreds of thousands of times.

An ultracapacitor can be viewed as two nonreactive porous plates, or collectors, suspended within an electrolyte, with a voltage potential applied across the collectors. In an individual ultracapacitor cell, the applied potential on the positive electrode attracts the negative ions in the electrolyte, while the potential on the negative electrode attracts the positive ions. A dielectric separator between the two electrodes prevents the charge from moving between the two electrodes.

Once the ultracapacitor is charged and energy stored, a load can use this energy. The amount of energy stored is vast compared to a standard capacitor because of the enormous surface area created by the porous carbon electrodes and the slight separation (10 Å) made by the dielectric separation. However, it stores a much smaller amount of energy than a battery. Because the charge and discharge rates are determined solely by its physical properties, the ultracapacitor can release energy much faster (with more power) than a battery that relies on slow chemical reactions.

2.7.7 Hydrogen Energy Storage

An alternative to storing energy in batteries is to store energy in hydrogen, which is produced using electricity through water electrolysis. Hydrogen storage is a key technology for advancing hydrogen and fuel cell technologies in applications, including stationary, portable, and transportation. After production, the hydrogen is stored for subsequent use as fuel to provide heat through combustion in homes and industry or electricity via a fuel cell or as a feedstock for other uses. Note that hydrogen has the highest energy per mass of any fuel; however, its low ambient temperature density results in a low energy per unit volume. This means developing advanced storage methods with the potential for higher energy density is required. Hydrogen can be stored physically either as a gas or a liquid. It can also be stored on the surfaces of solids (by adsorption) or within solids (by absorption).

DID YOU KNOW?

Hydrogen is an invisible gas with a color code, but you can't visually differentiate one color of hydrogen from another. So, within the energy industry, hydrogen is assigned color codes or nicknames to distinguish between the types of hydrogen. A guide to unlocking the current hydrogen color code is provided as follows:

- **Green hydrogen** is produced by clean (renewable) electricity sources.
- **Blue hydrogen** is low carbon hydrogen.
- **Grey hydrogen** is the most common form generated from natural gas.
- **Black/brown hydrogen** is derived from lignite and bituminous coal.
- **Pink hydrogen** is generated through electrolysis by nuclear energy.
- **Turquoise hydrogen** is produced using methane pyrolysis.
- **Yellow hydrogen** is made through electrolysis using solar power.
- **White hydrogen** is created through fracking.

2.7.8 Pumped Hydro Energy Storage (PHES)

Depending on whom you are talking to, one might say it is a "gravity battery," and others might call it a "water battery," but it is commonly known as pumped energy storage hydropower. These two monikers are correct when used to describe what this author calls nicknames for mechanical pumped hydro energy storage (PHES). And the nicknames are applicable because PHES acts similarly to a giant battery. After all, it can store energy and then release it when needed. Between 1960 and 1980, pumped storage plants for hydroelectric power in the United States were built. It is interesting to note that PHESs are the largest source of electricity storage technology in the United States—both in terms of capacity and number of plants. All remaining commercial-sized electrical storage uses thermal energy or batteries.

Simply put, PHES is made of two large pools of water, one higher than the other, acting like an hourglass to provide energy—the system stores energy in the form of the gravitational potential energy of water. The process involves pumping water into an elevated pool or storage reservoir during low electricity demand and low electricity prices, such as at night. When electricity demand is high, water flows downhill (gravity flow = water or gravity battery stored energy) from the reservoir through piping (penstock) to hydroelectric generators at a dam. Compared to conventional hydro plants, PHESs are more expensive in terms of construction costs; moreover, operating costs are costly because the water must be pumped uphill.

There are two types of pumped storage hydropower systems. An *open-loop system* has a continuous source of downstream water pumped uphill to an upper storage reservoir. More often or not, an open-loop plant pumps water that has already passed through the dam up to the storage reservoir above the dam. By contrast, *closed-loop systems* pump water from a lower storage reservoir, which is not continuously filled with water and is not connected to a flowing source.

How important and influential are these water batteries or gravity batteries? First, because of these units, grid overload is rare. When other energy sources like wind and solar power make more electricity than nearby industries and homes, the extra energy or power pushes water up into the top pool of water, where it sits, waiting and charging the water battery. Then, during a grid-overloading heat wave, the water is released from the top pool and flows down, acting as a prime mover (like a motor or diesel engine, etc.), spinning a turbine that creates electricity to energize lights, refrigerators, and other essentials.

We commonly think or feel that renewable energy is crucial for a clean energy future, but our needs go beyond environmental needs; the fact is, we need electrical power. This is not rocket science, mind-boggling, or a puzzle like finding the cure for cancer; it is essential in allowing and maintaining a good life, whatever that means. Water batteries can fill energy gaps when the weather limits the production of wind and solar energy.

Pumped storage hydropower provides more than 90% of U.S. energy storage. These projects are some of the most significant long-term energy storage systems today. These projects are placed in remote locations, and people living nearby generally know little about their importance until the lights go out.

EERE (2021) reports that in 2021, 18 states and all major regions of the country will use pumped-storage hydropower to store energy. Currently, South Carolina, Virginia, and California use most of these clean energy pools, and other states have projects to help grow the presence and usage of these water batteries.

In the United States, pumped storage hydropower can store more than 550 GWh of energy. To give an example of the power generated, consider that American video gamers use about as much energy as 85 million refrigerators or 5 million cars yearly. The 550 GWh of pumped storage hydropower in the U.S. provides this energy.

Presently, pumped hydro storage is the most efficient extensive energy storage system currently available, with a capacity of up to 80% and the ability to be turned on quickly. However, remember that energy storage, even the typical lead-acid storage battery, is not 100% efficient. Pumping water into a water battery's top reservoir requires a surge of energy. Another factor affecting efficiency in pumped hydro storage is evaporation.

Pumped hydropower storage sites provide more benefits to their locations than just storing water energy as a prime mover of electricity turbines. For instance, during a driving rainstorm that typically causes severe flooding in certain areas, the water pools accept excessive rainwater, which absorbs excess water and helps keep flooding to a minimum. This excess rainwater also adds to the pools' total water content, which reduces the need to use energy and pumps to pump up water levels in the pools. The pooled water can also provide water to fight wildfires and protect the area from extensive damage. During severe drought conditions, the pools can provide clean water for irrigating crops and a backup supply of safe drinking water.

In the United States, retrofitting existing pumped storage hydropower allows growth and increased energy output capacity without extensive new construction. This means that within the capacity of U.S. pumped storage—without any new construction—it can grow pumped storage as much as all other types of energy storage combined.

The bottom line of pumped hydro storage systems is that they can respond continuously to fluctuations in demand and sudden surges and provide energy at a low cost.

2.7.9 Compressed Air Energy Storage

Another mechanical energy storage option using cheap off-peak electricity is compressed air energy storage. An attractive feature of this technology is its relatively simple process. It can store electric energy as potential (compressed air) and be deployed near central power plants and distribution centers. In the process, a compressor is powered by available electricity to charge the air (compress), which is then stored in a chamber until the energy is needed. In response to demand, the stored energy can be discharged by expanding the stored air with a turboexpander generator.

2.7.10 Flywheel Energy Storage (FES)

Another mechanical energy storage system is the flywheel that, for short periods, can provide high power repeatedly. This system works by accelerating a flywheel (rotor) to an extremely high speed and maintaining energy in the system as rotational kinetic energy. The principle of energy conservation comes into play when energy is extracted from the system, reducing the flywheel's rotational speed. To add energy to the flywheel system, the speed of the flywheel must be increased. Safety is an issue with flywheels; they can disintegrate while in operation, causing substantial damage in areas where they are located.

2.8 PUMPED HYDRO ENERGY STORAGE AND AI

AI can be utilized in planning, feasibility study, discharge prediction, energy generation prediction, and planned maintenance. The various sub-sections of AI, including machine learning (ML), deep learning, artificial neural network (ANN), and fuzzy logic (converts an input to an output), are used for performance optimization, forecasting of parameters, monitoring and control optimization, policy and feature selection, feasibility studies, evaluation and capability assessment. AI and its sub-sections can forecast energy demand using historical consumption data and environmental variables such as temperature and humidity.

2.9 THE BOTTOM LINE

Energy comes in many different forms: kinetic energy, thermal energy, potential energy, chemical energy, and nuclear energy. Mechanical, electrical, electromagnetic (capacitors), and electrical chemical (battery) energy storage sources have been discussed. Other energy storage devices or systems exist, such as fireless locomotives, solid mass gravitational, hydraulic accumulators, and springs. Still, these are not discussed herein because, although they are storage devices or systems, they are small energy providers.

NOTE

1 Much of the information in this section is from Spellman, F.R., 2014. *Environmental Impacts of Renewable Energy.* Boca Raton, FL: CRC Press.

REFERENCES

EERE, 2013. *Severn Energy Literacy Principles*. Washington, DC: US Department of Energy.

EERE, 2021. *Pumped Energy Storage*. Washington, DC: Office of Energy Efficiency & Renewable Energy.

EIA, 2009. *What is energy*. Washington, DC: Energy Information Administration.

EIA, 2023. *U.S Energy Facts Explained.* Accessed 11/21/24 @ https://www.eia.gov/energy explained/US-energy-facts/.

NREL, 2009. *Cadmium is a toxic heavy Metal.* Golden, CO: National Renewable energy Laboratory.

3 Fossil Energy Sources

3.1 INTRODUCTION

When prehistoric plants and animals died and were gradually buried by layers of rock and soil, the result was the formation of nonrenewable fossil energy sources, including oil, coal, and natural gas. These different types of fossil fuels have been around for millions of years. Their formation depended on the combination of organic matter present, how long they were buried, and what temperature and pressure conditions existed as time passed.

Nealy, three-fourths of human-caused emissions came from the burning of fossil fuels. These fossil fuels are drilled or mined to produce electricity or refined for use as fuel for heating or transportation.

As I begin authoring this book, several significant events and situations warrant our attention. The promise that shifting from coal-generating power to renewable sources will create thousands of jobs has proven untrue. California is a good example: Promises were made that shifting to renewable energy sources would create 11,000 new jobs, but only 1,700 were delivered. Of the 1,700 jobs created, most went to consultants who served on a temporary, short-term basis.

The present state of the U.S. economy is second on the list of significant events or situations. The actual U.S. unemployment rate is 9–10+%, with increasing numbers of workers losing their jobs daily. For those still working in factory jobs, their future in these same types of jobs is threatened and, in some cases, doomed—these jobs are being done away, automated with, or outsourced overseas. Many of these jobs are unlikely to return to the U.S. Third on our list is the nominal world population clock that just posted a world population figure of 7.3 billion and counting—pointing to the increasing need for more resources, including energy.

How are these significant events and others related? What do they have to do with energy? What do they have to do with the so-called pending oil crisis (peak oil)? Finally, what do they have to do with renewable energy? I will discuss the interrelationships between oil and renewable energy and the significance of the events/situations described in the following.

SIDEBAR 3.1—HAITI

Let us discuss Haiti for a moment. Haiti is the poorest country in the Americas. Not only is it impoverished, but it is also the least developed. The 2010 earthquake only exasperated an already desperate situation in Haiti. Moreover, Haiti has consistently ranked among the most corrupt countries globally, and the ruthless criminal gangs pour salt in the wounds while roaming the country. Except for a few crops, it has no exports. Haiti has only survived because of the foreign aid it receives, mainly from the United States and lesser amounts from other countries.

DOI: 10.1201/9781003605430-4

Haiti's energy supplies must be imported; it has no indigenous supply of oil-derived energy. Haiti does have extensive exposure to sunlight, wind, tide, and wave energy. These sources can be employed in developing and producing renewable energy, all of which would lessen the need and expense for Haiti to import expensive hydrocarbon fuel products such as gasoline. The development of renewable energy sources in Haiti could go a long way in helping it lessen its dependence on worldwide generosity and loans. Indeed, economic growth could benefit if tied to decreased corruption, a shoring up of stable political footing, and increased human health and medical care.

Another real-world situation discussed here is interrelated to petroleum products and renewable energy production deals with the present and future state of the U.S. economy, which is the lack of viable jobs and lackluster progress in technological innovation. The United States has been the leader in technological development worldwide for some time. However, recent reports have surfaced pointing out that from 1999 to the present, the U.S. population grew by 32 million, or 1%, yet we had zero net job creation. Why? Take your pick; there are several reasons for our present economic condition. To begin with, we develop technology or technological advances in products such as iPads, computers, computer software, automobiles, farm machinery, furniture, textiles, wide-ranging applications of artificial intelligence (AI), and several other products. But most clothing, cell phones, computers, TVs, and many of our construction products (e.g., Chinese drywall) are produced in a foreign country. After we have spent significant amounts of capital, time, and volumes of expertise developing U.S. innovations, they are mimicked (copied) by foreign industries and produced at about nine times lower cost than it would cost to be made in the United States. These foreign producers have much cheaper labor costs and no regulatory pressure, and many receive financial assistance from their government. When these technological innovations leave the country, it creates an uneven playing field that simply does not teeter-totter in a direction favorable to us.

In addition to labor, regulatory, and related costs, the United States is no longer self-sufficient in oil production (although hydraulic fracturing has impacted). Our dependence on foreign energy sources is humiliating, shameful, and quite costly to us in many ways. For example, we have failed to grow jobs because we are no longer self-sufficient as a nation, except for producing agricultural products for the country's needs. The advent of extensive hydraulic fracturing will change this ratio; we will become self-sufficient. That is, until the fracked well runs dry.

Complicating any economic recovery we may achieve (in this writing, this seems doubtful and may be wishful thinking) is our dependence on foreign oil and the price of oil. One thing is sure: when economic conditions turn around and begin to climb in a positive direction, the cost of a gallon of gasoline ($3.18 and similar prices for diesel fuel at this writing) will increase, and increase, and increase again. The foreign energy producers are just sitting back, waiting for us to recover so they can turn the price increase lever one more level higher … and higher…, etc. These gasoline and diesel fuel price increases will quickly put a lid on any economic recovery; we will descend that slippery oil-soaked slope back to our present condition of stagnant malaise, or worse.

The continuing population growth impacts future fuel requirements and many other resources. Consider, for example, and as previously mentioned, at this very moment, the nominal world population clock reads 7.3 billion and counting. Population growth on this globe directly affects energy availability and use, as well as the pollution effects associated with using hydrocarbon fuels. The United States Census Bureau (USCB, 2008) predicts that the world's population in 2030 will be almost double that of 1980. Savinar (2010) predicts that oil production in 2030 will decline back to 1980 as worldwide demand significantly out-paces production. Bartlett (2004) claims that the rate of oil production per capita is falling and that the decline has gone undiscussed because a politically incorrect form of population control may be implied by mitigation.

It is essential to point out that one factor that has helped to lessen the effect of population growth on demand is the decline of the population growth rate since the 1970s. However, this is offset by increasing average longevity in developed nations. In 1970, the population grew to 2.1. By 2007, the growth rate had declined to 1.167% (CIA, 2008). However, the fact remains that the world's population is still growing, and pressure is continuing to be put on our limited energy supplies.

As mentioned, energy is only one of several issues concerning an increasing human population. The basic requirements of any population include air, water, food, shelter, and clothing. All these human needs are tied directly or indirectly to energy.

Regarding the air we breathe, the air itself is ubiquitous and a vital need naturally supplied to each of us. There are, however, issues related to the air we breathe that have been affected or tainted by energy use, namely, the air pollution that commonly accompanies the use of some forms of energy generation or use. There is continuing debate on hydrocarbon energy use and its impact on global climate change. While it is likely that we are undergoing global climate change, a general warming trend is not that apparent, nor has it been proven that air pollution caused by energy use is the primary causal factor of climate change. There can be little doubt, however, that human-caused (anthropogenic) pollution contributes to the current warming trend. Still, some critics argue that in a discussion about oil, oil resources, peak oil, and oil shortages, air and oil do not mix, and the use of hydrocarbons is not a primary cause of air pollution; they say there is no connection. In countering this argument, we suggest that the critic drive an automobile from any out-of-town location into downtown Los Angeles on one of those hot days in August or September, park their car, step out of their car, and take a deep breath of Los Angeles Basin air. When this occurs, the critic will get a mouthful and lungs full of more than just air and will readily realize that oil-derived vapors and air make a nauseous mix, making them intolerable or uncomfortable to inhale.

Along with good, clean air being one of those substances we cannot live without, we also need water. Along with its nutritional benefits, water was the primary energy producer in the early days of humans' existence. Water is still used in hydropower applications and, as such, is a renewable energy source (discussed later in the text) and a significant contributor to electricity generation. The problem with water, as with hydrocarbons, is that it is a finite natural resource. Today, we have the same amount of water on earth as we have always had. We cannot add to or delete from our present total water supply to any significant degree. The problem with our potable water is location; it is sparsely located or unavailable in many parts of the globe. Water is heavy, cumbersome, takes up a lot of space, and is not easily transported from location to location. Much of the globe's freshwater is polluted, not exclusively by energy use, but by the old mantra: We do not want it anymore, so throw it away. Often, the receiver of our throwaways (i.e.,) is and has been our ponds, lakes, streams, rivers, and oceans. Beyond the use of water to produce hydro-power energy, the interface between energy and water is often ugly, unsightly, and unhealthy in the form of polluted ponds, lakes, streams, and rivers from hydrocarbon-derived products such as pesticides, herbicides, rodenticides, and fertilizers. It seems odd that we carelessly pollute our air and water supplies, the two substances without which we can't live.

Water also interfaces with energy use and production in that it is not only used as an industrial cooling medium for machinery used in the refining, production, or use of hydrocarbon products but is also transported either by truck, tanker car, pipeline, or distributed via irrigation networks through power derived (e.g., pumps) from the use of hydrocarbon fuels such as gasoline and diesel fuel.

The connection between energy use and food has always existed and always will. From time immemorial to the present, we have expended one form of energy or another to hunt, gather, sow, and harvest our foodstuffs. In the current era, the mechanized farm machinery we are familiar with could not function without oil-derived fuels and lubricants. Moreover, many of the chemicals used to fertilize and protect crops from insects and diseases are oil-derived. Ours is an oil-based agriculture. And it is this oil-based agriculture that is driving the world's exploding population growth. As oil production increased, so did food production. As food production increased, so did the population. With population growth, the demand for food increased, escalating oil demand. Today, all the energy the agricultural industry uses to produce, store, and transport food comes from oil. Oil-derived products power tractors, harvesters, irrigation pumps, and energy for storage, trucking, railing, and flying those foods to the market.

Because oil-derived products such as diesel fuel and gas are essential to modern agriculture techniques, a fall in global oil supplies could cause spiking food prices and unprecedented famine in the coming decades (Goodchild, 2007). Pfeiffer (2004) contends that current population levels are unsustainable and that to achieve a sustainable economy and avert disaster, the U.S. population would have to be reduced by at least one-third and the world population by two-thirds.

DID YOU KNOW?

The largest consumer of fossil fuels in modern agriculture is ammonia production (for fertilizer) via the Haber process, which is essential for high-yielding intensive agriculture. The specific fossil fuel input to fertilizer production is primarily natural gas, to provide hydrogen via steam reforming.

The bottom line is that global population growth is only sustainable through our dependency on the infrastructure provided by oil. Whenever I make this statement to college-level environmental students, some accept my conclusion without much thought, and others give me that surprised and doubtful look. Early in my environmental engineering and health lecture presentations, I discovered that many students and others do not know much about the importance of oil and other nonrenewable energy sources and their impact on our present-day society.

"Well, we simply need to outlaw the nonrenewables and shift to wind, water, solar, and maybe even nukes and do away with oil in general…period."

Have you ever heard this statement or something close to it about doing away with oil … period?

Because I write and lecture about energy and the benefits, disadvantages, and ugly aspects of using fossil fuels, I often hear that we should shut down fossil fuels, especially oil and coal production and use.

So, I have developed a standard reply when critics, students, and the people on the street raise the topic of eliminating fossil fuels.

My reply. I straightforwardly point out that more than 500,000 products we are accustomed to using daily are manufactured using oil as a material or ingredient. To illustrate my point, I distribute a copy of and discuss Table 3.1.

TABLE 3.1
Products Manufactured using Oil or Oil Byproducts

Plastic Cups	Water Skies	Plastic Forks	Dresses
Transparent tape	Lens cleaner	Lip balm	Balloons
Water pipes	Rubber boots	Pantyhose	Shoes
Hair dryers	Oil filters	Bandages	Pens
Upholstery	Insulation	Running shoes	Bubble gum
Vaporizers	Trash bags	Chewing gum	Glue
Lipstick	Floor wax	Footballs	Basketballs
Caulking	Shoe polishes	Hair coloring	Crayons
Umbrellas	Dice	Golf balls	Mops
Computers	Fertilizers	Rubber bands	Tires
Petroleum jelly	Volley balls	Hearing aids	Perfumes
Nail polish	Synthetic clothing	House paint	Carpet
Loudspeakers	Movie film	Combs/brushes	Stereos
Electrician's tape	Air conditioners	Fishing lures	Car bodies
Electric blankets	Toothpaste	Denture adhesives	Floor wax
Motorcycle helmets	Fishing rods	Tape recorders	Antiseptics
Shower curtains	Roofing paper	Heart valves	Ice chests
Artificial turf	Paint brushes	Rubber cement	Toilet seats
Boats	Cold cream	Contacts	Telephones
Ammonia	Candles	Bras	Cameras
Fiberglass	Solvents	Purses	Cortisone
Car batteries	Eyeglasses	Parachutes	Plastic wrap
Insect repellent	Deodorant	Antihistamine	Compact disks
Inks of all types	Plywood adhesive	Hair curlers	Plastic chairs
Dishwasher liquid Shirts (non-cotton)	Refrigerator seals	Thermos bottle	Plastic toys

Does the information in Table 3.1 assuage the feelings and understanding of those who oppose using fossil fuels?

It depends.

Depending on what?

Depending on the mindset of those opposing fossil fuels, I have received a few responses in which the fossil fuel opponents and the advocates for using renewable energy sources had yet to learn about the vast number of everyday products that result from oil and oil byproducts. When they read the list of products, they morph into the silent mode, realizing that they do not want to live without many of the products listed in Table 3.1.

I also heard from a few respondents who have no problem using oil and its byproducts if fossil fuel energy sources can be developed more efficiently with little to no collateral pollution effects.

I have also heard this refrain from a few others, but with the caveat that AI should monitor, operate, and control energy production and usage.

Then, there is the most common response or statement about fossil fuels, in which respondents state that we have a limited supply of fossil fuels and may have reached **peak oil**; thus, we must advance the development and usage of renewable energy sources.

Let us discuss peak oil and its relevance in this discussion.

3.2 PEAK OIL

According to USDOE (2007), the *peak oil theory* states that the world's oil production rate will reach a maximum rate and then enter a terminal decline. This concept is based on the observed production rates of individual oil wells and the combined production rate of a field or related oil wells. This is not to say that the world is running out of oil, but it does signal the end of abundant and cheap oil on which all industrial nations depend (Campbell & Laherrere, 1998). Peak oil, also known as the Hubbert peak theory, concerns the long-term rate of conventional oil (and other fossil fuels) extraction and depletion. In a 1956 paper he presented at the American Petroleum Institute meeting, M.K. Hubbert proposed that oil production in the United States would peak between 1965 and 1970 (Hubbert, 1956). U.S. oil production peaked in 1971 and has been decreasing since then. Hubbert's theory is subject to continued discussion because of the potential effects of lowered oil production and the ongoing debate over aspects of energy policy. Opinions on the impact of passing Hubbert's peak range from the faith that the market economy will produce a solution, assuming significant investments in alternatives will occur before a crisis, to the prediction of doomsday scenarios of a global economy unable to meet its energy needs (Gwyn, 2004). This thesis postulates that oil prices will escalate and retreat as other fuel and energy sources are developed and used (CERA, 2006). Pessimistic predictions of future oil production operate on the thesis that either the peak has already occurred (Deffeyes, 2007), oil production is at the peak, or it will appear shortly (Koppelaar, 2006).

Although many experts were skeptical of his prediction, Hubbert proved correct when U.S. oil production peaked in 1971 (Brandt, 2007). Hubbert's prediction remained accurate even after discovering the Prudhoe Bay oil field (after 1971), with its significant volume of oil still not enough to bring U.S. oil production out of a long-term decline (USDOE, 2007).

Peak oil production and subsequent global production decline (based on optimistic assessments) are forecast to begin by about 2020 or later. Remember that the exact date of world oil peaks is a best guess; the actual date is uncertain, complicating the decision-making process. The decision-making process refers to deciding what or which alternative sources of energy to pursue to prevent the economic and social impacts of oil peaking from being chaotic. A fundamental problem in predicting oil peaking is uncertain and politically biased oil reserve claims from many oil-producing countries. In some instances, the veracity of such claims may be questionable (Hirsch, 2005).

If we accept predictions about peaking oil, the possible effects and consequences cannot be drawn and displayed without using an extensive brush. Few would argue against the fact that the widespread use of fossil fuels has been one of the most important agents of economic growth and prosperity since the Industrial Revolution. Fossil fuels have allowed humans to participate in energy consumption at a greater rate than it is being replaced. Some believe that mankind is in for a momentous paradigm shift when oil production decreases. This shift will cause humans and our modern technological society to change drastically. The actual impact of peak oil will depend heavily on the rate of decline and the development and adoption of effective alternatives. If alternatives are not forthcoming, the products produced with oil (including fertilizers, detergents, adhesives, solvents, and most plastics) will become scarce and expensive. In 2005, the USDOE published a report (known as the Hirsch report) with a stark warning: "The peaking of world oil production presents the U.S. and the world with an unprecedented risk management problem. As peaking approaches, liquid fuel prices and price volatility will increase dramatically, and without timely mitigation, the economic, social, and political costs will be unprecedented. Viable mitigation options exist on both the supply and demand sides, but to have a substantial impact, they must be initiated more than a decade before peaking (USDOE, 2005 Intro.)."

3.2.1 Hirsch Report

After careful analysis, the Hirsch report came to several conclusions and three scenarios (DOE, 2005):

3.2.1.1 Conclusions

1. World oil peak will happen. Some forecasters predict it within a decade, others later. For example, Simmons (2006) predicts global peak oil production during 2007–2009, Deffeyes (2008) predicts global peaking before 2009, and Laherrere (1998) predicts global peaking sometime between 2010 and 2020.
2. Oil peaking could cost economies dearly—particularly that of the United States.
3. Oil peaking presents a unique challenge. Previous transitions were gradual and evolutionary, while oil peaking will be abrupt and revolutionary.
4. The real problem is that liquid fuels for transportation—motor vehicles, aircraft, trains, and ships—are not readily available substitutes.
5. Mitigation efforts will require substantial time—an intense effort over decades.
6. Both supply and demand will require attention—higher efficiency can reduce demand, but large amounts of substitute fuels must be produced.

7. It is a matter of risk management—early mitigation will be less damaging than delayed mitigation.
8. Government intervention will be required—otherwise, the economic and social implications would be chaotic.
9. Economic upheaval is not inevitable—without mitigation, peaking will cause significant upheaval, but given enough lead time, the problems are soluble.
10. More information is needed; effective action requires a better understanding of several issues.

3.2.1.2 Scenarios

The Hirsch report listed three possible scenarios: waiting until world oil production peaks before taking crash program action leaves the world with a significant liquid fuel deficit for more than two decades; initiating a mitigation crash program 10 years before world oil peaking helps considerably but still leaves a liquid fuels shortfall roughly a decade after the time that oil would have peaked; or initiating a mitigation crash program 20 years before peaking appears to offer the possibility of avoiding a world liquid-fuels shortfall for the forecast period.

3.3 A MALTHUSIAN CATASTROPHE?

The peaking of oil presents several questions and a stark reality. Questions include: How will shortages and prices of oil affect you and your family? What about the effects on your community? How will oil shortages and prices affect industry and the economy? The stark reality is that the decline of oil production is not temporary. For those sitting back and waiting for some White Knight to come charging into the country on his great steed and then suddenly throwing and sinking his great platinum-clad spear into the ground to free a gigantic black stream of crude oil that shoots upward and forms a giant shaft reaching toward the clouds, it just isn't going to happen; it's literally a pipedream of mythical proportions.

The day is fast approaching when the inability to meet demand will raise fuel costs. Concurrently, the price of every sector of the economy is going to be affected in some way, whether it is a shortage of new material in manufacturing goods such as those listed in Table 3.1, the costs of transporting those goods to the market, storage in climate-controlled facilities, the machinery used in the construction of those facilities or structures of any kind, most fertilizer and pesticides used to maintain production yields in agriculture, and a list of other dependencies too long to list here. However, that is only the beginning of the impact of such a situation on our society. The resulting "domino effects" will generate enormous price increases across the board for all products and services, making many economically impractical. While the lack of oil supplies will affect the manufacture or availability of all these products and services, food will most likely be affected. An adequate food supply is essential not only to maintain world peace and prevent famine but also to ensure the health and well-being of the populace. This populace is increasing and will continue to grow even though the liquid fuel spigots are running dry. In related terms and meaning, this deadly doom and gloom scenario has been described as a Malthusian catastrophe. This term denotes a forced return to subsistence-level conditions once energy use has outpaced production. The tragic irony is that if this occurs and we are not prepared, for the first time in history, third-world subsistence economies will have the one-up on many of the rest of us who, in our lifetimes, have never planted, tended, and harvested any crop.

3.3.1 Blame or Tame?

Whenever anything goes wrong on this globe of ours, it is a human tendency for someone, anyone, or all of us to point a long finger of blame at someone or anyone besides ourselves. This finger-pointing is undoubtedly the case regarding energy shortages and energy-derived pollution problems. Many critics point the fingers of blame for our present and pending shortage of liquid fuels on the environmentalists. However, as USEPA (1974) points out, pointing the finger of blame at the environmental movement is misdirected. Although it is true that environmental regulations have caused some increased energy demand and have restricted supply, it is essential to recognize that other factors have been more significant. As mentioned, these include rapidly escalating demand for energy, energy pricing policies, oil import quotas, lack of incentives to invest in domestic energy facilities, and depletion of domestic oil and gas reserves.

Now we realize that the price of energy has increased. The summer of 2008, with its $4 plus gas prices, got many people's attention. The higher prices for heating oil, natural gas, and other oil-derived products have also garnered our attention. The problem is we tend to blame the higher costs of petroleum-based products on those who sell the products; we think they are greedy, and that is about the extent of our concern. We fail to understand that policymakers fail to understand the concept of exponential growth and formulate and pass a sensible long-term energy strategy. The result of these failures, and others, is to create an escalating energy crisis as time continues to run out. Unfortunately, most people are either oblivious to this or believe that some new energy source will magically appear—again, some White Knight will ride into town with the grand solution.

As noted, we always need to blame someone or anyone for things we do not like. However, blame is never the solution to any problem. Instead, we must get a hold of the issue and tame it. We need to think it through. Before doing that, however, we need to know the extent of the issue—the liquid fuel crisis.

3.3.2 Weathering the Storm

With the energy crisis pending, we may have three choices: We can mine for more oil, aggressively pursue energy conservation, or develop renewable energies. Renewable energy sources are needed because of limited fossil fuel resources, increased demand, and environmental impacts. Natural processes can constantly regenerate renewable energy at reasonable rates. Renewable energy technologies turn these fuels into usable forms of energy—most often electricity, but also heat, chemicals, or mechanical power. Some possibilities include:

- Biomass
- Wind Energy
- Hydro Power
- Tidal Energy

- Wave Energy
- Solar Energy
- Geothermal Power
- Hydrogen Fuel Cells

3.4 ALTERNATIVE/RENEWABLE ENERGY[1]

The worldwide use of liquid fossil fuels, their decreasing availability, politics, fears about climate change, and other economic forces are pushing for substitute, alternate, and renewable fuel sources. This is the case, of course, because of the current and future financial problems that \$4+/gal gasoline has generated (especially in the United States) and the perceived crisis developing with high carbon dioxide emissions, the major contributing factor of global climate change.

Before proceeding with an introductory discussion of alternative/renewable energy sources, it is essential to clearly distinguish between the two terms, the current buzzwords, alternative and renewable energy. *Alternative energy* is an umbrella term that refers to any usable energy source intended to replace fuel sources without the undesired consequences of the replaced fuels. The term "alternative" presupposes an undesirable connotation (for many people, fossil fuel has joined that endless list of four-letter words)—that is, alternative energy is fueled energy that does not use up natural resources or harm the environment. Examples of alternate fuels include petroleum as an alternative to whale oil, coal as an alternative to wood, alcohol as an alternative to fossil fuels, and coal gasification as an alternative to petroleum. The key point in understanding the term alternative energy is that these fuels need not be renewable.

Renewable energy is generated from natural resources—such as sunlight, wind, water (hydro), ocean thermal, wave and tide action, biomass, and geothermal heat—which are naturally replenished (and thus renewable). Renewable energy resources are inexhaustible—replenished at the same rate as they are used—but limited in the amount of energy available per unit of time. If we have not come full circle in our cycling from renewable to nonrenewable back to renewable, we are getting close to that. Consider, for example, that in 1850, about 90% of the energy consumed in the United States was from renewable energy resources (hydro power, wind, burning wood, etc.). However, the United States heavily relies on nonrenewable fossil fuels, natural gas, oil, and coal.

RENEWABLE ENERGY—A JEKYLL AND HYDE PERSONALITY?

Dr. Jekyll Side (Good)	Mr. Hyde's Side (Bad)
Sustainable	Intermittent in time (some)
Fuel is free	Upfront costs
Inexhaustible	Distant from populations (some)
Non-polluting (to a point)	May involve environmental issues
Easily applied to off-grid/distributed power use	Large footprint

Most renewable energy is used for electricity generation, heat in industrial processes, heating and cooling buildings, and transportation fuels. In the United States, electricity producers (utilities, independent producers, and combined heat and power plants) consumed 22% of the world's total renewable energy in 2012 for producing electricity. Most of the remaining 78% was biomass consumed for industrial applications (principally paper-making) by plants producing only heat and steam. Biomass is also used for transportation fuels (ethanol) and residential and commercial space heating. In 2023, the consumption of U.S. renewable consumption by source was (EIA, 2023):

- Hydropower 10%
- Biomass Wood 23%
- Biomass Waste 5%
- Biomass Biofuels 32%
- Wind 18%
- Geothermal 1%
- Solar 11%

From this listing, it is evident that there are (currently) five primary forms of renewable energy: solar, wind, biomass, geothermal, and hydroelectric energy sources. Each holds promise and poses challenges regarding future development.

The United States imports approximately 50% of its oil. Replacing some of our petroleum with fuels provided by solar power or made from plant matter, for example, could save money and strengthen our energy security.

Renewable energy is plentiful, and the technologies are improving all the time. There are many ways to use renewable energy. Our focus should be to find and develop a renewable source of liquid fuels (e.g., produced from biomass) because our economy runs on liquid fuels. Most non-liquid renewable energies will not (at the present time) provide power for airplanes and heavy trucks; that is, solar, wind, hydro, geothermal, wave and tidal, and fuel cells cannot power the primary vehicles we use today. Note that we did not list trains. Remember that today, many trains are powered by diesel-power or diesel-electric systems. Trains could be retrofitted to steam power developed by burning coal and wood products (a step back into the past); however, we cannot do this to power heavy trucks and airplanes.

Most Americans still do not understand that we are running short of the fuels we use daily—the fuels that made us what we are today. We built the world's greatest economy on oil that sold for $3–4 per barrel. That is no longer the case. To maintain our living level, the so-called good life, we must find and develop renewable energy sources to power our homes and secure our future.

3.5 THE BOTTOM LINE

Fossil fuels are ubiquitous in global society, and there is considerable pressure to switch to renewable energy sources because of their environmental impacts, including global warming and pollution, and their inability to be renewable. Both fossil fuel and renewable energy can be made more efficient by employing AI. In renewable energy, for example, AI can assist in weather forecasting and optimum times for maximizing solar, wind, and tidal energy sources.

NOTE

1 Much of the information and data in this section is from EIA's *Renewable Energy Trends* 2004. Accessed 06/12/24 @ http:www.eia.doe.gov/cneaf/solar.renewables/page/trends/rentrends04.html. EIA's, 2007. *How Much Renewable Energy Do We Use?* Accessed 06/12/24 @ http://tonto.eia.doe.gov/energy_in_brief/renewable_energy.cfm.

REFERENCES

Bartlett, A.A., 2004. Thoughts on long-term energy supplies: Scientists and the silent lie. *Physic Today* 57(7): 53–57.

Brandt, A.R., 2007. Testing hubbert. *Energy Policy* 35(5): 3074–3088.

Campbell, C.J., and Laherrere, J.H., 1998. The end of cheap oil. Scientific American xi: 21–26.

CERA, 2006. CERA says the peak oil theory is faulty. *Energy Bulletin*. Accessed 01/16/24 @ https://www.energybulletinnet/22381.html.

CIA, 2008. *The World Factbook*. Accessed 01/17/24 @ https://www.cia.gov/library/publications/the-world-factbook/geos/X.html.

Deffeyes, K.S., 2007. *Current Events-Join Us as We Watch the Crisis Unfolding*. Princeton University: Beyond Oil. Accessed 01/16/24 @ https://www.princeton.edu/hubbert/current-events.html.

Deffeyes, K.S., 2008. *Hubbert's Peak: The Impending World Oil Shortage*. Princeton: Princeton University Press.

DOE, 2005. *Hirch Report*. Washington, DC: US Department of Energy.

EIA 2007. *Renewable Energy Trends in Consumption and Electricity 2007*. Washington, DC: Energy Information Agency.

EIA, 2023. *Monthly Energy Review*, April 2015. Washington, DC: Energy Information Agency. Accessed 11/11/24 @ https://www.eia.doe.gov/emeu/mer/pdf/pages/sec10_3/pdf.

Goodchild, P., 2007. *Peak Oil and Famine: Four Billion Deaths*. Accessed 01/17/24 @ https://www.countercurrents.org/goodchild291007.htm.

Gwyn, R., 2004. *Demand for Oil Outstripping Supply*. Accessed 1/15/24 @ https://www.commondreams.org/views04/0128-10.htm.

Hirsch, R.L., 2005. *The Mitigation of the Peaking of World Oil Production Summary of an Analysis*. Accessed 01/16/24 @ https://www.peakoil.net/USDOE.html.

Hubbert, M.K., 1956. *Nuclear Energy and the Fossil Fuels 'Drilling and Production Practice'*. Accessed 01/08/24 @ https://www.hubbertpeak.com/hubbert/1956/1956.pdf.

Koppelaar, R.H.E.M., 2006. *World Production and Peaking Outlook*. Accessed 01/12/24 @ https://peakoil.nl/wp-content/upoads/2006/09/aspnl_2005_report.pdf.

Laherrere, J.H., 1998. The end of cheap oil. Scientific American. xxii: 166–174.

Pfeiffer, D.A., 2004. *Eating Fossil Fuels*. Accessed 10/11/24 @ https://www.fromthewilderness.com/free.ww3/100303_eating_oil.html.

Savinar, M., 2010. *Are We Running Out? I Thought There Was 40 Years of the Stuff Left*. Accessed 01/17/24 @ https://www.lifeaftertheoilcrash.net.

Simmons, M., 2006. *Twilight in the Desert: The Coming Saudi Oil Shock and the World Economy* 2005. New York: Wiley.

USCB, 2008. *Total Midyear Population for the World:* 1950–2050. Accessed 01/17/24 @ https://www.census.gov/compendia/statab/tables/09s1285.xls.

USDOE, 2005. *Peaking of World Oil Production: Impacts, Mitigation, & Risk Management*. Accessed 01/05/24 @ https://www.netl.doe.gov/publications.others/pdf/Oil_Peaking_NETL.pdf.

USDOE, 2007. *Peak Oil—The Turning Point*. Washington, DC: United States Department of Energy.

USEPA, 1974. *Liquid Fuels or Environment*. Washington, DC: US Environmental Protection Agency.

4 AI for Energy

4.1 WHAT IS ARTIFICIAL INTELLIGENCE?

Artificial intelligence (AI) is a transformative branch of computer science dedicated to creating intelligent agents capable of representing human-like thought functions. This book defines AI as the simulation of human intelligence by mimicking the human brain for analysis, environmental system modeling, and decision-making.

AI has emerged as a transformative force in various industries, and energy production and use is no exception. Moreover, AI is growing in use and applications, so you can hardly avoid hearing about AI today. For example, most of us have heard about or have used Alexa or Siri, the voice-controlled virtual assistants. However, not everyone using these virtual assistants understands their source, what wake word detection is all about, how voice processing works, and how the machine learning (ML) models use custom triggers and generation AI—however, the user proceeds without asking any questions. Whatever system is being used, the user finds that it either operates or does not operate, which is about the extent of many users' AI knowledge. This mindset, a way of thinking and operating, is a typical modus operandi in our daily ride through life. For example, most people have no idea how a vehicle motor can power their vehicle. Still, they turn the ignition key or push the start button anyway, start the car, and move along without knowing anything about vehicle engines. Today, many of those using AI or ML, Deep Learning (DL), or the Internet of Things (IoT) do so without thinking about it; it's just like breathing, eating, sleeping, and sky diving, just not thought about until we run out of electrical power in the first case or the other cases when we need oxygen, food, the ability to sleep and realize that when we jump, we forget our parachute.

DID YOU KNOW?

Another factor that users of AI must be aware of is that much of the data contained in many applications are in the "Black Box Mode," meaning that various sources contain data that is obscure and difficult to analyze, whereas, in contrast, in the "White Box Mode" is where data is transparent and can be analyzed by the user.

In addition to AI-assisted triggers used in virtual assistants, AI is everywhere, reaching for more in our homes, cars, and phones. For example, AI makes social media content recommendations and detects finance fraud, and it is employed in transportation, manufacturing, healthcare, and education.

Those involved with energy production, storage, and use should (must) learn AI. Why?

DOI: 10.1201/9781003605430-5

It is the future, like the IoT.

Like ML.

Like DL.

All sorts of digitation are occurring now and more so in the future.

It is the future. This is the future. This is now!

Get aboard the digital train, and do not worry about a caboose—riding at the end of the train is not an option. No, not for anyone, not for technology, like energy.

I have been lucky. I have lived during the advancement of the digital age, and the passage of time is all about adapting; learning to adapt is how we move on and survive. Moreover, adapting is learning.

How about you? Are you ready for the AI revolution?

Can you adapt?

Digital enlightenment, adaptation, and switchover from typewriter to personal computer began in 1984. I remember this year well because I was introduced to the Apple Macintosh computer. The advent of an affordable personal computer available for use in preparing my college environmental engineering and health and for a few medical nurses courses I taught was beneficial, especially in fashioning my lecture notes, handouts, and examinations.

As the years passed and my lecture attendance continued to grow—the students felt I was a somewhat new hat (a bit different) and quickly advertised that I was a "professor easy" in favor of a wide range of students; another item I noticed is that by 2020, almost every student possessed a laptop computer, smartphone, or smartwatch.

This 2020 awakening, for me, also was my first exposure to AI available at the fingertip. That Superbowl ad was what some called a "resource" for humanity from the computer giant IBM. The ad conveyed that there was now a computer for the rest of us. I quickly became a fan of the Mac but did not express my satisfaction with Mac tattoos, a "MacQuarium," Apple log haircut, paper Macs, Mac-inspired much is, and more—no, for me, the old hat was the call of the age, to a degree, at least.

One drastic change I noticed in the classroom, delivered via AI, is that the quality of students' reports, papers, theses, and dissertations was primarily well-written and grammatically correct—AI-assisted work.

Does this AI-assisted student work bother me? No, it is smooth sailing when I must read their work. And, besides, who wants to ride in the caboose in this digital age?

I currently use Microsoft Word and other associated applications in my work—it is just a personal choice—but Apple hooked me into the computer age and Microsoft into the AI age … and hopefully, more to come.

4.2 THE BOTTOM LINE

The future is occurring faster than expected. AI and computer applications, both those we use daily and those yet to be discovered and applied, are part of the changing times. AI is being increasingly applied to energy production, use, and storage to enhance efficiency, improve process operations, and optimize resource utilization. We can ride the AI train or get left behind, standing in a forest without trees.

5 The Digital Energy Professional

5.1 DIGITALIZATION OF ENERGY PRACTICE[1]

When artificial intelligence (AI) is included, energy systems become more dependable, efficient, and sustainable. The key is that we must adapt. Do not ride the caboose; ride the AI locomotive instead.

The point is that digitization "greases" many skids. We need to apply the grease.

The advantages and disadvantages of incorporating AI, Machine Learning (ML), and Deep Learning (DL) into industrial and domestic operations are not to ignore Internet of Things (IoT), which several industries use. Considering this ongoing incorporation of learning digitation, evolving technology, and implications, it is filled with many disadvantages if one does not know, read, and understand the language—the "new" language.

I always recall what Voltaire said so long ago, "If you wish to converse with me, please define your terms?"

AI and its associates have created a new language we must define, understand, grasp, and use. This language is in addition to the language associated with energy engineering.

New language?

Yes, indeed—for sure. Let me present a real-world digital-based employment example. Let us say that you are the chief engineer of a large engineering firm. You need to hire a new engineer-in-training (EIT). You have sorted through a half dozen job applicants' resumes and company job application forms. You are impressed with the two candidates' resumes and other information. You schedule personal interviews with the two candidates. In the morning, during your interview with Candidate 1, you were pleased with her extensive education and experience and her answers to all your queries, especially when she could correctly answer and define all the digitation terms you threw at her.

So, after the Candidate 1 interview, you were convinced that if her references were positive, she would be the one to hire.

Later in the day, during the Candidate 2 interview, you were just as impressed with this young woman's background, presentation, appearance, attitude, and the rest. You could feel the onset of a puzzling question: which candidate should I hire?

Before the interview ended, Candidate 2 answered this question for you. After preliminary greetings and a review of Candidate 2's background, resume, college studies and course grades, and some general small talk, you decided it was time to present technical questions—the same questions you posed with Candidate 1; the difference between the experience is that it only took you one question and the

DOI: 10.1201/9781003605430-6

response to it to tell you all you needed to know about Candidate 2 and your determination of whom to hire.

You stated that in filling the open position, it was important that the hired person be utterly literate in computer operations and have a strong background in AI, robotics, and digital systems, all of which are increasingly used in environmental engineering.

So, your first and only queries were as follows:

"Ok, in a database management system, better known as DBMS, what is the difference between indexing and hashing?"

There was a moment of silence.

There was more than a moment of silence.

You looked at Candidate 2 and noticed the lines on her forehead, her eyes wide, and her head turned slightly to the left. She did try to answer your question but stumbled, her voice changing. Spasmodic dysphonia (shaky voice) was apparent. She fumbled for an answer and instead kept looking away from you.

So, after a short period, you stood up, reached your hand out to shake hands, and waited for her to stand, shake hands, and depart.

She slowly got up from the chair and gave you a weak handshake, often described as a limp experience. Then she turned and left the room, never to be seen again.

You hired Candidate 1 the following day and have not looked back since.

DID YOU KNOW?

The difference between indexing and hashing is as follows:

Indexing is a technique that quickly retrieves records from database files and is generally used to optimize database performance.

Hashing is a faster and more flexible technique that allows searching for the desired data location on a disk without using an index structure.

5.2 SINK OR SWIM

Many humans are under extreme emotional stress when their job environment or potential new job environment shifts from "old school" operations to the latest digital wave. They have trouble with the wave and cannot surf it; they are lost somewhere in the "Cloud," so to speak. In this digital age, we must quickly adapt, get laid off, or change careers, a classic sink-or-swim scenario.

Only some of the news could be better. We must accept that we live in a time of excitement (hopefully, this is the case for most of us), and technological innovations and the digital sector drive transformative change. Business paradigms are changing and shifting. Digital technologies are reshaping the traditional workplace, enabling us to work from home rather than at an office or other worksite. The latest developments in AI, generative AI, and related advances are growing the frontiers of the digital revolution. Also, you might think or suppose that the COVID-19 pandemic has slowed innovation and transformation, but the opposite is true; digital conversion is accelerating. The future is here, and you are riding the wave or left to ride the caboose in this digital age—you must adapt, or otherwise ....

5.3 DIGITAL METAMORPHOSIS

To ride the digital wave (aka the digital train), we must, as stated, adapt, adapt, or otherwise. In this adaptation process, we need to learn the nuts and bolts and the applications of AI and its subsets, including ML, DL, Natural Language Processing (NLP), Expert Systems, Computer Vision, and the IoT. While traditional web technologies (i.e., the programming language, database, server administration, servers, markup language, style rules, and cloud-side codes) remain at the top of the ladder, in the majority of usage at present, the engine in digital service is a new wave of cloud-based services, following the engine (aka the Flagship part of the digital train, at present) in the digital train is the mobile internet, Big Data, design thinking (DL), AI, robotics, advanced neural machine learning, augmented-reality technology, and additive manufacturing, in that order, which are moving (the movers) up the train of usage, rails of popularity, and innovation.

DID YOU KNOW?

Natural Language Processing (NLP) focuses on understanding and generating human language. Expert Systems use rules and knowledge to make decisions in specific domains. Computer vision involves analyzing visual information to understand and interpret images and videos.

Let's look at the current components that make up the digital train and list their function:

- **Cloud-based services**—Traditionally, organizations had to purchase and maintain their physical servers. Now, cloud-based servers run in the cloud on infrastructure owned by the cloud service provider; they remove the need to buy, run, and manage physical servers.
- **Mobile internet**—It is also known as *mobile web* for handheld devices, such as smartphones or feature phones (i.e., a pre-wireless smartphone) through a wireless mobile network.
- **Big data architecture**—It is a systemic design and structure of a data management system.
- **The Internet of Things (IoT)**—It is a network of physical objects embedded with sensors, software, and other technologies to connect and exchange data with other devices and systems via the Internet.
- **Digital thinking (DT)**—It is a non-linear, iterative aspect or process that challenges assumptions, redefines complications, and creates groundbreaking models for testing.
- **AI technology**—It enables computers and machines to simulate human learning, understanding, problem-solving, decision-making, originality, and independence.
- **Robotics**—Robotics follows the digital train. Within environmental engineering and science, it focuses on robotic automation algorithms used in ecological functions such as performing vital tasks related to air pollution monitoring, water, soil, wildlife monitoring, and waste cleanup operations.

- **Advanced neural ML**—The "name neural" is based on a model of how the human brain functions and comprises a collection of processing units called *nodes*. Advanced Neural ML is learning without definite instructions and is primarily used in deep learning.
- **Augmented-reality (AR) technology**—Augmented-reality (AR) technology is following in the digital train. AR is popular because it presents an enhanced real-world version via computer-generated digital information, including visual, sound, and other sensory elements.
- **Additive manufacturing**—The final section of the digital train, not the caboose presented in this discussion, is known as additive manufacturing (aka 3D printing). In operation, additive manufacturing constructs a three-dimensional object from a CAD or digital 3D model.

Note: To be successful in AI-based environmental engineering, you must know the terminology, live it, write it, and use it in speech. Therefore, the keywords and abbreviations used in this book are presented and defined in the following section. This is in addition to the glossary at the end of the book.

5.4 KEY TERMS AND ABBREVIATIONS

5.4.1 Terms and Definitions

Algorithm—It is a specification of how to solve a class of problems by performing calculation, data processing, and automated reason tasks.

Artificial intelligence (AI)—AI refers to the intelligence generated by machines. AI is applied when a machine mimics cognitive functions that humans associate with learning and problem-solving.

Chatbot—An AI that conducts a conversation via auditory or word-based methods.

Generative AI—Generative AI refers to models or algorithms that generate new output, such as text, photos, videos, data code, and 3D representations, from vast amounts of data they were prepared on. Understanding the differences between AI and Generative AI is crucial in today's technology-driven world, as these two concepts offer unique capabilities and applications in several fields while married at the hip. Traditional AI systems are designed to understand, interpret, and make decisions based on the data they have been supplied. They can answer questions, recommend purchasing products, and drive motor vehicles autonomously. This is accomplished through machine learning algorithms that allow these systems to gain knowledge from experience and improve over time. On the other hand, Generative AI goes beyond understanding and interpreting data. It generates entirely new content from the very beginning. It can create human-like text through advanced deep learning, compose music, and create images indistinguishable from those made by humans.

Generative pre-trained transformers (GPT)—GPT denotes a family of AI models. These models, primarily larger language models such as those developed by OpenAI (founded in 2015 in San Francisco by Elon Musk and others), are based on a transformer architecture capable of understanding and generating human-like language.

Machine learning (ML)—The scientific study of statistical models and algorithms that computer systems use to perform specific tasks.

The following AI terminology and industry abbreviations are relatively new—as foreign as hieroglyphics—and are head-scratchers without definition.

AE—auto recorder
AI—artificial intelligence
ANN—artificial neural network
CFD—computational fluid dynamics
CUSUM—Shewhart cumulative sum
DBN—Deep Belief Network
DL—Deep Learning
DRL—Deep Reinforcement Learning
GPU—Graphical Processing Unit
GRU—Gated Recurrent Unit
LSTM—Long Short-Term Memory
MLP—Multi-Layer Perception
ML—Machine Learning
NN—Neural Network
TL—Transfer Learning

AI's deep learning technique creates a model of a complex non-linear system by learning from examples of the system's inputs and outputs. Deep Learning extracts essential features or representations directly from the raw input data by organizing its computational processing units (aka artificial neurons) into a hierarchy of layers.

Hierarchy of layers?

Yes.

As a subset of machine learning, methods based on neural networks with representation learning are what the hierarchy of layers is all about. "Deep" means using multiple layers in the network. The organization in these layers supplies the view that the computational depth of the network inspires the name "deep learning." When trained for a particular task, whatever task might be, the model can be used (and is used) to generate outputs. The function of these outputs is based on the neural network's architecture.

DID YOU KNOW?

What exactly is AI? Artificial intelligence allows a computer program to think and learn independently. It is synthetic because it simulates human intelligence into machines that do things humans usually do. This technological marvel extends beyond mere automation, incorporating a broad spectrum of AI skills, including enabling machines to understand, reason, learn, and interact in a human-like manner. AI is designed to complement humans and not to replace them—at least for now.

5.5 THE MACHINE LEARNING INTERFACE

In 2022, I learned a lot while watching the installation of computerized automatic sampling at water and wastewater facilities. Moreover, I watched the integration of AI, the IoT, and ML into large plant operations, which was (and is) revolutionizing water and wastewater treatment (and also the other sectors of environmental engineering).

The critical facets of ML and its co-technologies that I observed and learned include the following points:

- **Real-time monitoring**—Devices, such as IoT sensors, are embedded wholesale, blanket-fashion. Real-time monitoring delivers updated data about events, processes, and systems. This capability is essential because it provides real-time data on energy generation and storage parameters. Another plus of this type of monitoring is that it enables continuous monitoring and immediate recognition of glitches and abnormalities.
- **Predictive modeling**—This technique uses statistics to predict outcomes (most commonly future outcomes). The algorithms analyze the data collected in ML to predict potential issues and optimize energy processes. Technically, predictive modeling begins with data processing, data splitting, and model tuning (meaning that deep learning models can be trained with limited data). From my perspective, the huge plus is that data-driven decisions enhance efficiency and reduce operational costs, especially in energy conservation and water quality management.
- **Computerization and manipulation (automation and control)**—When IoT and ML are combined, they enable an automated control system that dynamically adjusts procedures based on real-time data. This ensures that energy system processes are continuously operating at peak levels.
- **Viability**—These technologies contribute to more sustainable energy management by improving process accuracy and efficiency, minimizing environmental impact, and reducing energy consumption. Finding cost savings in energy usage is often tricky, but it is possible with planning and competent supervision.
- **Blockchain integration**—Blockchain is an emerging technology that creates a secure and transparent framework for managing energy systems data and works to monitor efficiency. Blockchain's primary value is providing mathematical proof about the state of energy data so that different parties to transactions can agree on the outcome.

5.6 GOOD AND BAD OF AI IN ENERGY

After training in AI, ML, and the IoT, most come to realize that instead of being a job killer in energy production, storage, and use, it is straightforward to see the potential advantages of incorporating AI into energy operations. Based on Duggal (2024), the good of using AI is listed as follows:

- **Reduction of human error**—AI can reduce errors to null when appropriately programmed.

- **Human risk reduction**—The robotics involved in AI, ML, and IoT have revolutionized many different fields and allow hazardous processes to be accomplished without human hands-on involvement. Typical applications of AI in risk management include threat intelligence analysis, event management, security information, fraud detection, risk reduction in the workplace, and data monitoring and classification.
- **On-call 24 × 7**—Twenty-four by seven service means AI, ML, and IoT are productive continuously, enhancing plant productivity, and they can handle repetitive jobs easily, with near-seamless operations around the clock.
- **Digital assistants**—They deliver operational content upon user request.
- **New discoveries**—AI, ML, and IoT are driving forces behind numerous innovations and discoveries that are aiding humans in resolving the most challenging issues.
- **Unemotional operation**—AI, ML, and IoT are devoid of emotions and are highly practical and rational in their approach. Whenever human bias is removed from operations, it is a plus.
- **Repetition, repetition, repetition, and more repetition**—This is the old "here we go again" repetition. Based on experience and observation, little is as dull as repetition. The problem is that we all do a lot of repetitive tasks as part of our daily work, our work-related tasks. AI, ML, and IoT may efficiently automate these menial chores and even remove the "boring" tasks for people. When we are not bored, we are more creative.
- **Daily uses**—While teaching upper-grad and graduate students one of my college topics on the Internet and Homeland Security, I always injected the following in my lectures: "Where do you think we would be today without the Internet?" We utilize a variety of apps, including Alexa, Windows, Google, Siri, and Google Maps, to make calls, take selfies, and respond to emails. With all this available technology, we give little thought to the enormous technological advancements for our daily use.
- **AI in perilous situations**—The most common uses of AI, ML, and IoT to support risk management and mitigation processes are essential. Having a robot perform perilous tasks lets us avoid many dangerous situations.
- **Medical uses**—AI-powered tools have applications ranging from diagnosis and treatment to drug discovery and clinical trials. They help medical professionals keep employees in top condition.

5.6.1 The Other Side

We all know that a coin has two sides. AI, ML, and IoT have the advantages listed above. Still, like the coin, these critical technologies have another side or downside, depending on how you view them or their applications.

Let us now look at AI's main disadvantages, again depending on one's point of view.

- **Significant costs**—"A significant cost" because having the ability to create a robot, a machine that can simulate human intelligence, is no small feat. This achievement is not cheap because enormous amounts of time and resources can cost a lot of money. Also, AI is constantly changing, adapting, and advancing with the latest hardware and software; staying updated is a critical requirement, thus making it quite expensive.

- **Imagination restrained**—This points to immobilized or held-back creativity. AI cannot learn to think outside the digital box. However, it can learn over time with pre-fed data and past experiences but cannot be imaginative or creative in its approach. AI can only do what it is programmed or told to do.
- **Makes humans lethargic and lazy—**AI is addictive, and it has rapidly integrated into many aspects of our daily lives. Whether used as personal assistants on our smartphones or the algorithms underpinning AI and its messaging, AI does the work humans used to do. AI technologies allow us to use our brains less and less, and future generations are at risk.
- **There is no right from wrong—**Ethics and morality are essential to humans, but ethics and morality in AI differ. These human features are challenging to incorporate into AI. With no concern for ethics and morality, this moment is known as AI singularity, referring to when AI surpasses human intelligence, leading to machines that can outthink and outnumber us. AI can grow uncontrollably and eventually take over or wipe out humanity.
- **Lack of feelings**—We have been taught or advised that machines, including computers, do not have feelings. Even though robots are superior to humans when functioning effectively, humans function as a team and cannot be replaced by digital technology.
- **Machines are programmed**—AI, ML, and IoT machines can only be used to complete tasks they have been programmed for; they cannot do anything else. If adjustments are needed, we must manually alter the codes.
- **Robot it!**—The most obvious (perceived) disadvantage of AI, ML, and IoT is their use in robots and chatbots that take human jobs. Unemployment is a situation most of us attempt to avoid. The bottom line for workers worrying about digital processes or systems taking their jobs is to join the chorus. In this digitized world, we must adapt and become computer literate. Moreover, while disruption caused by advanced technology is a fact of life, AI, ML, and IoT do not cause widespread loss of jobs (presently). However, they change the demand for specific skills and contribute to more flexible contingent work arrangements.

Throwing one's hands in the air and accepting defeat and loss of employment due to chatbots and robots in this digital age is not the answer. The answer is, again, adapting to change by working with and on implementing and maintaining the changes and not being against them.

An offshoot of the robot is the Internet bot. One of the questions I often hear is, "What is a bot?" Now, my students do not ask this question because I have found that they have become experts at or with bots—the question usually arises from those who are older and from a different time and place and not all that up on computers, the Internet, bots, or anything digital other than their TV remotes.

Different types of bots, including *Chatbots*, simulate human conversation and provide customer service and information. *Web crawlers* index content for search engines. *Social bots* operate on social media platforms to respond to prompts and interact with users. Finally, there are the *malicious bots* we all need to be aware of and their potential to conduct cyber-attacks.

5.7 THE BOTTOM LINE

We need humans behind the digital revolution. To qualify and succeed in efficient energy production and usage or any other technology (even in industry, business, finance, banking, and so forth), we need to adapt to the skills required for the job. Times are changing fast, and roles are being redefined and recast. There is a definite need to board the digital train and hold on while the industry advances to levels yet to be identified. AI is pivotal in advancing energy efficiency and sustainability initiatives.

NOTE

1 This section is adapted from Spellman, F.R., 2025. *Science of AI in Environmental Engineering*. Boca Raton, FL: CRC Press.

REFERENCE

Duggal, N., 2024. *Advantages and Disadvantages of Artificial Intelligence (AI)*. Accessed 8/11/24 @ https://muchrach.com.

Part 2

Energy Sources

6 The Wind Industry and AI

6.1 INTRODUCTION[1]

Do not expect Don Quixote, mounted in armor on his old nag, Rocinante, with or without his trusty partner Sancho Panza, to charge today's windmills. Instead, expect—you can count on it, bet on it, and rely on it—that the rest of us will make the charge to build those windmills to satisfy our growing, relentless need for renewable energy. What other choice do we have?

Wind energy or power is all about wind. In simple terms, wind is the atmosphere's response to uneven heating conditions. Regarding the Earth's atmosphere, and again, to state the obvious, it is constantly in motion. Anyone observing the constant weather changes and cloud movement around them knows this phenomenon. Although its physical manifestations are apparent, the importance of the dynamic state of our atmosphere is much less noticeable.

The constant motion of Earth's atmosphere (air movement) consists of both horizontal (*wind*) and vertical (*air currents*) dimensions—the atmosphere's motion results from thermal energy heating the Earth's surface and the air molecules above. Energy flows from the equator poleward because of differential heating of the Earth's surface.

Hanson (2004) points out that the energy resources contained in the wind in the United States are well-known and mapped in detail. Even though this is the case, air movement plays a critical role in transporting the energy of the lower atmosphere, bringing the warming influences of spring and summer and the cold chill of winter. Generally, wind and air currents are fundamental to how nature functions, and the effects of air movements on our environment are often overlooked. All life on earth has evolved or has been sustained with mechanisms dependent on air movement: pollen is carried by the wind for plant reproduction; animals sniff the wind for essential information; wind power was the motive force that began the earliest stages of the Industrial Revolution. Now we see the effects of winds in other ways: Wind causes weathering (erosion) of the earth's surface; wind influences ocean currents; air pollutants and contaminants such as radioactive particles transported by the wind impact our environment.

DID YOU KNOW?

Wind speed is generally measured in m/s, but Americans usually think in mph. A good rule of thumb to keep at hand is that to convert m/s to mph double the value in m/s and add 10%.

DOI: 10.1201/9781003605430-8

6.2 AIR IN MOTION

In all dynamic situations, forces are necessary to produce motion and changes in motion—winds and air currents. The atmosphere's air (comprised of various gases) is subject to two primary forces: gravity and pressure differences from temperature variations.

Gravity (gravitational forces) holds the atmosphere close to the Earth's surface. Newton's law of universal gravitation states that everybody in the universe attracts another body with a force equal to:

$$F = G\frac{m_1 m_2}{r^2} \tag{6.1}$$

where

F = force
m_1 and m_2 = the masses of the two bodies
G = universal constant of 6.67×10^{-11} N $\times$ m^2/kg^2
r = distance between the two bodies

The force of gravity decreases as an inverse square of the distance between them. Thermal conditions affect density, which in turn causes gravity to affect vertical air motion and planetary air circulation. This affects how air pollution is naturally removed from the atmosphere.

Gravity is consistently a downward force, although it can be overridden by forces in other directions. It acts on each gas molecule, accounting for the more significant air density near the Earth.

Atmospheric air is a mixture of gases, so the gas laws and other physical principles govern its behavior. The pressure of a gas is directly proportional to its temperature. Pressure is force per unit area ($P = F/A$), so temperature variation in air gives rise to a difference in pressure of force. This difference in pressure resulting from temperature differences in the atmosphere creates air movement on both large and local scales. This pressure difference corresponds to an unbalanced force, and when a pressure difference occurs, the air moves from a high- to a low-pressure region.

In other words, horizontal air movements (called *advective winds*) result from temperature gradients, which give rise to density gradients and, subsequently, pressure gradients. The force associated with these pressure variations (*pressure gradient force*) is directed at right angles to (perpendicular to) lines of equal pressure (called *isobars*).

Figure 6.1 maps the pressures over a region by taking barometric readings at different locations. Lines drawn through the points (locations) of equal pressure are called isobars. All points on an isobar are of equal pressure, which means no air movement along the isobar. The wind direction is at right angles to the isobar in the direction of the lower pressure. In Figure 6.1, air moves down a pressure gradient toward a lower isobar like a ball rolling down a hill. If the isobars are close together, the pressure gradient force is significant, and high wind speeds characterize such areas. If isobars are widely spaced (see Figure 6.1), the winds are light because the pressure gradient is slight.

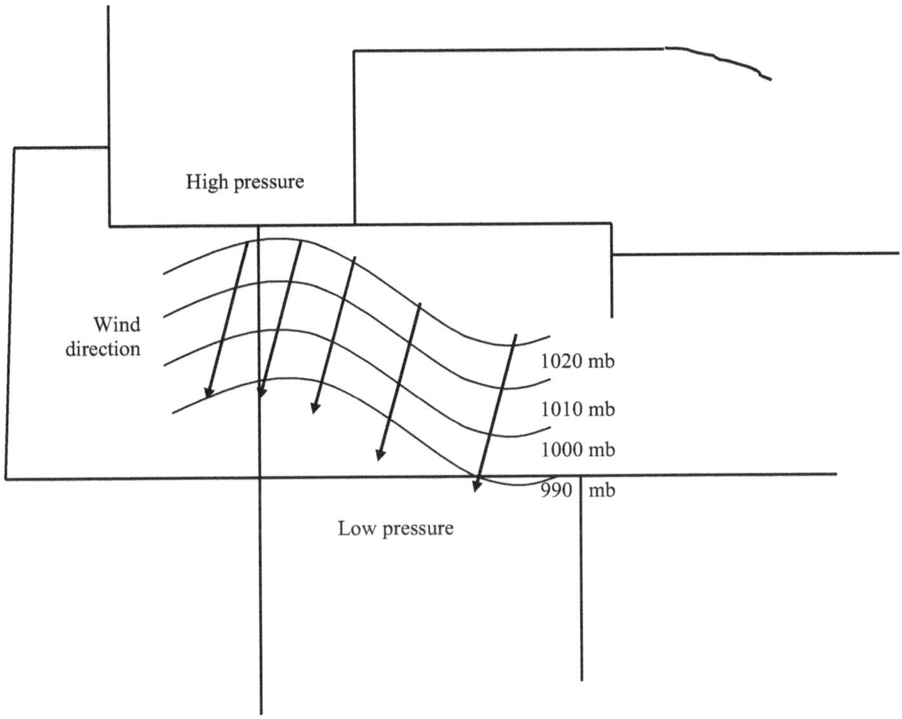

FIGURE 6.1 Isobars are drawn through locations having equal atmospheric pressures. The air motion, or wind direction, is at right angles to the isobars and moves from a region of high pressure to an area of low pressure.

Localized air circulation gives rise to *thermal circulation* (a result of the relationship based on a law of physics whereby the pressure and volume of a gas are related to its temperature). A change in temperature causes a change in gas pressure and volume. With a shift in volume comes a change in density since P = m/V, so regions of the atmosphere with different temperatures may have different air pressures and densities. As a result, localized heating sets up air motion and increases *thermal circulation.* To gain an understanding of this phenomenon, consider Figure 6.2.

Once the air has been set into motion, secondary (velocity-dependent forces) act. The Earth's rotation causes these secondary forces (Coriolis force) and contact with the rotating Earth (friction). The *Coriolis force,* named after its discoverer, French mathematician Gaspard Coriolis (1772–1843), is the effect of rotation on the atmosphere and all objects on the Earth's surface. In the Northern Hemisphere, it causes moving objects and currents to be deflected to the right; in the Southern Hemisphere, it causes deflection to the left because of the earth's rotation. Air is deflected from its expected path in large-scale north or south movements. That is, air moving poleward in the Northern Hemisphere seems to be deflected toward the East; air moving southward appears deflected toward the West.

Figure 6.3 illustrates the Coriolis effect on a propelled particle (analogous to the apparent effect of an air mass flowing from Point A to Point B). The action of the

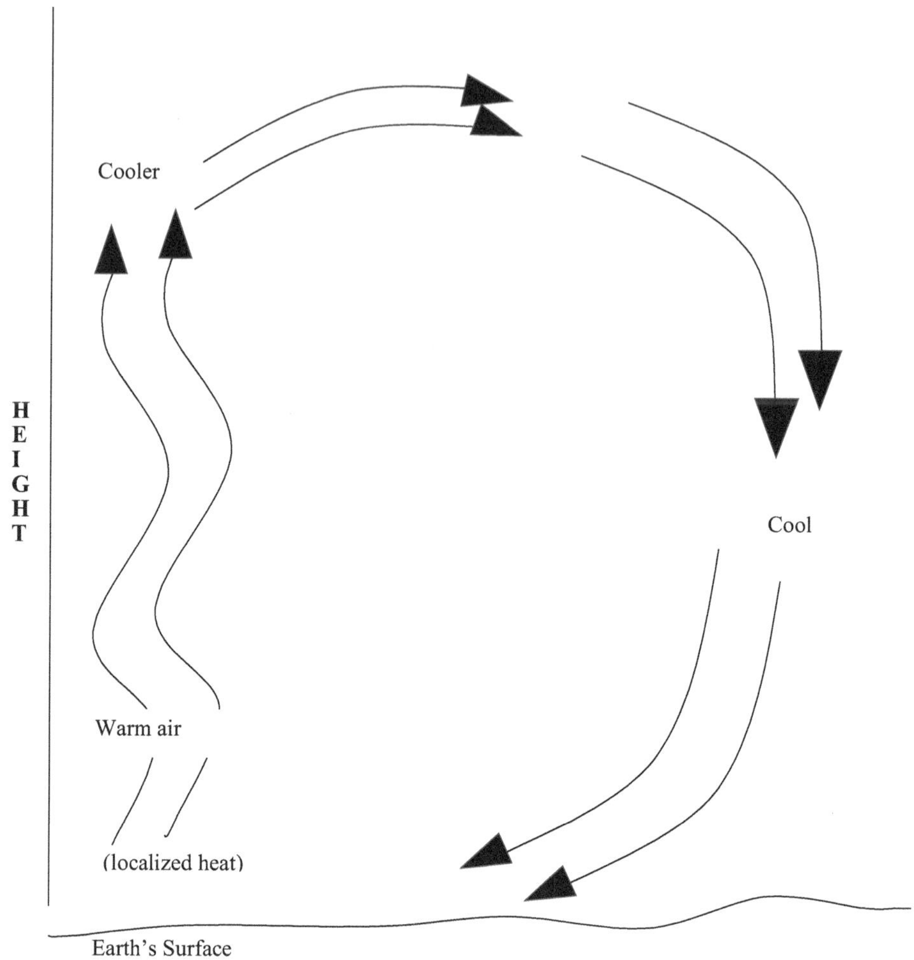

FIGURE 6.2 Thermal circulation of air. Localized heating, which causes air in the region to rise, initiates the circulation. As the warm air rises and cools, cool air near the surface moves horizontally into the region vacated by the rising air. The upper, still cooler air then descends to occupy the area vacated by the cool air.

earth's rotation on the air particle as it travels north over the earth's surface as it rotates beneath it from east to west can be seen. Projected from point A to Point B, the particle will reach Point B as it moves in a straight line (deflected). The earth rotates east to west beneath it.

Friction (drag) can also cause the deflection of air movements. This friction (resistance) is both internal and external. The friction of its molecules generates internal friction. Friction is also generated when air molecules run into each other. External friction is caused by contact with terrestrial surfaces. The frictional force's magnitude along a surface depends on the air's magnitude and speed, and the opposing frictional force is in the opposite direction of the air's motion.

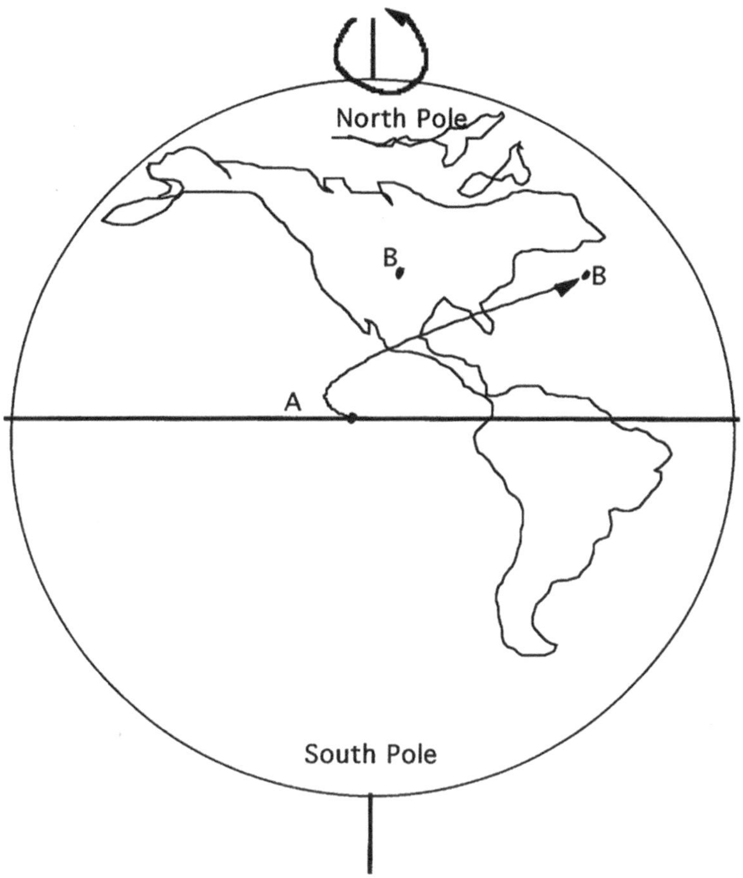

FIGURE 6.3 The effect of the earth's rotation on the trajectory of a propelled particle.

6.3 WIND ENERGY[2]

Wind energy is the movement of the wind to create power. Since early recorded history, people have been harnessing the energy of the wind for milling grain, pumping water, and other mechanical power applications. Wind energy propelled boats along the Nile River as early as 5,000 B.C. By 200 B.C., simple windmills in China were pumping water, while vertical-axis windmills with woven reed sails were grinding grain in Persia and the Middle East.

Wind energy spread worldwide, and by the 11th century, people in the Middle East were using windmills extensively for food production; returning merchants and crusaders carried this idea back to Europe. The Dutch refined the windmill and adapted it for draining lakes and marshes in the Rhine River Delta. When settlers took this technology to the New World in the later 19th century, they began using windmills to pump water for farms and ranches and later to generate electricity for homes and industry. The first known wind turbine for producing electricity was by Charles F. Brush in

Cleveland, Ohio, in 1888; it was a 12-kW unit used to charge batteries in the cellar of a mansion. The first wind turbine outside of the United States to generate electricity was built by Poul la Cour in 1891 in Denmark. He used electricity from his wind turbines to electrolyze water to make hydrogen for the gas lights at the schoolhouse. By the 1930s and 1940s in the United States, hundreds of thousands were used in rural areas not yet served by the grid. The Oil crisis in the 1970s created renewed interest in the wind until the U.S. government stopped giving tax credits. Today, several hundred thousand windmills operate worldwide, and many are used for water pumping. However, the use of wind energy as a pollution-free means of generating electricity on a significant scale is attracting the most current interest in the subject. As a matter of fact, with the present and pending shortage and high cost of fossil fuels to generate electricity and the green movement toward the use of cleaner fuels, wind energy is the world's fastest-growing energy source. It will power industries, businesses, and homes with clean, renewable electricity for many years. In the United States since 1970, wind-based electricity generating capacity has increased markedly, although (at present) it remains a small fraction of total electric capacity and consumption. But this trend is beginning to change—with the advent of \$4/gal of gasoline, high heating, and cooling costs, and subsequent increases in the price of electricity, worldwide political unrest or uncertainty in oil-supplying countries, one only needs to travel the "wind corridors" of the United States encompassing parts of Arizona, New Mexico, Texas, Missouri and north through the Great Plains to the Pembina Escarpment and Turtle Mountains of North Dakota and elsewhere witness the considerable activity the seemingly exponential increase in wind energy development and wind turbine installations; these machines are being installed to produce and provide electricity to the grid.

Total annual U.S. electricity generation from wind energy increased from 6 billion kilowatt-hours (kWh) in 2000 to about 434 billion kWh in 2022 (EIA, 2023).

DID YOU KNOW?

We can classify wind energy as a form of solar energy. As mentioned, winds are caused by uneven heating of the atmosphere by the sun, irregularities of the Earth's surface, and the rotation of the Earth. As a result, winds are strongly influenced and modified by local terrain, bodies of water, weather patterns, vegetative cover, and other factors. The wind flow, or motion of energy when harvested by wind turbines, can be used to generate electricity.

6.4 WIND POWER BASICS

Wind energy or wind power describes the process by which the wind is used to generate mechanical power or electricity. Wind turbines convert the wind's kinetic energy into mechanical power. This mechanical power can be used for specific tasks (such as grinding grain or pumping water), or a generator can convert it into electricity (EERE, 2006a).

As mentioned, we have been harnessing the wind's energy for hundreds of years. Windmills have been used to pump water or grind grain from old Holland to farms in the United States. Today, the windmill's modern equivalent—a wind turbine—can use the wind's energy to generate electricity. They accomplish this by turning aircraft propeller blades into moving air and powering an electrical generator that supplies an electrical current. Unlike fans that use electricity to make wind, wind turbines use wind to make electricity. The wind turns the blades, which spin a shaft that connects to a generator and makes electricity (WINDEIS, 2009).

DID YOU KNOW?

Whenever wind energy is considered a possible renewable energy source, the amount of land area required, accessibility to generators, and aesthetics must be considered.

6.4.1 Wind Turbine Types

Whether called wind-driven generators, wind generators, wind turbines, wind-turbine generators (WTG), or wind energy conversion systems (WECS), modern wind turbines fall into two primary groups: the horizontal-axis (HAWT) variety, like the traditional farm windmills used for water pumping, and the vertical-axis (VAWT) design, like the eggbeater-style Darrieus rotor-model, named after its French inventor, the only vertical axis machine with any commercial success. Wind hitting the vertical blades, called aerofoils, generates a lift to create rotation. No yaw (rotation about vertical axis) control is needed to keep them facing into the wind. The heavy machinery in the nacelle (cowling) is located on the ground. Blades are closer to the ground, where wind speeds are lower. Most large modern wind turbines are horizontal-axis turbines, highlighted and described in detail in this text.

DID YOU KNOW?

Groups of wind turbines are in what is called a "wind farm" or a "wind park."

6.4.2 Horizontal-Axis Wind Turbines (HAWTs)

Wind turbines are available in various sizes and, therefore, power ratings. Utility-scale turbines range from 100 kW to as large as several megawatts. Horizontal-axis wind turbines (HAWTs) typically have either two or three blades. "Downwind" HAWTs have a turbine with the blades behind (downwind from) the tower. No yaw control is needed because they naturally orient themselves in line with the wind. However, these downwind HAWTs undergo a shadowing effect in that when a blade swings behind the tower, the wind it encounters is briefly reduced, and the blade flexes. "Upwind" HAWTs usually have three blades in front of (upwind of) the tower. These upwind wind turbines require a complex yaw control to keep them facing the wind. They operate more smoothly and deliver more power and, thus, are the most used modern wind turbines. The largest machine has blades that span more than the length of a football field, stands 20 building stories high, and produces enough electricity to power 1,400 homes.

6.4.2.1 Inside the HAWT

A HAWT consists of three main parts: a turbine, a nacelle, and a tower. Several other essential parts are contained within the tower and nacelle, including an anemometer, blades, brake, controller, gearbox, generator, high-speed shaft, low-speed shaft, pitch, rotor, tower, wind direction, wind vane, yaw drive, and yaw motor (see Figure 6.4; Alternative Energy News, 2010).

6.4.2.1.1 Turbine Components/Parameters

- **Anemometer**—It measures the wind speed and transmits wind speed data to the controller.
- **Blades**—Most turbines have either two or three blades. Wind blowing over the blades causes the blades to "lift" and rotate; they capture the wind's kinetic energy and help the turbine rotate.

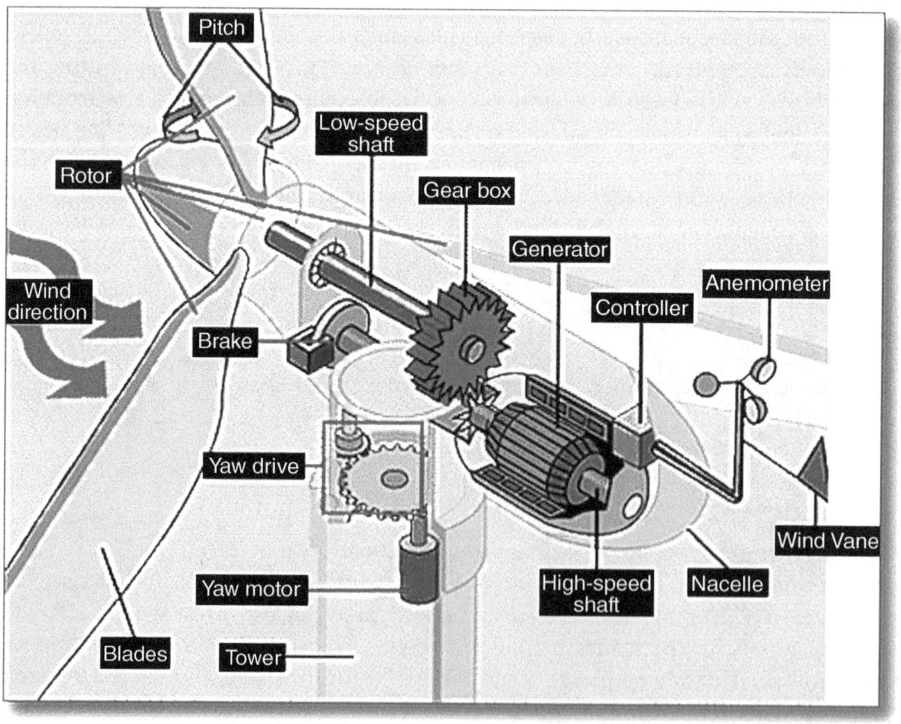

FIGURE 6.4 Horizontal-axis wind turbine components.

- **Brake**—A disk brake, which can be applied mechanically, electrically, or hydraulically to stop the rotor in emergencies.
- **Controller**—It starts the machine at wind speeds of about 8–16 miles per hour (mph) and shuts off at about 55 mph. Turbines do not operate at wind speeds above 55 mph because the high winds might damage them.
- **Gearbox—**Gears connect the low-speed shaft to the high-speed shaft and increase the rotational speeds from about 30–60 rotations per minute (rpm) to about 1,000–1,800 rpm, the rotational speed required by most generators to produce electricity. The gearbox is a costly (and heavy) part of the wind turbine, and engineers are exploring "direct-drive" generators that operate at lower rotational speeds and don't need gearboxes.
- **Generator**—It is usually an off-the-shelf induction generator that produces 60-Hz AC electricity.
- **High-speed shaft**—It drives the generator.
- **Low-speed shaft**—The rotor turns the low-speed shaft at about 30–60 rpm.
- **Nacelle**—It sits atop the tower and contains the gearbox, generator, low- and high-speed shafts, controller, and brake.
- **Pitch**—Blades are turned, or pitched, out of the wind to control the rotor speed and keep the rotor turning in winds that are too high or too low to produce electricity.
- **Rotor**—The blades and the hub together are called the rotor.
- **Tower**—It is made from tubular steel, concrete, or steel lattice. Because wind speed increases with height, taller towers enable turbines to capture more energy and generate more electricity.
- **Wind direction**—This is an "upwind" turbine, so-called because it operates facing into the wind. Other turbines are designed to run "downwind," facing away from the wind.
- **Wind vane**—It measures wind direction and communicates with the yaw drive to orient the turbine properly to the wind.
- **Yaw drive**—Upwind turbines face the wind; the yaw drive keeps the entire nacelle and thus the rotor facing into the wind as the wind direction changes. Downwind turbines don't require a yaw drive, and the wind blows the rotor downwind.
- **Yaw motor**—It powers the yaw drive.

DID YOU KNOW?

During the nacelle's rotation, the cables inside the tower can be twisted; if the turbine keeps turning in the same direction, which can happen if the wind keeps changing in the same direction, the cables can become increasingly twisted. The wind turbine is, therefore, equipped with a cable twist counter, which notifies the controller that it is time to straighten the cables.

—Khaligh and Onar (2010)

6.5 WIND ENERGY/POWER CALCULATIONS

A wind turbine is a machine that converts the kinetic energy in the wind into the mechanical energy of a shaft. Calculating the energy and power available in the wind relies on fundamental physics and geometry knowledge. The kinetic energy of an object is the extra energy it possesses because of its motion. It is defined as the work needed to accelerate the body of a given mass from rest to its current velocity. Once in motion, a body maintains its kinetic energy unless its speed changes. The equation provides the kinetic energy of a body:

$$KE = 1/2mv^2 \tag{6.2}$$

where

KE = kinetic energy
m = mass
v = velocity

Example 6.1: Determining Power in the Wind

Step 1:

To find the kinetic energy of moving air (wind), let's say we have a large packet of wind [i.e., a geometrical package of air passing through the plain of a wind turbine's blades (which sweep out a cross-sectional area A), with a thickness (B); see Figure 6.5] passing through the plane over a given time.

Step 2:

To determine power in the wind, we begin by considering the kinetic energy of the packet of air shown in Figure 6.5, along with mass m moving at velocity v, as shown in equation (6.2). Next, we need to divide by time to get power:

$$\text{Power Though Area A} = -\left(\frac{1}{2}\,\frac{\text{m passing through A}}{\text{t}}\right)v^2 \tag{6.3}$$

Step 3:

The mass flow rate is

$$m = \frac{\text{m passing through A}}{\text{t}}\rho Av \tag{6.4}$$

where ρ is the air density—the mass per unit volume of the Earth's atmosphere.

Step 4:

To obtain the power in the wind equation, we must combine equations (6.3) and (6.4):

$$\text{Power through area A} = \tfrac{1}{2}(\rho Av)v^2$$

$$\text{Power in the wind} = P_W = \tfrac{1}{2}\rho Av^3 \tag{6.5}$$

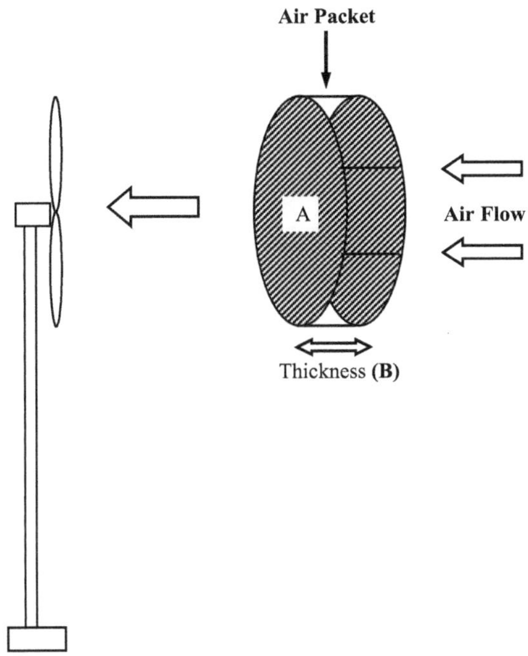

FIGURE 6.5 A packet of air passing through the plane of a wind turbine's blades, with thickness (B) passing through the blades over a given time.

where
P_W (W) = power in the wind
ρ (kg/m³) = air density (2.225 kg/m³ at 15°C and 1 atm)
A (m²) = cross-sectional area that wind passes through perpendicular to the wind
v (m/s) = windspeed normal to A (1 m/s = 2.237 mph)

DID YOU KNOW?

Air density and pressure decrease with increasing altitude. They also change with variance in temperature or humidity.

From equation (6.5), it is evident that power increases like the cube of wind speed. That is, doubling the wind speed increases the power by eight. Moreover, the energy in 1 hour of 20 mph winds is the same as in 8 hours of 10 mph winds. We cannot use average wind speed because the relationship is non-linear regarding power (W/m²) and windspeed (mph). Power in the wind is also proportional to A. $A = (\pi/4)D^2$ for a conventional HAWT, so wind power is proportional to the blade diameter squared. Large giant wind turbines are more cost-effective because the cost is approximately proportional to the blade diameter.

6.5.1 Air Density Correction Factors

Earlier, we pointed out that air density is affected by different temperatures and pressures. Air density correction factors can correct air density for temperature and altitude. Correction factors for temperature and altitude correction can be found in standardized tables. The following equation determines air density for different temperatures and pressures:

$$\rho = \frac{P \times MW \times 10^{-3}}{RT} \tag{6.6}$$

where

P = absolute pressure (atm) that starts at zero gas pressure (no molecules—a vacuum) and has no practical maximum limit.
MW = molecular weight of air (g/mol) = 28.97 g/mol
T = absolute temperature (K) [The temperature in Kelvin (273 + °C) or Rankine (absolute zero = −460°F, or 0°R).]
R = ideal gas constant = $8.2056 \times 10^{-5} \times m^3 \times atm \times K^{-1} \times mol^{-1}$

DID YOU KNOW?

Air density is more significant at lower temperatures.

6.5.2 Elevation and Earth's Roughness

Windspeed is affected by its elevation above the Earth and the roughness of the Earth. Because power increases like the cube of windspeed, we can expect a significant economic impact from even a moderate increase in windspeed; thus, windspeed is an important parameter in the operation of wind turbines. Earth's surface features cannot be ignored when deciding where to place wind turbines and in the calculation of their productivity output. Natural obstructions such as mountains and forests and human-made obstructions such as buildings provide friction as winds flow over them. The point is there is a lot of friction in the first few hundred meters above ground—smooth surfaces (like water) are better. Thus, the greater the elevation above the Earth, the greater the windspeed; tall wind turbine towers are better.

When actual measurements are not possible or available, it is possible to characterize or approximate the impact of rough surfaces and height on wind speed using the following equation:

$$\frac{v}{v_0} = \left(\frac{H}{H_0}\right)^{\alpha} \tag{6.7}$$

where

α = friction coefficient—obtained from standardized tables; typical value of α in open terrain is 1/7; for a large city, $\alpha = 0.4$; for calm water, $\alpha = 0.1$
v = windspeed at height H
v_0 = windspeed at height H_0 (H_0 is usually 10 m)

DID YOU KNOW?

As pointed out, wind energy is a cubed function of wind speed, which means that if wind speed doubles, there is eight times as much available energy, not twice as much as one might expect.

6.5.3 Wind Turbine Rotor Efficiency

Generally, when we think about efficiency, we think about input versus output and know that if we put 100% into something and get 100% output, we have a very efficient machine, operation, or process. We can approximate efficiency input versus output in engineering by performing mass balance calculations. Using these calculations, we know that according to conservation laws, materials cannot disappear or be destroyed; thus, they must exist and be somewhere. However, in this input versus output concept, there are a couple of views on maximum rotor efficiency that the ill-informed or misinformed have taken, and neither makes sense, but it is stated here to point out what makes sense. The two wrong assumptions: First, it can be assumed that downwind velocity is zero; the turbine extracted all the power from the wind; second, downwind velocity is the same as the upwind velocity—the turbine extracted no power. Again, in a word: Wrong! Albert Betz, in 1919, set the record straight when he theorized that there must be some ideal slowing of the wind so that the turbine extracts the maximum power. Betz theorized in his wind turbine efficiency theory (i.e., to their power input versus power output and overall efficiency in general) that the efficiency rule or laws of conservation, as stated above, do not apply. Betz law states the maximum possible energy that can be derived from a wind turbine. We have provided the following explanation to understand Betz's law and its derivation.

6.5.3.1 Derivation of Betz's Law

To understand Betz's law, we must first understand the constraint on wind's ability to convert kinetic energy into mechanical power. Visualize wind passing through a turbine (see Figure 6.6). It slows down, and the pressure is reduced, so it expands.

The following equation is used to determine the power extracted by the blades, another step in the derivation of Betz's law:

$$P_b = \frac{1}{2} m \left(v^2 - v_d^2\right) \tag{6.8}$$

where

m = mass flow rate of air within stream tube
v = upwind undisturbed windspeed
v_d = downwind windspeed

Using equation (6.9), the mass flow rate is determined as the next step in Betz's law derivation. In making the determination, it is easiest to use cross-sectional area A at the plane of the rotor because we know what this value is. The mass flow rate is

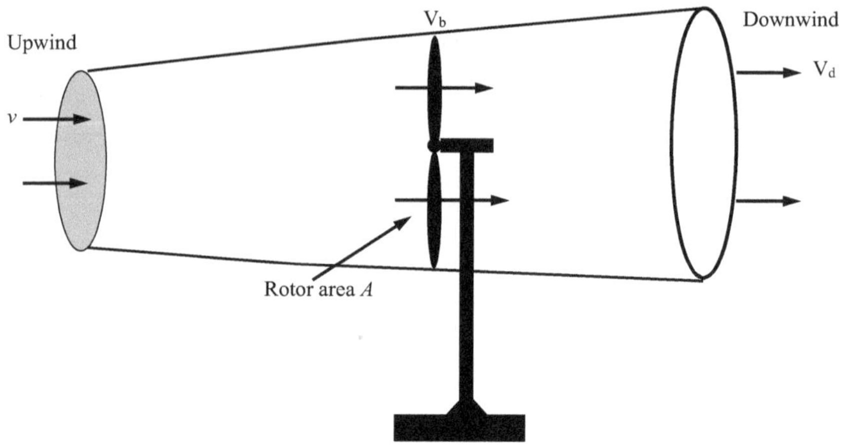

FIGURE 6.6 Wind passing through a turbine.

$$m = p\ Av_b \tag{6.9}$$

Assume that the velocity through the rotor v_b is the average of upwind velocity v and downwind velocity v_d:

$$v_b = \frac{v + v_d}{2} \quad \Longrightarrow \quad m = \rho\ A\left(\frac{v + v_d}{2}\right)$$

Equation (6.8) becomes

$$\begin{aligned} P_b &= \tfrac{1}{2}\rho A\left(\frac{V + v_d}{2}\right)\left(v^2 - v_d^2\right) \\ &= \frac{v^3}{2}[(1 + \lambda)(1 - \lambda^2)] \end{aligned} \tag{6.10}$$

$$\Longrightarrow \quad P_b = (\tfrac{1}{2}\rho Av^3)\ \tfrac{1}{2}[(1 + \lambda)(1 - \lambda^2)]$$

P_W = Power in the Wind C_P = Rotor Efficiency

The next step is to find the speed windspeed ratio λ, which maximizes the rotor efficiency, C_P:

$$C_P = \tfrac{1}{2}\left[(1 + \lambda)(1 - \lambda^2)\right] = \tfrac{1}{2} - \frac{\lambda^2}{2} + \frac{\lambda}{2} - \frac{\lambda^3}{2}$$

Set the derivative of rotor efficiency to zero and solve for λ:

$$\frac{\partial C_P}{\partial \lambda} = -2\lambda + 1 - 3\lambda^2 = 0$$

$$\frac{\partial C_P}{\partial \lambda} = 3\lambda^2 + 2\lambda - 1 = 0$$

$$\frac{\partial C_P}{\partial \lambda} = (3\lambda - 1)(\lambda + 1) = 0 \quad \Rightarrow \quad \lambda = 1/3$$

Maximizes rotor efficiency

When we plug the optimal value for λ back into C_P to find the maximum rotor efficiency:

$$C_P = \tfrac{1}{2}\left[(1+1/3)(1-1/3)\right] = 16/27 = 59.3\% \tag{6.11}$$

The maximum efficiency of 59.3% occurs when air is slowed to one-third of its upstream rate. Again, this factor and value is called the "Betz efficiency" or "Betz's law" (Betz, 1966). In plain English, Betz's law states that the rotor cannot capture all wind power; otherwise, air would be completely still behind the rotor and not allow more wind to pass. For illustrative purposes, in Table 6.1, we list wind speed, power of the wind, and power of the wind based on Betz limit (59.3%).

TABLE 6.1
Betz Limit for 80 M Rotor Turbine

Wind Speed mph/ms	Power (kW) of Wind	Power (kW) Betz Limit
5/2.2	36	21
10/4.5	285	169
15/6.7	962	570
20/8.9	2,280	1,352
25/11.2	4,453	2,641
28/12.5	6,257	3,710
30/13.4	7,695	4,563
35/15.6	12,220	7,246
40/17.9	18,241	10,817
45/20.1	25,972	15,401
50/22.4	35,626	21,126
55/24.6	47,419	28,119
56/25.0	50,053	29,681
60/26.8	61,563	36,507

Source: Adapted from Devlin (2007).

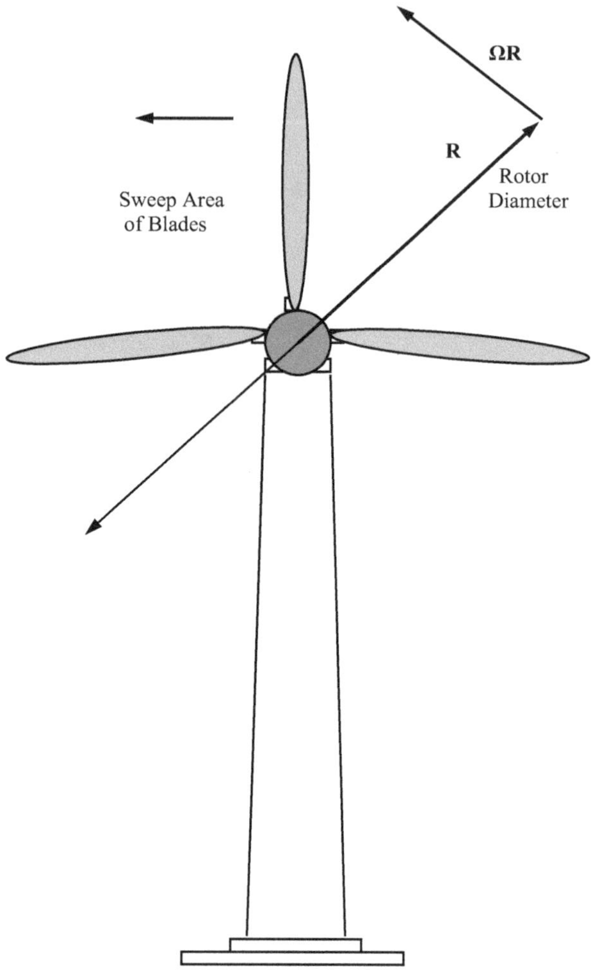

FIGURE 6.7 Tip-speed ratio (TSR).

6.5.4 Tip-Speed Ratio (TSR)

Efficiency is a function of how fast the rotor turns. The tip-speed ratio (TSR) is a significant factor in wind turbine design. The tip-speed ratio is the rotating blade tip's speed ratio to the free stream wind (see Figure 6.7). Stated differently, TSR is the speed of the outer tip of the blade divided by wind speed. There is an optimum angle of attack that creates the highest lift-to-drag ratio. If the wind turbine rotor spins too slowly, most of the wind will pass straight through the gap between the blades, therefore giving it no power! But if the rotor spins too fast, the blades will blur and act like a solid wall to the wind. Moreover, rotor blades create turbulence as they spin through the air. If the next blade arrives too quickly, it will hit that turbulent air. Thus, it is better to slow down the blades. Because the angle of attack is dependent on wind speed, there is an optimum tip-speed ratio

$$TSR = \Omega R / V \tag{6.1}$$

where

Ω = rotational speed in radians/sec
R = rotor radius
V = wind "free stream" velocity

6.6 SMALL-SCALE WIND POWER

A small home-sized wind machine may be the answer to meeting your electricity needs. A residential-scale wind turbine typically features rotors with diameters ranging from 8 to 25 ft and a height extending beyond 30 ft. It can supply the power needs of an all-electric home or small business. Utility-scale turbines range in size from 50 to 750 kW. Small turbines below 50 kW are used for homes, telecommunication dishes, or water pumping.

According to EERE (2006b), potential users of small home-size wind turbines should ask and answer many questions before installing a small wind system. This section provides some of the answers and explanations EERE delivers to address these questions.

We begin by asking the obvious question: What are the benefits to homeowners using wind turbines? Wind energy systems provide a cushion against electricity price increases. Wind energy systems reduce U.S. dependence on fossil fuels and don't emit greenhouse gases. If you are building a home in a remote location, a small wind energy system can help you avoid the high cost of extending utility power lines to your site.

Although wind energy systems involve a significant initial investment, they can be competitive with conventional energy sources when you account for a lifetime of reduced or altogether avoided utility costs. The length of payback time—the time before the savings resulting from your system equal the system cost—depends on the system you choose, the wind resource in your site, electric utility rates in your area, and how you use your wind system.

DID YOU KNOW?

The difference between power and energy is that power (kilowatts [kW]) is the rate at which electricity is consumed, while energy (kilowatt-hours [kWh]) is the quantity consumed.

Another frequently asked question: Is wind power practical for me? Small wind energy systems can be used with an electricity transmission and distribution system (called grid-connected systems) or in stand-alone applications not connected to the utility grid. A grid-connected wind turbine can reduce your consumption of utility-supplied electricity for lighting, appliances, and electric heat. The utility makes up the difference if the turbine cannot deliver the energy you need. When the wind system produces more electricity than the household requires, the excess can be sold to the utility. With the interconnections available today, switching takes place automatically. Stand-alone wind energy systems can be appropriate for homes, farms, or even entire communities (a co-housing project, for example) far from the nearest utility lines.

Stand-alone systems (not connected to the utility grid) require batteries to store excess power generated when the wind is calm. They also need a charge controller to keep the batteries from overcharging. As mentioned earlier in the text, deep-cycle batteries, such as those used for golf carts, can discharge and recharge 80% of their capacity hundreds of times, which makes them a good option for remote renewable energy systems. Automotive batteries are shallow cycle and should not be used in renewable energy systems because of their short life in deep-cycling operations.

Small wind turbines generate direct current (DC) electricity. In exceedingly small systems, DC appliances operate directly off the batteries. If you want to use standard appliances that use conventional household alternating current (AC), you must install an inverter to convert DC electricity from the batteries to AC (see Figure 6.8).

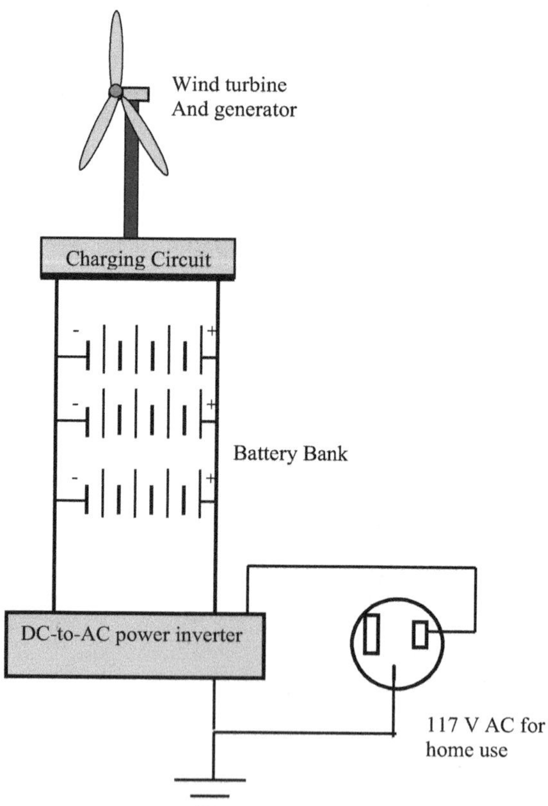

FIGURE 6.8 Stand-alone small-scale wind-power system.

Although the inverter slightly lowers the system's overall efficiency, it allows the home to be wired for AC, a definite plus with lenders, electrical code officials, and future homebuyers.

> **Safety Note**
>
> Batteries should be isolated from living areas and electronics because they contain corrosive and explosive substances. Lead-acid batteries also require protection from temperature extremes.

In grid-connected systems (or interactive systems), the only additional equipment required is a power condition unit (inverter) that makes the turbine output electrically compatible with the utility grid. Usually, batteries are not needed.

The type of system can be practical if the following conditions exist.

6.6.1 Conditions for Stand-Alone Systems

- You live in an area with average annual wind speeds of at least 4.0 m/s (9 mph)
- A grid connection is unavailable or can only be made through an expensive extension. The cost of running a power line to a remote site to connect with the utility grid can be prohibitive, ranging from $15,000 to more than $50,000 per mile, depending on terrain.
- You have an interest in gaining energy independence from the utility.
- You would like to reduce the environmental impact of electricity production.
- You acknowledge the intermittent nature of wind power and have a strategy for using intermittent resources to meet your power needs.

6.6.2 Conditions for Grid-Connected Systems

- You live in an area with average annual wind speeds of at least 4.5 m/s (10 mph).
- Utility-supplied electricity is expensive in your area (about 10–15 cents per kWh).
- The utility's requirements for connecting your system to its grid are not prohibitively expensive.
- Local building codes or covenants allow you to erect a wind turbine on your property legally.
- You are comfortable with long-term investments.

After comparing stand-alone and grid-connected systems and determining which is best suited for your circumstance, the next question is whether your location is the right site to install a small-scale wind turbine system. Is there enough wind on your site, is it legal to install the system on your property, and are there environmental and/or economic issues?

In determining if your site is right for wind turbine installation, you must decide if the wind blows hard enough to make a small wind turbine system economically worthwhile. That is a key question and not always quickly answered. The wind resource can vary significantly over just a few miles because local terrain influences the wind flow. Yet, there are steps you can take that will go a long way toward answering the above question.

As a first step, wind resource maps like the ones included in DOE's *Wind Energy Resource Atlas of the United States* (RREDC, 2010) can be used to estimate the wind resource in your region. The highest average wind speeds in the United States are generally found along seacoasts, on ridgelines, and the Great Plains; however, many areas have wind resources strong enough to power a small wind turbine economically. The wind resource estimates on the Wind Energy Resource Atlas maps generally apply to terrain well exposed to the wind, such as plains, hilltops, and ridge crests. Local terrain features may cause the wind resources at a specific site to differ considerably from these estimates.

Average wind speed information can be obtained from a nearby airport. However, caution should be used because local terrain influences and other factors may cause the wind speed recorded at an airport to differ from your location. Airport wind data are generally measured at about 20–33 ft (6–10 m) above ground. Average wind speeds increase at height and may be 15%–25% greater at a typical wind turbine hub height of 80 ft (24 m) than those measured at airport anemometer heights. The Wind Energy Resource Atlas contains data from airports in the United States and makes wind data summaries available.

Again, it is essential to have site-specific data to determine the wind resource at your exact location. If you do not have on-site data and want a more transparent, predictable picture of your wind resource, you may wish to measure wind speeds at your location for a year. You can do this with a recording anemometer, which generally costs $500–$1,500. The most accurate readings are taken at "hub height" (i.e., the elevation at the top of the wind turbine tower). This requires placing the anemometer high enough to avoid turbulence created by trees, buildings, and other obstructions. The standard wind sensor height used to obtain data for the DOE maps is 10 m (33 ft).

It is not unusual to have varied wind resources within the same property. If you live in complex terrain, select the installation site. If you locate your wind turbine on the top of the windy side of a hill, for example, you will have more access to prevailing wind than in a gully or on the leeward (sheltered) side of a mountain on the same property. Consider existing obstacles and plan for future obstructions, including trees and buildings, which could block the wind. Also, recall that the power in the wind is proportional to its speed (velocity) cubed (v^3). This means that the amount of power you get from your generator increases exponentially as the wind speed increases. For example, if your site has an annual average wind speed of about 5.6 m/s (12.6 mph), it has twice the energy available as a site with a 4.5 m/s (10 mph) average (12.6/103).

Another proper indirect measurement of the wind resource is the observation of an area's vegetation. Strong winds can permanently deform trees, especially conifers or evergreens. The deformity, known as "flagging," has been used to estimate the average wind speed for an area (see Figure 6.9).

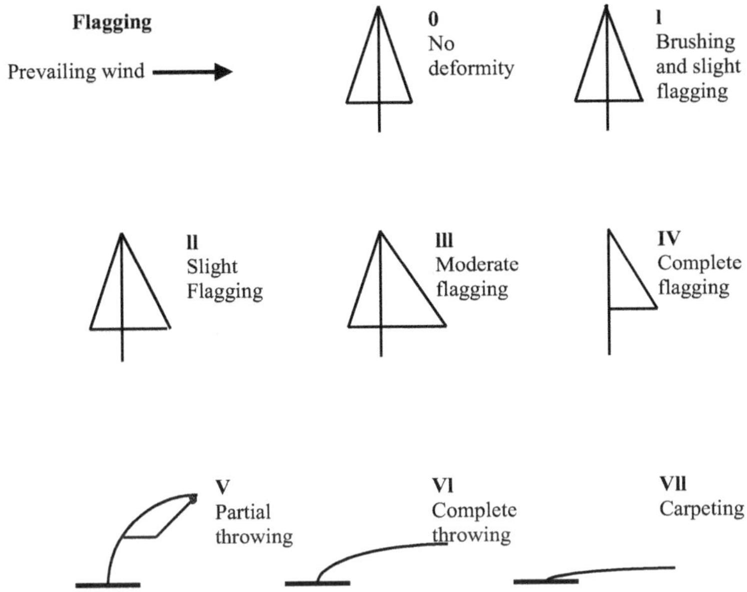

Griggs-Putnam Index of Deformity

Index	1	11	111	1V	V	V1	V11
Wind mph	7-9	9-11	11-13	13-16	15-18	16-21	22+
Speed m/s	3-4	4-5	5-6	6-7	7-8	8-9	10

FIGURE 6.9 A crude method of approximating average annual wind speed from the deformation of trees (and other foliage).

In addition to ensuring the proper siting of your small wind turbine system, legal, environmental, and economic issues must be addressed. For example, you should also

- research potential legal and environmental obstacles;
- obtain cost and performance information from manufacturers;
- perform a complete economic analysis that accounts for a multitude of factors;
- understand the basics of a small wind system; and
- review possibilities for combining your system with other energy sources, backups, and efficiency improvements.

Regarding economic issues, because energy efficiency is usually less expensive than energy production, making your house more energy efficient will likely result in being able to spend less money since you may need a smaller wind turbine to meet your needs. But a word of caution: before you spend any money, research potential legal and environmental obstacles to installing a wind system. Some jurisdictions, for example, restrict the height of the structures permitted in residentially zoned areas, although variances are often obtainable. Your neighbors might object to a wind machine that blocks their view, or they might be concerned about noise. Consider obstacles that might block the wind in the future (significant, planned developments or saplings, for example). Saplings that will grow into large trees can be a problem in the future. As mentioned, trees can affect wind speed (see Figure 6.9). If you plan to connect the wind generator to your local utility company's grid, determine its requirements for interconnections and buy electricity from small independent power producers.

When you are convinced that a small wind turbine is what you want and no obstructions are restricting its installation, approach buying a wind system as you would any major purchase.

6.7 AI IN WIND POWER

Wind energy is an efficient and sustainable renewable energy source that is expanding exponentially globally. However, the sector has efficiency, scaling, and cost challenges. Artificial intelligence helps address these challenges by providing immediate improvements and the incentive to explore additional enhancements.

The wind energy production industry uses AI to provide environmental data, including weather forecasting and wind analysis. The benefits include accurate predictions and adjustments to weather and wind conditions, which lead to operational efficiency and reduce the chance of system damage due to severe weather conditions. AI also aids in site selection, preconstruction planning, optimization, turbine-planned maintenance, monitoring, scheduling, and inspection.

6.8 THE BOTTOM LINE

According to Archer and Jacobson (2004), technology is much more advanced today in utilizing our wind resources, and the United States is home to one of the best wind resource areas in the world, the Midwest states of North and South Dakota, Nebraska, Kansas, Montana, Iowa, and Oklahoma. However, as with any other source of energy, nonrenewable or renewable, there are advantages and disadvantages associated with their use. On the positive side, wind energy is a free, renewable resource, so no matter how much it is used today, there will still be the same supply. Wind energy is also a source of clean, non-polluting electricity. Wind turbines can be installed on farms or ranches, thus benefiting the economy in rural areas, where most of the best wind sites are found. Moreover, farmers and ranchers can continue to work the land because the wind turbines use only a fraction of the land—the height and distance between turbines means that land used for wind turbines can also be used for agriculture and grazing, i.e., only about 5% of the land in a wind farm is occupied by the turbines themselves. One huge advantage of wind energy is that it is a domestic energy source produced in the United States or the country where it is installed and where wind is abundant. Again, in the United States, the wind supply is plentiful.

On the other hand, wind energy does have a few negatives, such as wind projects facing opposition. Wind power must compete with conventional generation sources on a cost basis. Even though wind power costs have decreased dramatically in the past 10 years, technology requires a higher initial investment than fossil-fueled generators. The challenge of using wind as a power source is that the wind is intermittent and does not always blow when electricity is needed. Wind energy cannot be stored (unless batteries are being used), and not all winds can be harnessed to meet the timing of electricity demands. Another problem is that good sites are often located in remote locations, far from cities where electricity is needed. Moreover, wind resource development may compete with other uses for the land, and those alternative uses may be more highly valued than electricity generation. Finally, regarding the environment, wind power plants have relatively little impact on the environment compared to other conventional power plants; there is some concern over the noise produced by the rotor blades, aesthetic (visual) impacts, and sometimes birds being killed by flying into the rotors. Most problems have been resolved or reduced through technological development or adequately locating wind plants. But keep in mind that the "not in my backyard" (NIMBY) point of view is still alive and strong in the United States and has succeeded in killing many projects, even renewable, "clean" energy projects.

We understand some of the issues related to the NIMBY point of view, although we disagree with them. While it is questionable and arguable that the United States has reached its apogee in attaining the so-called good life for most of its citizenry and that many feel that our best days are behind us, it is not questionable about our future if we do not innovate and free ourselves from the bondage of fossil fuels and the accompanying economics, politics, and potential turmoil. With fossil fuel supplies dwindling and world political pressures for accessibility ramping up to dangerous levels, we need to find reliable alternative energy sources for the future. We may never be able to improve on the so-called "good life," but it would be beneficial for most of us if we can maintain the level of living that we have obtained to this point while at the same time reducing oil imports and carbon emissions.

One thing is sure: wind energy production continues to be one of the fastest-growing energy technologies. It looks set to become a significant generator of electricity worldwide.

NOTES

1 Parts of this section are adapted from Spellman, F.R., 2006. *Environmental Science and Technology*. Lanham, MD: Government Institutes Press.
2 Much of the information in this section is from USDOE-EERE, 2005. *History of Wind Energy*. Accessed 06/14/09 @ http://www1/eere/emergu/gpv/womdamdjudrp/printable_versions/wind_hisotry.htm.

REFERENCES

Alternative Energy News, 2010. *Wind Generator Technologies*. Accessed 02/20/24 @ https://www.alternative-energy-news.info/technology/wind-pworer/widn-turbines/.

Archer, C., and Jacobson, M.Z., 2004. *Evaluation of Global Wind Power*. Stanford, CA: Department of Civil and Environmental Engineering, Stanford University.

Betz, A., 1966. *Introduction to the Theory of Flow Machines*. (D.G. Randall, Trans.) Oxford: Pergamon Press.

Devlin, L, 2007. Wind Turbine Efficiency. Accessed @ http://k0lee.com/2007/11/wind-turbine-efficiency/.

EERE, 2006a. *Wind & Hydropower Technologies Program*. Accessed 02/20/24 @ https://www1.3343.energy.gov/windandhydro/wind_how.html.

EERE, 2006b. *How Wind Turbines Work*. Accessed 02/20/24 @ https://www1.eere.energy.gov/Windandhydro/wind_how.html.

EIA, 2023. *Wind Explained*. Accessed 11/21/24 @ https://www.eia.gov/energyexplained/wind/electricity-generation-from-wind.php.

Hanson, B.J., 2004. *Energy Power Shift*. Maple, WI: Lakota Scientific Press.

Khaligh, A., and Onar, O.C., 2010. *Energy Harvesting*. Boca Raton, FL: Taylor and Francis.

RREDC, 2010. *Wind Energy Resource Atlas of the U.S.* Accessed 02/25/24 @ rredc.nrel.gov/Wind/pubs/atlas.

WINDEIS, 2009. *Wind Energy Guide*. Accessed 06/13/2024 @ https://windeis.anl.gov/guide/basics/index.cfm.

7 Solar Energy and AI

7.1 INTRODUCTION

It is fitting to begin our discussion of the various kinds of renewable energy with the sun—the medium-sized star that symbolizes life, power, strength, force, clarity, and, yes, energy. The sun nourishes our planet. Solar energy is a renewable resource because it is continuously supplied to the earth by the sun. When we first consider the sun and solar energy, we quickly realize there is nothing new about renewable energy. The sun was the first energy source; it has been around for 4.5 billion years if anything else we know. On Earth, without the sun, there is nothing—absolutely nothing, zilch. There is the old 5-5-5-thumb rule that guides the essence of life on Earth; that is, we commonly state that one cannot survive a lack of breath (oxygen) after 5 minutes; we must have water available to drink in 5 or less days; and we need food to eat in 5 weeks or less to stay alive. All this is true, of course, but without the sun, the 5-5-5 rule has no validity—without the sun, nothing related to life has validity—without the sun, there is nothing: no air, no water, no food, and no life.

The sun, the ultimate energy source, provided light and heat to the first humans. During the day, people searched for food. They hunted and gathered and probably stayed together for their safety. When nightfall arrived, we can only imagine that they huddled together for warmth and reassurance in the light of the stars and moon, waiting for the sun and its life-giving and life-sustaining light to return.

Solar energy (a term used interchangeably with solar power) uses the sun's power, using various technologies, "directly or indirectly" to produce energy. Solar energy is one of the best renewable sources available because it is one of the cleanest. Direct solar radiation absorbed in solar collectors can provide space heating and hot water. Passive solar energy can enhance use in buildings for space heating and lighting requirements. Solar energy can also be used to produce electricity, and this is the renewable energy area that is the focus of attention in this section.

Radiant energy from the sun, in the form of photons, strikes the surface of the earth with the average equivalent of about 168 kWh of energy (equal to 575,000 Btus thermal energy) per square foot per year; it varies, of course, with location, cloud cover and orientation with the surface (Hanson, 2004). The question then becomes, how much of this energy is used in the United States by various consumers? According to EERE (2008a), approximately 11% of renewable energy is solar power used by U.S. citizens, which is 1.2 quadrillion Btu.

In its Solar Energy Technology Program, USDOE (2009) points out that solar technologies diversify the energy supply and reduce the country's dependence on imported fuel. Moreover, solar energy technologies provide a beneficial environmental impact by helping to improve air quality and offset greenhouse gas emissions. An additional benefit of the solar technology industry's growing benefit is that it stimulates our economy by creating solar manufacturing and installation jobs.

DOI: 10.1201/9781003605430-9

According to USDOE (2009), the two solar electric technologies with the most significant potential (based on cost-effectiveness) are concentrating solar power (CSP) and photovoltaics (PV).

7.2 CONCENTRATING SOLAR POWER (CSP)[1]

CSP offers a utility-scale, dependable, firm, dispatchable renewable energy option that can help meet a nation's demand for electricity. EERE (2008a) points out that nine trough plants producing more than 400 megawatts (MW) of electricity have operated reliably in the California Mojave Desert since the 1980s. CSP plants produce power by using mirrors to focus sunlight to heat a working fluid—a thermal receiver like a boiler tube. The receiver absorbs and converts sunlight into heat. This high-temperature fluid is used to spin a turbine or power an engine that drives an electricity generator.

Concentrating solar power systems can be classified by how they collect solar energy: linear concentrators, dish/engines, or power tower systems (NREL, 2009). All three are typically engineered with tracking devices that follow the sun seasonally and throughout the day to maximize the system's electrical output (Chiras, 2002).

7.2.1 Linear Concentrators

As mentioned, linear CSP concentrators or collectors are one of the three types of CSP systems in use today. Linear CSP collectors capture or collect the sun's energy using long rectangular, curved (U-shaped) mirrors. The mirrors are tilted toward the sun, focusing sunlight on tubes (or receivers) that run the length of the mirrors. The reflected sunlight heats a fluid flowing through the tubes. The hot fluid is then used to boil water to create superheated steam that spins a conventional steam turbine generator to produce electricity. Alternatively, steam can be generated directly in the solar field, eliminating the need for costly heat exchangers.

Linear concentrating collector fields consist of many collectors in parallel rows typically aligned in a north-south orientation to maximize annual and summertime energy collection. With a single-axis sun-tracking system, this configuration enables the mirrors to track the sun from east to west during the day, ensuring that the sun reflects continuously onto the receiver tubes (EERE, 2008b).

There are two major types of linear concentrator systems: parabolic trough systems, where receiver tubes are positioned along the focal line of each parabolic mirror, and linear Fresnel reflector systems, where one receiver tube is positioned above several mirrors to allow the mirrors greater mobility in tracking the sun.

7.2.1.1 Parabolic Trough Systems

EERE (2008b) points out that the predominant CSP systems currently operating in the United States are linear concentrators using parabolic trough collectors. All rays of light that enter parallel to the axis of a parabolic-shaped mirror will be reflected to one point, the focus. Hence, it is possible to concentrate virtually all the radiation incident upon the mirror in a relatively small area in the focus (Perez-Blanco, 2009). Parabolic trough systems form a long-curved array of mirrors (usually coated silver or polished aluminum). Trough systems have been developed in the United States, Spain, and Japan (Chiras, 2002). The receiver tube (Dewar tube) runs its length and is positioned along the focal line of each parabola-shaped reflector (Duffie & Beckman, 1991; Patel, 1999). The tube is fixed to the mirror structure. The heated fluid—either a heat-transfer fluid (usually oil) or water/stream—flows through and out of the field of solar mirrors to where it is used to create steam [or, in the case of a water/steam receiver, it is sent directly to the turbine (i.e., the turbine is the prime mover of the electrical generator that produces electrical current flow in an accompanying electrical distribution system)]. Temperatures in these systems range from 150°F to 750°F (80°C–400°C).

Currently, the largest individual trough systems generate 80 MW of electricity. However, individual systems being developed will generate 250 MW. In addition, individual systems can be collocated in power parks. This capacity would be constrained only by transmission capacity and availability of contiguous land area.

Trough designs can incorporate thermal storage. Thermal storage systems are discussed later, but for informative purposes, we can see that in such a system, the collector field is oversized to heat a storage system during the day that can be used in the evening or during cloudy weather to generate additional steam to produce electricity. Parabolic trough plants can be designed as hybrids, meaning that they use fossil fuel to supplement the solar output during periods of low solar radiation. A natural gas-fired heater or gas-steam boiler/reheater is used in this design. Troughs may be integrated into existing or new combined-cycle natural gas and coal-fired plants (EERE, 2008b).

7.2.1.2 Fresnel Reflector Systems

The linear Fresnel reflector system is a second linear concentrator technology, a solar power collector. Flat or slightly curved mirrors mounted on trackers on the ground are configured to reflect sunlight onto a water-filled receiver tube fixed in space above these mirrors. A small parabolic mirror is sometimes added atop the receiver to focus the sunlight further. The critical advantage of Fresnel reflector systems is their simplicity compared to other systems.

7.3 DISH/ENGINE SYSTEM

Dish/engine systems use a mirrored dish-shaped parabolic mirror like a giant satellite dish. The dish-shaped surface directs and concentrates sunlight onto a thermal receiver (the interface between the dish and the engine/generator), which absorbs and collects the heat and transfers it to the engine generator. Solar dish/engine systems convert the energy from the sun into electricity at a very high efficiency. The power conversion unit includes the thermal receiver and the engine/generator. It absorbs the concentrated solar energy beams, converts them to heat, and transfers them to the engine/generator. A thermal receiver can be a bank of tubes with a cooling fluid—usually hydrogen or helium—typically the heat-transfer medium and an engine's working fluid. Alternate thermal receivers are heat pipes, where the boiling and condensing of an intermediate fluid transfers the heat to the engine. The most common type of heat engine used today in dish/engine systems is the Stirling engine (conceived in 1816 and mentioned earlier). This system uses the fluid heated by the receiver to move pistons and create mechanical power. Mechanical energy runs a generator or alternator to produce electricity (USDOE, 1998).

Disk/engine systems use dual-axis collectors to track the sun. As mentioned, the ideal concentrator shape is parabolic, created by a single reflective surface, multiple reflectors, or facets. Many options exist for receiver and engine type, including the Stirling cycle, microturbine, and concentration PV modules.

Because of the high concentration ratios achievable with parabolic dishes and the small size of the receiver, solar dishes are efficient (as high as 30% efficient) at collecting solar energy at very high temperatures (900°F–2,700°F; 500°C–1,500°C).

USDOE (1998) points out that solar dish/engine systems have environmental, operational, and potential economic advantages over more conventional power generation options because they:

- produce zero emissions when operating on solar energy;
- operate more quietly than diesel or gasoline engines;
- are more accessible to operate and maintain than conventional engines;
- start up and shut down automatically; and
- operate for long periods with minimal maintenance.

Because of their size and durability, solar dish/engine systems are well-suited and often used for non-traditional power generation. Individual units range in size from 10 to 25 kW. They can operate independently of power grids in remote, sunny locations, pumping water and providing power to people living in isolated places.

As mentioned, dish/engine systems produce high temperatures, which in turn can be used to produce steam that can be used either for electricity production or select high-temperature industrial processes (Chiras, 2002). Dish/engine systems can also be linked to provide utility-scale power to a transmission grid. Such systems could be located near consumers, substantially reducing the need for building or upgrading transmission capacity. Mainly because of their high efficiency, the cost of these systems is expected to be lower than that of other solar systems for these applications (USDOE, 1998).

7.4 POWER TOWER SYSTEM

The power tower or the central receiver system consists of a large field of flat, sun-tracking mirrors known as heliostats (silver-laminated acrylic membranes) focused on concentrating sunlight onto a receiver on the top of a tower. A heat-transfer fluid heated in the receiver generates steam, which, in turn, is used in a conventional turbine generator to produce electricity. Some power towers use water/steam as the heat-transfer fluid. Other advanced designs are experimenting with molten (liquefied) nitrate salt because of its superior heat transfer and energy-storage (heat retention) capabilities, allowing continued electricity production for several consecutive cloudy days (NREL 2010). The energy-storage capability, or thermal storage, allows the system to continue dispatching electricity during cloudy weather or at night. Therefore, a power tower system comprises five main components: heliostats, receivers, heat transport and exchange, thermal storage, and controls (see Figure 7.1). Individual commercial plants can be sized to produce up to 200 MW of electricity.

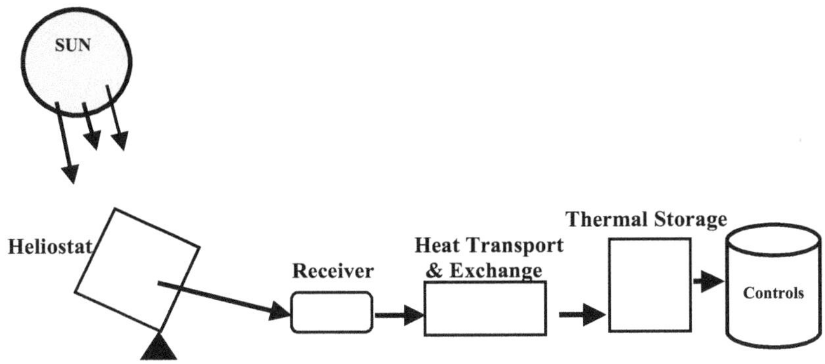

FIGURE 7.1 Solar power tower main components.

DID YOU KNOW?

Smaller CSP systems can be located directly where the power is needed. For example, a single dish/engine system can produce 3–25 kW of power and is well-suited for distributed applications. Larger, utility-scale CSP applications provide hundreds of megawatts of electricity for the power grid. Both linear concentrator and power tower systems can be easily integrated with thermal storage, helping to generate electricity during cloudy periods or at night. Alternatively, these systems can be combined with natural gas, and the resulting hybrid power plants can provide high-value, dispatchable power throughout the day.

DID YOU KNOW?

Concentrated solar power plants use water not only for lens cleaning but also for stream cycle production. Water consumption is an issue because they need water and are most effective in locations where the sun is most intense, which in turn often corresponds to places like the Mohave Desert, where there is little water (USDOE, 2008).

7.5 THERMAL STORAGE (EERE, 2008c)

Thermal energy storage (TES) has become critical to any CSP system deployed today. One challenge facing the widespread use of solar energy is the reduced or curtailed energy production when the sun sets or is blocked by clouds. Thermal energy storage provides a workable solution to this challenge. In a CSP system, as mentioned, the sun's rays are reflected onto a receiver, creating heat that is then used to generate electricity. The thermal energy can be stored later if the receiver contains oil or molten salt as the heat transfer medium. This allows CSP systems to be the cost-competitive option for providing clean, renewable energy. Presently, steam-based receivers cannot store thermal energy for later use.

An essential criterion in selecting a storage medium, such as the oil or molten salt referred to above, is the *substance's specific heat.* Specific heat (c) measures the heat energy required to increase a unit quantity of a substance by a unit of temperature. For example, at a temperature of 15°C, the heat energy needed to raise water's temperature to 1 K (equal to 1°C) is 4,186 J/kg. Different materials absorb different amounts of heat when undergoing the same temperature increase. The relationship between a temperature change ΔT and the amount of heat Q added (or subtracted) is given by

$$Q = mc\Delta T \tag{7.1}$$

where c is the specific heat and m is the mass of the substance. Table 7.1 lists specific heats for different materials per pound and cubic foot. Another important parameter related to specific heat and thermal energy storage materials is *heat capacity,* which is defined as the ratio of the heat energy absorbed by a substance to the substance's increase in temperature and is given by

$$\text{Heat Capacity} = \text{Specific Heat} \times \text{Density} \tag{7.2}$$

Another medium for thermal storage is *phase change materials*, classified as latent heat storage units. Specifically, a phase material is a substance with a high heat of fusion (i.e., the amount of thermal energy that must be absorbed or evolved for 1 mole of a substance to change states from a solid to a liquid or vice versa without a temperature change) which, melting and solidifying at a specific temperature, can store and release large amounts of energy. Heat is absorbed or released when the material changes from solid to liquid. One joint group of substances being used as phase change materials for solar applications (and elsewhere) are eutectic salts; salts are combined with water, such as sodium sulfate decahydrate, also known as Glauber's Salt. This salt melts at 91°F with the addition of 108 Btu per pound. Conversely, when the temperature drops below 91°F, 108 Btu per pound of heat energy is released as the slat solidifies (Hinrichs & Kleinbach, 2006).

TABLE 7.1
Thermal Energy Storage Materials

	Specific Heat	Density		Heat Capacity	
Material	**Btu/lb/°F**	**(lb/ft³)**	**kg/m³**	**(Btu/ft³-°F)**	**(kJ/m³-°C)**
Water	1.00	62	1,000	62	4,186
Iron	0.12	490	7,860	59	3,521
Copper	0.09	555	8,920	50	3,420
Aluminum	0.22	170	2,700	37	2,430
Concrete	0.23	140	2,250	32	2,160
Stone	0.21	170	2,700	36	2,270
White pine	0.67	27	435	18	1,220
Sand	0.19	95	1,530	18	1,540
Air	0.24	0.075	1.29	0.02	1.3

Several TES technologies have been tested and implemented since 1985. These include the two-tank direct, two-tank indirect, and single-tank thermocline systems.

7.5.1 Two-Tank Direct System

Solar thermal energy in this system is stored in the same fluid used to collect it. The fluid is stored in two tanks—one at high and the other at low temperatures. Fluid from the low-temperature tank flows through the solar collector or receiver, where solar energy heats it to a high temperature, and it then flows back to the high-temperature tank for storage. Fluid from the high-temperature tank flows through a heat exchanger, generating steam for electricity production. The fluid exits the heat exchanger at a low temperature and returns to the low-temperature tank. Two-tank direct storage was used in early parabolic trough power plants and at California's Solar Two power tower. The trough plants used mineral oil as the heat transfer and storage fluid; Solar Two used molten salt.

7.5.2 Two-Tank Indirect System

This system functions the same way as the two-tank direct system, except different fluids are used as the heat-transfer and storage fluids. This system is used in plants where the heat-transfer fluid is too expensive or not suited for use as the storage fluid. The storage fluid from the low-temperature tank flows through an extra heat exchanger, which is heated by the high-temperature heat-transfer fluid. The high-temperature storage fluid then flows back to the high-temperature storage tank. The fluid exits this heat exchanger at a low temperature and returns to the solar collector or receiver, where it is heated back to the high temperature. Storage fluid from the high-temperature tank generates steam like the two-tank direct system. The indirect system requires an extra heat exchanger, which adds cost. This system will be used in many of the parabolic power plants in Spain and has also been proposed for several U.S. parabolic plants. The plants will use organic oil as the heat-transfer fluid and motor salt as the storage fluid.

7.5.3 Single-Tank Thermocline System

This system stores thermal energy in a single tank in a solid medium, most commonly silica sand. At any time during operation, a portion of the medium is at a high temperature and a portion is at a lower temperature. A temperature gradient or thermocline separates the hot- and cold-temperature regions (see Figure 7.2). High-temperature heat-transfer fluid flows into the top of the thermocline and exits the bottom at low temperature. This process moves the thermocline downward and adds thermal energy to the system. Reversing the flow pushes the thermocline upward and removes thermal energy from the system to generate steam and electricity. Buoyancy effects create thermal stratification of the fluid within the tank, which helps to stabilize and maintain the thermocline.

Note: Using a solid storage medium and only needing one tank reduces the cost of the single-tank thermocline system relative to the two-tank systems.

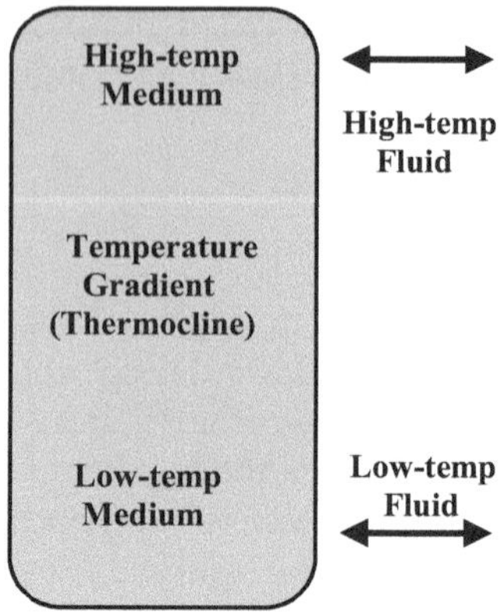

FIGURE 7.2 Single-tank thermocline thermal energy storage system.

7.6 PHOTOVOLTAICS (PV)

Hinrichs and Kleinbach (2006) point out that photovoltaic (PV) has been and will continue to be one of the more glamorous technologies in the energy field. Photovoltaic (Gr. *photo* light and *volt*, electricity pioneer Alessandro Volta) technology uses abundant energy in the sun, and it has little impact on our environment. *Photovoltaics* is the direct conversion of light (photons) into electricity (voltage) at the atomic level. Some materials exhibit a property known as the *photoelectric effect* (discovered and described by Becquerel in 1839) that causes them to absorb photons of light and release electrons. When these free electrons are captured, an electric current results (i.e., electricity is the flow of free electrons) that can be used as electricity. The first photovoltaic module (billed as a solar battery) was built by Bell Laboratories in 1954. In the 1960s, the space program began to make the first serious use of technology to provide power aboard spacecrafts. Space program use helped this technology make giant advancements in reliability and helped to lower costs. However, the oil embargo of the 1970s (the so-called energy crisis) propelled photovoltaic technology to the forefront of recognition for use other than space applications only. Photovoltaics can be used in many products, from small consumer items to large commercial solar electric systems.

Figure 7.3 illustrates the photoelectric effect. When light shines on the negative plate, electrons are emitted with kinetic energy inversely proportional to the wavelength of the incident light. Figure 7.4 illustrates the operation of a basic *photovoltaic cell*, also called a *solar cell*. Solar cells are made of silicon and other semiconductor materials such as germanium, gallium arsenide, and silicon carbide, and they are used in the microelectronics industry. For solar cells, a thin semiconductor wafer is specially treated to form an electric field, positive on one side and negative on the other. When light energy strikes the solar cell, electrons are jarred loose from the atoms in the semiconductor material (see Figures 7.3 and 7.4). Suppose electrical conductors are attached to the positive and negative sides, forming an electrical circuit. In that case, the electrons can be captured in the form of an electrical current—that is, again, recall that electron flow is electricity. This electricity can power a load, such as a light, tool, toaster, and other electrical appliances and apparatus.

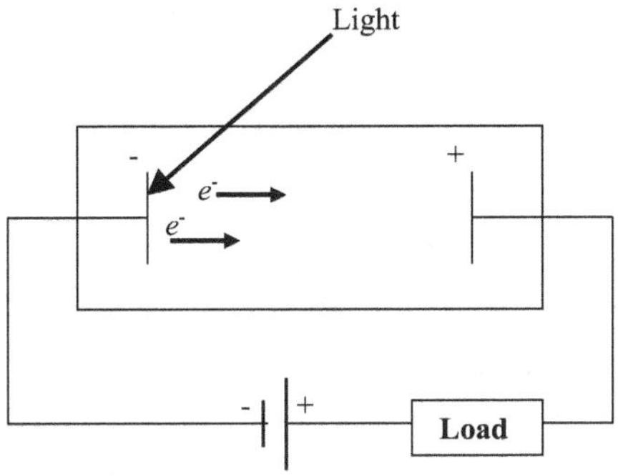

FIGURE 7.3 Photoelectric effect.

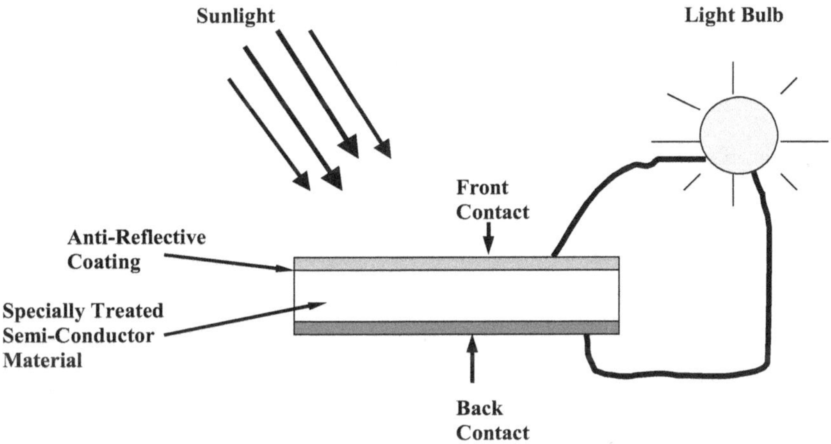

FIGURE 7.4 Operation of primary photovoltaic cell.

Several solar cells are electrically connected and mounted on a support panel or frame, called a photovoltaic module (see Figure 7.5). Solar panels used to power homes and businesses are typically made from solar cells combined into modules that hold about 40 cells. Modules are designed to supply electricity at a specific voltage, such as a standard 12-V system. The current produced is directly dependent on how much light strikes the module.

Multiple modules can be wired together to form an array. Generally, the larger the area of a module or array, the more electricity it will produce. Photovoltaic modules and arrays produce direct-current (DC) electricity. They can be connected in series and parallel electrical arrangements for the required voltage and current combination.

7.7 SOLAR ENERGY AND DOMESTIC/INDUSTRIAL USE

When considering a potential renewable energy source, we first want to know if it is reliable, clean, and affordable. This is especially true for consumers and business owners contemplating solar applications for home and business use. Until recently,

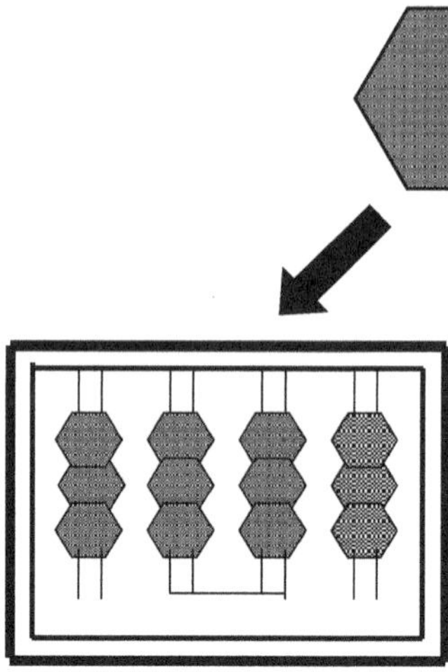

FIGURE 7.5 Single solar cell and solar cell module.

solar energy has been used in home and industrial locations but only on a somewhat limited basis. This trend is changing, especially in those locations where sunlight is prevalent throughout the year. With the soaring costs of fossil fuel supplies and their pending decreased availability, solar power is beginning to receive more attention. For domestic use, solar technologies include photovoltaics, passive heating, daylighting, and water heating. Industrial and commercial facilities may use the same solar technologies for residential buildings. These nonresidential buildings can also use solar energy technologies that would be impractical for a home. These technologies include ventilation, air preheating, solar process heating, and solar cooling.

7.7.1 Solar Hot Water

Just as the sun heats the surface layers of exposed bodies of water—ponds, lakes, streams, and oceans—it can also heat water used in buildings and swimming pools. Two main parts are added to harness the sun's ability to heat water: a solar collector and a storage tank. The most common collector is a *flat-plate collector*. The flat-plate collector system is the most used collector and is ideal for applications that require water temperatures under 140°F (60°C). Mounted on the roof, it consists of a thin, flat, rectangular, insulated box with a transparent glass cover that faces the sun. Small copper tubes run through the box and carry the fluid—either water or other fluid, such as an antifreeze solution—to be heated. The tubes, typically arranged in series (i.e., water flows in one end and out the other), are attached to an absorber plate, which, along with the tubes, is painted black to absorb the heat. As heat builds up in the collector, it heats the fluid (water or propylene glycol) passing through the tubes. The motive force for passing fluid through the tubes can be accomplished by active or passive means. In an active system, water heaters, the most common type of system, rely on an electric pump and controller to circulate water or other heat-transfer fluids through the collectors. Passive solar water-heater systems rely on

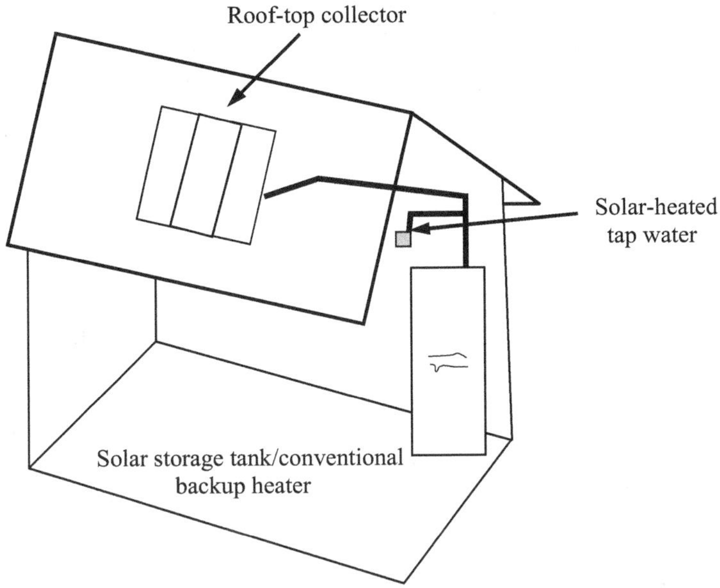

FIGURE 7.6 Simplified representation of a home solar water heating system.

gravity and the tendency for water to circulate as it is heated naturally. Because passive systems have no "moving parts," they are more reliable, easier to maintain, and often have a longer life than active systems.

The storage tank holds the hot liquid (see Figure 7.6). It is usually a large, well-insulated tank, but modified water heaters can store hot liquid. When the fluid is other than water, it is heated by passing through a coil of tubing in the tank, which is full of hot fluid. Active solar systems usually include a storage tank and a conventional water heater. In two-tank systems, the solar water heater preheats water before it enters the traditional water heater.

7.7.2 Solar Process Heat

In addition to conventional domestic water heating systems, commercial and industrial buildings may use the same solar technologies—photovoltaics, passive heating, daylighting, and water heating. These nonresidential buildings can also use solar energy technologies that would be impractical for a home. These technologies include ventilation air preheating, solar process heating, and solar space cooling.

7.7.2.1 Space Heating

Many large buildings need ventilated air to maintain indoor air quality (IAQ). Although all that fresh air is excellent for IAQ, heating this air can use large amounts of energy and be very expensive in cold climates. But an elegantly simple solar ventilation system can preheat the air, saving energy and money. This type of system typically uses a transpired collector.

Transpired air collector systems essentially consist of a dark-colored, perforated façade installed on a building's south-facing wall. An added fan or the building's existing ventilation system draws ventilation air into the building through the perforated absorber plate on the facade and up the plenum (the air space between the absorber and the south wall). According to USDOE (2006), solar energy absorbed by the dark absorber and transferred to the air flowing through it can preheat the intake air by as much as 40°F (22°C). Reduced heating costs will pay for the systems in 3–12 years.

In addition to meeting a portion of a building's heating load with clean, free solar energy, the transpired collector helps save energy and money in other ways. It recaptures heat loss through a building's south-facing wall; heat that escapes through the south wall is captured in the air between the structural wall and the transpired collector and returned to the interior. Also, by introducing make-up air through ceiling-mounted ducts, the system eliminates the wasteful air stratification that often plagues high-ceiling buildings (USDOE, 2006).

Using a working system in an industrial application is best to illustrate an active, real-world example of transpired collectors (used as solar preheaters for outdoor ventilation air). The following case study is an example; experts have monitored it extensively.

Case Study 7.1—General Motors Battery Plant[2]

The General Motors battery Plant in Oshawa, Canada, is a 100,000-ft² facility where automotive batteries are manufactured. Built in the 1970s, it has an open shop floor and a 298-foot-high ceiling. GM operates two full-time production shifts within the plant and conducts maintenance activities at night and on weekends, so the building is continuously occupied.

Until the early 1990s, GM relied solely on a team-operated fan coil system for space heating. However, the system could not provide the necessary quantities of heated outdoor air, resulting in inadequate plant ventilation.

In 1991, plant management installed a transpired collector to correct the ventilation problems. During the following two years, the transpired collector system was modified slightly to improve airflow; the original fans and motors were replaced with axial fans and high-efficiency motors, and the original ducting was replaced with upgraded fabric ducting.

The GM plant collector comprises 4,520 ft² of absorber sheeting. The lower 21 ft of the transpired collector is a black, perforated, aluminum wall cladding with 1.6-mm holes totaling 2% porosity. The average depth of the plenum between the transpired collector and the plant's structural fence is 6 inches. The canopy at the top of the wall is a manifold for the airflow and a solar heat collection device. The canopy face is made of a perforated plate with 1% porosity. The transpired collector covers about 50% of the plant's south-facing wall; the remainder of the south facade has shipping doors and other obstructions that make it unsuitable for mounting collector cladding.

The GM transpired collector hast to fan/distribution systems, each consisting of a constant-speed fan, a recirculation damper system, and a fabric distribution duct.

The total airflow delivered by the system's fans is 40,000 cfm. Recirculated air and air drawn through the solar collectors make up this flow; the percentages of each depend on the temperature of the air coming from the collector.

The GM Battery Plant's transpiring collector has been monitored extensively since its installation. The data in this case study reflect the system's performance during the 1993/1994 heating season. An in-depth report on the monitoring program is available (Enermodal Engineering Limited., 1995).

The data shows that the annual energy savings for the 4,520-ft^2 collector was 940 million Btu/year: 678 MBtu resulted from the thermal energy gained directly from the outside air as it passed through the absorber, and 262 MBtu resulted from heat loss recaptured by the wall form inside the building. Other possible energy-saving mechanisms—such as destratification and heat recapture—likely contributed to improved system performance; however, these effects are highly structure-specific and have not been incorporated into the savings reported here.

The cost of the transpired collector system at the GM plant was \$66,530, or \$14.72/ft^2 of the installed collector. The cost per square foot is higher than typical installations for two reasons: (1) this system was installed soon after the technology was introduced before design and installation procedures had been streamlined, and (2) the cost includes the fan and ducting modifications that were implemented during the first 2 years of operation.

It is essential to point out that the GM transpired that the collector had several operational problems. After the system was installed, employees complained about fan noise and cold drafts, and the system was occasionally disabled. The fan and duct upgrades described previously eliminated the problems on one of the fan systems; the other fan continues to generate noise, and employees still disable it when working in the immediate vicinity. The manufacturer (Conserval) has addressed these complaints by specifying smaller but more numerous fans in subsequent installations.

Also, both bypass dampers and a recirculation damper required additional maintenance. The recirculation damper became stuck in full recirculation mode, and a new modulating motor was installed to fix the problem. The bypass dampers occasionally become bound, which leads to unacceptably high leakage rates. These dampers were kept closed manually through the 1993–1994 heating season.

7.7.2.2 Water Heating

Solar water-heating systems are designed to provide large quantities of hot water for nonresidential buildings. A typical system includes solar collectors that work along with a pump, heat exchanger, and/or one or more large storage tanks. The two main types of solar collectors used for nonresidential buildings—an evacuated-tube collector and a linear concentrator—can operate at high temperatures and have high efficiency. An evacuated-tube collector is a set of double-walled glass tubes and reflectors to heat the fluid inside the tubes. A vacuum between the two walls insulates the inner tube, retaining heat. Linear concentrators use long, rectangular, curved (U-shaped) mirrors tilted to focus sunlight on tubes that run along the length of the mirrors. The concentrated sunlight heats the fluid within the tubes.

7.7.2.3 Space Cooling

Space cooling can be accomplished using thermally activated cooling systems (TACS) driven by solar energy. Because of a high initial cost, TACS is not widespread. The two systems currently in operation are solar absorption and desiccant systems. Solar absorption systems use thermal energy to evaporate a refrigerant fluid to cool the air. In contrast, solar desiccant systems use thermal energy to regenerate desiccants that dry the air, thereby cooling the air. These systems also work well with evaporative coolers in more humid climates.

7.8 LET THERE BE NATURAL LIGHT[3]

Lighting is an essential element of any building. Proper lighting improves the aesthetics of indoor spaces and provides illumination for tasks and activities. An efficient lighting strategy, including natural daylighting, can give appropriate illumination levels and reduce energy costs. *Daylighting* is a passive solar design that illuminates a living space, an industrial or commercial working space. In the Northern Hemisphere, southern exposure receives the most significant amount of sunlight, while the opposite is true in the Southern Hemisphere. In other words, large picture-type windows in the south-facing wall or roof allow natural light to penetrate and save on expensive electrical lighting.

Because the proper use of daylighting requires integrating natural and artificial lighting sources early in the building design process, it is important to understand the fundamentals of lighting and lighting technology. Accordingly, in this section, we first discuss lighting technology before discussing the principles and practices associated with daylighting.

7.8.1 Lighting Technologies

Lighting technologies include:

- **Lamps**—They include lighting sources like fluorescent and incandescent bulbs and solid-state lighting.
- **Ballasts**—These are used with electric discharge lamps such as fluorescent lamps, ballasts transform and control electrical power to the light.
- **Luminaries (fixtures)**—They are complete lighting units that contain the bulbs and, if necessary, the ballasts.
- **Lighting controls**—Devices such as timers and sensors that can save energy by turning lights off when not needed.
- **Daylighting**—It is the use of natural light in a building.
- **Solid-state lighting**—No other technology offers as much potential to save energy and enhance the quality of building environments, contributing to energy and climate change solutions.

7.8.1.1 Lamps

Commonly called lightbulbs, lamps produce light. When comparing lamps, it is essential to understand the following performance characteristics:

- **Color rendering index (CRI)**—A measurement of a light source's accuracy in rendering different colors compared to a reference light source with the same correlated color temperature. The highest attainable CRI is 100. Lamps with

CRIs above 70 are typically used in office and living environments.

- **Correlated color temperature (CCT)**—A measurement on the Kelvin (K) scale that indicates the warmth or coolness of a lamp's color appearance. The higher the color temperature, the cooler the color appearance. Typically, a CT rating below 3,200 K is considered warm, while above 4,000 K is considered excellent.
- **Efficacy**—The ratio of the light output to the power, measured in lumens per watt (lm/W). The higher the efficacy, the more efficient the lamp.

Lamp types include:

Incandescent lamps—A standard incandescent lamp consists of a large, thin, frosted glass envelope. Inside the glass is an inert gas such as argon and nitrogen. At the center of the lamp is a tungsten filament. Electricity heats the filament. The heated tungsten emits visible light in a process called incandescence. Most standard light bulbs are incandescent lamps. They have a CRI of 100 and CCTs between 2,600 and 3,000, making them attractive lighting sources for many applications. However, these bulbs are typically inefficient (10–15 lm/W), converting only about 10% of the energy into light while transforming the rest into heat. Another type of incandescent lamp is the halogen lamp. Halogen lamps also have a CRI of 100. However, they are slightly more energy efficient and maintain light output over time. A halogen lamp also uses a tungsten filament. However, the filament is encased inside a much smaller quartz envelope. The gas inside the envelope forms the halogen group. If the temperature is high enough, the halogen gas will combine with tungsten atoms as they evaporate and redeposit them on the filament. This recycling process lets the filament last much longer. In addition, it is now possible to run the filament hotter. This means more light per unit of energy is obtained. Because the quartz envelope is so close to the filament, it becomes about four times hotter than a standard incandescent lamp. As a result of this wasted heat energy, halogen lamps—popular in torchieres (i.e., portable electric lamps with a reflector bowl that directs light upward to given indirect illumination)—are not very energy efficient. The exposed heat from halogen torchieres can pose a serious fire risk, especially near flammable objects. Because of this inefficiency and risk, manufacturers have developed torchieres for other lamps, such as compact fluorescent lamps.

Fluorescent lamps—Fluorescent lamps (70–100 lm/W) consist of a sealed glass tube. The tube contains a small amount of mercury and an inert gas, like argon, kept under shallow pressure. In these electric-discharge lamps, a fluorescing coating on the glass, called phosphor powder, transforms some of the ultraviolet energy generated into light. Fluorescent lamps also require ballast to start and maintain their operation. Early fluorescent lamps were sometimes criticized for not producing enough warm colors, making them appear too white or uncomplimentary to skin tones, and a cool white fluorescent lamp had a CRI of 62. But today, lamps with CRIs of 80 and above simulate natural daylighting and incandescent light. They also are available in a variety of CCTs: 2,900–7,000. The "T" designation for fluorescent lamps is tubular—the lamp's shape. The number after the "T" gives the diameter of the lamp in eighths of an inch. The T8 lamp—available straight or U-shaped—has become the standard for new construction. It also commonly serves as a retrofit replacement for 40-watt T12 lamps, improving efficacy, CRI, and efficiency. In some cases, T10 lamps offer advantages over T12 and T8 lamps, including higher efficiency, higher CRI values, a more comprehensive selection of CCTs, and compatibility with several ballast types. Another lamp type is the T5FT fluorescent lamp. These lamps produce maximum light output at higher ambient temperatures than linear or U-shaped ones. Linear fluorescent lamps often are less expensive than compact fluorescent lamps. They can also make more light, are easier to dim, and last longer.

DID YOU KNOW?

Cold cathode fluorescent lamps are one of the latest technological advances in fluorescent technology. The "cold" in the cold cathode means no heating filament in the light to heat the gas, making it more efficient. Also, since there is no filament to break, they are ideal for use in rough service environments where a regular lamp may fail. They are often used as backlights and exit signs in LCLD monitors.

Compact fluorescent lamps (CFLs)—They are small-diameter fluorescent lamps folded for compactness, with an efficacy of 50–75 lm/W for 27–40 W. There are several styles of CFLs: two-, four-, and six-tube lamps, as well as circular lamps. Some CFLs have the tubes and ballast permanently connected. Others have separate tubes and ballasts. Some CFLs feature a round adaptor, allowing them to screw into common electrical sockets and making them ideal replacements for incandescent lamps. They last up to 10 times longer than incandescent lamps and use about one-fourth of the energy, producing 90 percent less heat. However, typical 60–100-W incandescent lamps are no more than 5.3 inches long, while standard CFLs are longer than 6 inches. Therefore, sub-CFLs have been developed. No more than 4.5 inches long, sub-CFLs fit into most incandescent fixtures.

DID YOU KNOW?

CFLs are also suitable replacements for halogen lamps in torchieres because of their energy efficiency, brightness, and low heat output.

High-Intensity Discharge (HID) *Lamps* Compared to fluorescent and incandescent lamps, high-intensity discharge (HID) lamps produce a large quantity of light in a small package. HID lamps produce light by striking an electrical arc across tungsten electrodes housed inside a specially designed inner glass tube. This tube is filled with both gas and metals. The gas aids in the starting of the lamps. Then, the metals produce light once they are heated to evaporation. Like fluorescent lamps, HID lamps require ballast to start and maintain their operation. Types of HID lamps include mercury vapor (CRI range 15–55), metal halide (CRI range 65–80), and high-pressure sodium (CRI range 22–75). The first commercially available HID lamps were Mercury vapor lamps (25–45 lm/W), which initially produced a bluish-green light. Today, they are also available in a color-corrected, whiter light. But the newer, more efficient high-pressure sodium and metal halide lamps often replace them. Standard high-pressure sodium lamps have the highest efficacy of all HID lamps but produce a yellowish light. High-pressure sodium lamps that produce whiter light are now available, but efficiency is somewhat sacrificed. Metal halide lamps are less efficient but produce an even whiter, more natural light. Colored metal halide lamps are also available.

DID YOU KNOW?

HID lamps are typically used when high light levels are required over large areas and when energy efficiency and/or long life are desired. These areas include gymnasiums, large public areas, warehouses, outdoor activity areas, roadways, parking lots, and pathways. However, HID sources—especially metal halide (45-100 lm/W)—have been used in small retail and residential environments.

Low-pressure sodium Lamps—They have the highest efficacy of all commercially available lighting sources, producing up to 180 lm/W. Even though they emit a yellow light, a low-pressure sodium lamp should not be confused with a standard high-pressure sodium lamp—a high-intensity discharge lamp. Low-pressure sodium lamps operate much like fluorescent lamps and require a ballast. The lamps are also physically large, about 4 ft long for a 180-W size—so light distribution from fixtures is less controllable. There is a brief warm-up period for the lamp to reach full brightness.

With a CRI of 0, low-pressure sodium lamps are used where color rendition is unnecessary, but energy efficiency is. They are commonly used for outdoors, roadways, parking lots, and pathway lighting. Low-pressure sodium lamps are preferred around astronomical observatories because the yellow light can be filtered out of the random light surrounding the telescope.

Solid-State Lighting (SSL)—Compared to incandescent and fluorescent lamps, solid-state lighting (SSL) creates light with less directed heat. A semiconducting material converts electricity directly into light, making light energy efficient. SSL includes a variety of light-producing semiconductor devices, including light-emitting diodes (LEDs) and organic light-emitting diodes (OLEDs). Warm white LEDs have an efficacy of 50 lm/W, while cool white LEDs can achieve an efficacy of up to 100 lm/W. Until recently, tiny light bulbs that fit easily into an electrical circuit were used as simple indicator lamps in electronics and toys. However, recent research has achieved efficiencies equal to those of fluorescent lamps. And the cost of semiconductor material, which used to be quite expensive, has lowered, making LEDs a more cost-effective lighting option. Ongoing research shows that LEDs have great potential as energy-efficient lighting for residential and commercial building use. New uses for LEDs include small area lighting, such as task and under-shelf fixtures, decorative lighting, and pathway and step marking. As LEDs become more powerful and effective, they will be used in more general illumination applications, with entire walls and ceilings becoming the lighting system. They are already being used successfully in many general illumination applications, including traffic signals and exit signs. OLEDs are currently used in fragile, flat display screens, such as portable televisions, vehicle dashboard readouts, and postage-stamp-sized data screens built into pilots' helmet visors. Because OLEDs emit their light and can be incorporated into arrays on fragile, flexible materials, they also could be used to fashion large, fragile panels for light sources in buildings.

DID YOU KNOW?

RBG designation—red, green, blue. One way to create white light with LEDs is to mix the three primary light colors.

PC designation—phosphor conversions. White light can be produced by a blue, violet, or near-UV LED coated with yellow or multi-chromatic phosphors. The Combined light emission appears white.

7.8.1.1.1 Ballasts

Ballasts consume, transform, and control electrical power for electric-discharge lamps, providing the necessary circuit conditions for starting and operating them. Electric-discharge lamps include fluorescent, high-intensity discharge, and low-pressure sodium. When comparing ballasts, it is important to understand the following performance characteristics:

- **Ballast factor (BF)**—It is the ratio of the light output of a lamp or lamps operated by a specific ballast to the light output of the same lamp(s) operated by a reference ballast. It can be used to calculate the actual light output of that specific lamp-ballast combination. BF is typically different for each lamp type. Ballasts with extremely high BFs could reduce lamp life and accelerate lumen deficiency because of high lamp current. Extremely low BFs also could reduce lamp life because they reduce lamp current.
- **Ballast efficacy factor (BEF)**—It is the ratio of ballast factor (as a percentage) to power (in watts). BEF comparisons should be made only among ballasts operating in the same type and number of lamps.
- **System efficacy**—It is the light output ratio to the power, measured in lumens per watt (lm/W), for a particular lamp ballast system.

There are three basic types of ballast: Magnetic, hybrid, and electronic.

1. **Magnetic ballasts**—These contain a magnetic core of several laminated steel plates wrapped with copper windings. These ballasts operate lamps at line frequency (60 Hz in North America). Magnetic ones are the least expensive and least efficient of all ballasts. They have more significant power losses than electronic ballasts. But today's magnetic ballasts are 10% more efficient than the older high-loss ones, which used aluminum windings. Magnetic ballasts are available with dimming capability. However, they cannot be dimmed below 20% and still use more electricity than electronic ballasts.
2. **Hybrid ballasts** (cathode-disconnect ballasts)—-They use a magnetic core-and-coil transformer and an electronic switch for the electrode-heating circuit. Like magnetic ballasts, they operate lamps at a frequency (60 Hz in North America). After they start the lamp, these ballasts discount the electrode-heating circuit. Hybrid ballasts cost more than magnetic ones but are more energy efficient.
3. **Electronic ballasts**—In the early 1980s, manufacturers began to replace the core-and-coil transformer with solid-state, electronic compounds that could operate lamps at 20–50 kHz. These electronic ballasts experience half the power loss of magnetic ballasts. Also, lamp efficacy increases by 10%–15% compared to 60-Hz operation. Electronic ballast is the most expensive but also the most efficient. Operating lamps with electronic ballast reduces electricity use by 10%–15% over magnetic ballasts of the same light output. They are also quieter and lighter, and they virtually eliminate lamp flicker. Electronic ballasts are also available as dimming ballasts. These ballasts allow the light level to be controlled between 1% and 100%. There are a variety of electronic ballast available for use with fluorescent lamps. Electronic ballast has been successfully used with lower-watt high-intensity discharge (HID) lamps (primarily 35–100 w/MH). These ballasts provide energy savings over magnetic ballasts of 8%–20%. Their lighter weight also helps with some HID applications, such as track lighting.

7.8.1.1.2 Luminaires (Lighting Fixtures)

A lamp or lamp-ballast combination may produce light very efficiently, but the overall system efficiency may still be poor if installed in an inefficient luminaire. The best luminaire manufacturers will design their fixtures around specific lamps to optimize the light delivered to the work area. For example, a luminaire designed specifically for a compact fluorescent lamp can deliver almost 10 times as much illumination as an incandescent fixture fitted with the same compact fluorescent lamp.

Luminaire components include reflectors, diffusers—which absorb some of the light from a lamp—and polarizing panels. Reflectors can direct more of the lamp's light out of the luminaire onto the work area, and polarizing panels can sometimes increase the contrast of a visual task.

When comparing luminaires, it is essential to understand the flowing performance characteristics:

- **Illuminance**—It refers to the amount of light that reaches a surface. It is measured in footcandles (lumens/square foot) or lux (lumens/square meter).
- **Luminaire efficacy/efficiency rating (LER)**—It refers to the light output (lumens) per watt of electricity use (lm/W).

It is a good idea to consult a certified lighting designer before selecting luminaires or fixtures for an office building, factory, warehouse, or parking lot. A lighting designer will not only help find the most energy-efficient luminaires but also provide lighting that creates a comfortable and more productive work environment.

Today, energy-efficient commercial lighting design includes more than just the ambient or general lighting of a workspace, such as ceiling luminaires. When designing or retrofitting the lighting, the general Illuminance can be reduced if task lighting is implemented correctly into the overall design. Task lighting can result in significant energy savings and improved visibility for workers.

DID YOU KNOW?

The most efficient light source technology for exit signs is LEDs. The most popular parking lot luminaires use energy-efficient, high-intensity discharge or low-pressure sodium lamps. However, the most efficient light source technology for outdoor use is outdoor photovoltaic lighting.

7.8.1.1.3 Lighting Controls

Lighting controls help conserve energy and make a lighting system more flexible. The most common light control is the on/off switch. Light control technologies include manual dimming, photosensors, occupancy sensors, clock switches or timers, and centralized controls.

Manual dimming—These controls allow occupants of a space to adjust the light output or luminance. This can result in energy savings through reductions in input power, as well as reductions in peak power demand and enhanced lighting flexibility. Slider switches allow the occupant to change the lighting over the complete output range. They are the simplest of manual controls. Preset scene controls change the dimming setting for various lights simultaneously with a button. It is possible to have different settings for the morning, afternoon, and evening. Remote control dimming is also available. This type of technology is well suited for retrofit projects, where it is helpful to minimize rewiring.

DID YOU KNOW?

Fluorescent lighting fixtures require special dimming ballasts and compatible control devices. Some dimming systems for high-intensity discharge lamps also require special dimming ballasts.

Photosensors—These devices automatically adjust the light output of a lighting system based on detected Illuminance. The technology behind photosensors is the photocell. A photocell is a light-responding silicon chip that converts incident radiant energy into electrical current. While some photosensors turn lights off and on, others can also dim lights. Automatic dimming can help with lumen maintenance. Lumen maintenance involves dimming luminaires when they are new, which minimizes the wasteful effects of over-design. The power supplied to them is gradually increased to compensate for light loss over the life of the lamp.

DID YOU KNOW?

All photosensors decrease the electric power used for lighting. In addition to lowering the electric power demand, dimming the lights also reduces the heat load on a building's cooling system. Any solar heat gain in a building during the day must be considered for a whole-building energy usage analysis.

Occupancy sensors—Turning lights on and off base does not detect motion within a space. Some sensors can also be used with dimming controls to keep the lights from turning completely off when space is unoccupied. This control scheme may be appropriate when occupancy sensors control separate zones in ample space, such as in a laboratory or an open office area. In these situations, the lights can be dimmed to a predetermined level when the space is unoccupied. Sensors can also enhance the efficiency of centralized controls by switching off lights in unoccupied areas during regular working hours and after hours. There are three basic types of occupancy sensors, and these include:

- **Passive infrared (PIR) sensors**—They react to the movement of a heat-emitting body through their field of view. Wall box-type PIR occupancy sensors are best suited for small, enclosed spaces such as private offices, where the sensor replaces the light switch on the wall, and no extra wiring is required. They should not be used where walls, partitions, or other objects might block the sensors' ability to detect motion.
- **Ultrasonic**—These sensors emit an inaudible sound pattern and re-read the reflection. They react to changes in the reflected sound pattern. These sensors detect very minor motion better than most infrared sensors. Therefore, they are good to use in spaces such as restrooms with stalls, which can block the field of view because the hard surfaces will reflect the sound pattern.
- **Dual-technology (hybrid)**—Occupancy sensors that use passive infrared and ultrasonic technologies to minimize the risk of false triggering (lights coming on when the space is unoccupied). They also tend to be more inexpensive.

Clock switches or timers—These control lighting for a preset period. They come equipped with an internal mechanical or digital clock that automatically adjusts for the time of year. The user determines when the lights should be turned on and when they should be turned off. Clock switches can be used in conjunction with photosensors.

Centralized controls are building controls or automation systems that automatically turn on, turn off, or dim electrical lights around a building. The centralized control system can turn on the lights in the morning before employees arrive. During the day, it can dim the lights during periods of high-power demand. And the lights can be turned off automatically. A centralized lighting control system can significantly reduce energy use in buildings where lights are left when not needed.

7.8.1.1.4 Daylighting

As mentioned earlier, daylighting places windows or other openings and reflective surfaces so that natural light provides adequate internal lighting during the day. When professionally designed and effectively integrated with the electric lighting system, daylighting can offer significant energy savings by offsetting a portion of the electric lighting load. A related benefit is reducing cooling capacity and use by lowering a substantial component of internal gains. In addition to energy savings, daylighting improves occupant satisfaction and comfort. Windows also provide visual relief, contact with nature, time orientation, the possibility of ventilation, and emergency egress.

The daylight zone has high potential, particularly in daytime-occupied spaces. Site solar analysis should assess the access to daylight by considering what is "seen" from the various potential window orientations. What proportion of the sky is seen from typical task locations in the room? What are the exterior obstructions and glare sources? Will the building design shade a neighboring building or landscape feature dependent on daylight or solar access? It is essential to establish which spaces will most benefit from daylight and which spaces have little or no need for daylight; within the spaces that can be used in daylight, place the most critical visual tasks in positions near the window. Try to group functions by similar lighting requirements and occupancy patterns. Avoid placing the windows in the direct line of sight of the occupant, as this can cause extreme contrast and glare. It is best to orient the occupant at 90° from the window. Where privacy is not a significant concern, consider interior glazing (relights or borrow lights) that allows light from one space to be shared with another. This can be achieved with transom lights, vision glass, or translucent panels if privacy is required. The floor plan configuration should maximize the perimeter daylight zone. This may result in a higher skin-to-volume ratio than a typical compact building design. A standard window can produce functional illumination to a depth of about 1.5 times the window's height. Light shelves or other reflector systems can increase this to 2.0 times or more. As a rule of thumb, the higher the window on the wall, the deeper the daylight penetration.

Window Design Considerations—The daylight that arrives at a work surface comes from tree sources:

1. **The exterior reflected component**—This includes ground surfaces, pavement, adjacent buildings, wide windowsills, and objects. Remember that excessive rough reflectance will result in glare.
2. **The direct sun/sky component**—The direct sun component is typically blocked from occupied spaces because of heat gain, glare, and UV degradation. The sky dome then becomes an essential contribution to daylighting the pace.
3. **The internal reflected component**—Once daylight enters the room, the surrounding wall, ceiling, and floor surfaces are essential to light reflectors. High reflectance surfaces will better bounce the daylight around the room and reduce extreme contrasts. The window frame material should be light-colored to reduce contrast with the view and have a non-specular finish to eliminate glare spots. Window jambs and sills can be beneficial light reflectors. Deep jambs should be splayed (angled toward the interior) to reduce the contrast around the perimeter of the window.

DID YOU KNOW?

The most critical interior light-reflecting surface is the ceiling. High-reflectance paints and ceiling tiles are now available with 0.90 or higher reflectance values. Titling the ceiling plant toward the daylight source increases the daylight reflected from this surface. In small rooms, there should also be a high-reflectance matte finish. The sidewalls, followed by the floor, have less impact on the reflected daylight in the space.

Major room furnishings such as office cubicles or partitions can significantly impact reflected light, so select light-colored materials. Suggested reflectance levels for various room surfaces are:

- Ceilings: > 80%
- Walls: 50%–70%
- Floors: 20%–40%
- Furnishings: 25%–45%

Because light has no scale for architectural purposes, the room's proportions are more important than the dimensions. A room with a higher ceiling than the room depth will have deeper daylight penetration, whether from side lighting (windows) or top lighting (skylights and clerestories). Raiding the window head height will also result in deeper penetration and even illumination. Punched window openings, such as small, square windows separated by wall area, result in uneven illumination and harsh contrast between the window and adjacent wall surfaces. A more even distribution is achieved with horizontal strip windows.

DID YOU KNOW?

There is no direct sunlight on the polar-side wall of a structure from the autumnal equinox to the spring equinox in parts of the globe north of the Tropic of Cancer and in parts of the globe south of the Tropic of Capricorn.

Effective aperture (EA)—One method of assessing the relationship between visible light and window size is the effective aperture method. The *effective aperture* (EA) is defined as the product of the visible transmittance and the window-to-wall ratio (WWR). The window-to-wall ratio is the proportion of window area compared to the total wall area where the window is located. For example, if a window covers 25 square feet in a 100-square-foot wall, the WWR is 25/100 or 0.25. A good starting target for EA ranges from 0.20 to 0.30. A higher WWR (larger window) results in a lower visible transmittance for a given EA number.

Example: WWR = 0.5 (half the wall in glazing)

VT = 0.6, EA = 0.3

or WWR = 0.75, VT = 0.4 for the same EA of 0.3

DID YOU KNOW?

Typically, lowering the visible transmittances will also lower the shading coefficient, but you must verify this with glazing manufacturer data since this is not always the case.

Light Shelves are effective horizontal light-reflecting overhangs placed above the eye level with a transom window. They enhance the lighting from windows on the equator-facing side of a building. Exterior shelves are more effective shading devices than interior shelves. A combination of exterior and interior shelves will provide an even illumination gradient.

Top lighting strategies—Top lighting can benefit large single-level floor areas and the top floors of multi-story buildings. The general types of top lighting include skylights, clerestories, monitors, and sawtooth roofs.

- **Skylights**—Horizontal skylights can be an energy problem because they tend to receive maximum solar gain at the day's peak. The daylight contribution also peaks midday and falls off severely in the morning and afternoon. High-performance skylight designs incorporate reflectors or prismatic lenses that reduce the peak daylight and heat gain while increasing early and late afternoon daylight contributions. Another option is light pipes, where a high reflectance duct channels the light from the skylight down to a diffusing lens in the room. These may be advantageous in deep roof constructions.
- **Clerestory window**—Vertical glazing high on an interior wall. South-facing clerestories can be effectively shaded from direct sunlight by a professionally designed horizontal overhand. In this design, the interior north wall can be sloped to reflect better the light coming down into the room. Use light-colored overhangs and adjacent roof surfaces to improve the reflected component. If exterior shading is impossible, consider interior vertical baffles to diffuse the light better. A south-facing clerestory will produce higher daylight illumination than a northern-facing clerestory. East- and west-facing clerestories have the same problems as east- and west-facing windows: they have complex shading and potentially high heat gains.
- **Roof monitor**––This design consists of a flat section raised above the adjacent roof with vertical glazing on all sides. It often results in excessive glazing area, which results in higher heat losses and gains than a clerestory design. The multiple orientations of glazing can also create shading problems.

DID YOU KNOW?

A building designed for daylighting without an integrated electrical lighting system will be a net energy loser because of the increased thermal loads. When the electric lighting load is reduced, there will be more than offsetting savings in electrical and cooling loads. The benefits from daylighting are maximized when both occupancy and lighting sensors control the electric lighting system.

- **Sawtooth roof**—It is an old design often seen in industrial buildings. Typically, one sloped surface is opaque and the other is glazed. A contemporary sawtooth roof may have solar collectors or photovoltaic cells on the south-facing slope and daylight glazing on the north-facing slope. Unprotected glazing on the south-facing sawtooth surface may result in high heat gains. In these applications, an insulated diffusing panel may be a good choice.

7.9 SOLAR ENERGY AND THE AI CONNECTION

To produce solar energy for use on Earth, it is necessary to have the Sun, and the solar receivers must be directed toward the Sun. The advantage of solar power with applied AI is that AI can store tremendous amounts of historical and current data related to local weather conditions. This is part of predictive analytics. This enables the solar cells to be better positioned to absorb the Sun's rays by informing the operators (human and the AI system) of the best times to get maximum energy from the Sun. The solar cells must be positioned optimally along with the Sun's rays to receive maximum solar input. AI also comes into play with predictive solar system maintenance, making the entire operation more efficient and detecting system faults and anomalies.

7.10 THE BOTTOM LINE

Solar energy has some obvious advantages in that the source is free; however, the initial investment in operating equipment is not free or inexpensive. Solar energy is also environmentally friendly, requires almost no maintenance, and reduces our dependence on foreign energy supplies. The most significant downside of solar energy use is that solar panels cannot capture enough energy to provide heat for homes or offices in areas without direct sunlight during certain times of the year. Geographically speaking, the higher latitudes do not receive as direct sunlight as tropical areas. Because of the sun's position in the sky, solar panels must be placed in sun-friendly locations such as the U.S. Desert Southwest and the Sahara region of northern Africa. Another downside of solar energy is the system's efficiency; it may be seriously affected by how well it was installed. AI in solar energy production uses neural networks to predict energy demand and produce futile demand. Deep learning algorithms can accurately predict future demand using historical consumption rates and environmental variables.

The real bottom line of solar energy is obvious: its advantages outnumber its disadvantages.

NOTES

1. Most of the information in this section is adapted from DOE, 2001. *Concentrating Solar Power: Energy from Mirrors*. Washington, DC: U.S. Department of Energy.
2. Case study is found in USDOE's Federal Technology Alert on Transpired Collectors. Accessed 02/10/10 @ www.eren.doe.gov/solarbuildings.
3. Much of the information in this section is from USDOE: EERE, 2008. *Energy Efficiency and Renewable Energy*. Accessed 02/13/10 @ www1.eere.energy.gov/buildings/commercial/printable_versions/lighting.html.

REFERENCES

Chiras, D.D., 2002. *The Solar House: Passive Heating and Cooling*. White River Junction, VT: Chelsea Green Publishing Company, p. 33.
Duffie, J., and Beckman, W., 1991. *Solar Engineering of Thermal Processes*. New York: John Wiley & Sons, Inc.
EERE, 2008a. *Solar Energy Technologies Program*. Accessed 02/04/24 @ www.eere.energy.gov.
EERE, 2008b. *Linear Concentrator Systems*. Accessed 02/04/24 @ https://www.EERE.energy.gov/solar/printable_versions/linear_concentrators.html.
EERE, 2008c. *Thermal Storage*. Accessed 02/08/24 @ https://www.eere.energy.gov/solar/printable_versions/thermal_storage.html.
Enermodal Engineering Limited, 1995. *Performance of the Perforated-Plate/Canopy Solarwall at GM Canada, Oshawa*. Ottawa, ON: Energy Technology Branch, Dept of Natural Resources.
Hanson, B.J., 2004. *Energy Power Shift: Benefiting from Today's New Technologies*. Maple, WI: Lakota Scientific Press.
NREL, 2009. *Concentrating Solar Power*. Accessed 06/13/24 @ https://www.nrel.gov/learning/re_csp.html.
NREL, 2010. *CSP Systems*. Accessed 02/07/24 @ https://www.nrel.gov/concentratingsolar.html.
Patel, M., 1999. *Wind and Solar Power Systems*. Boca Raton, FL: CRC Press.
Perez-Blanco, H., 2009. *The Dynamics of Energy: Supply, Conversion, and Utilization*. Boca Raton, FL: CRC Press.
USDOE, 1998. *Solar Dish/Engine Systems*. Washington, DC: US Department of Energy.
USDOE, 2006. *Solar Buildings: Transpired Air Collectors*. Accessed 02/11/24 @ www.eren.doe.Gov/solarbuildings.
USDOE, 2008. *Concentration Solar Power Commercial Application Study*. Washington, DC: US Dept. of Energy.
USDOE, 2009. *Solar Energy Technologies Program*. Accessed 06/13/24 @ https://www1.eere.Energy.gov/solar/printable_versions/about.html.

8 Bioenergy and AI

8.1 INTRODUCTION

The ongoing development of and need for an expanding use of artificial intelligence and machine learning (AI/ML) integrated with experimentation, genomics [a genome is the complete set of deoxyribonucleic acid (DNA) within a living cell], biosystem design, and bioprocessing technology are poised to revolutionize scientific investigation, particularly bioenergy research. Advances in data science, synthetic biology, and laboratory automation have increased interest in developing AI, ML, and autonomous experimentation for broader genomics-based research and biotechnology applications.

8.2 WHEN THE FAT LADY SINGS

In a recent New York Times article, Thomas Friedman (2010) stated, "The fat lady has sung." Specifically, Friedman spoke about America's transition from "The Greatest Generation" to what Kurt Anderson called "The Grasshopper Generation." According to Friedman, we are "going from the age of government handouts to the age of citizen givebacks, from the age of companions fly free to the age of paying for each bag." He goes on to say that we all accept that our parents were the greatest generation, but it is us that we are concerned about and that it is the "we" that comprise the Grasshopper Generation; "we are eating through the prosperity that was bequeathed us like hungry locusts."

As mentioned (and the central theme of this text), what we are "eating through," among other things, is our readily available, relatively cheap energy source. Like the grasshopper, we can gobble it until it is gone or find alternative renewable energy sources.

One of the promising sources of energy is bioenergy. Bioenergy is a general term for energy derived from materials such as straw, wood, or animal wastes, which, in contrast to fossil fuels, were living matter relatively recently. Such materials can be burned directly as solids (biomass) to produce heat or power but can also be converted into liquid biofuels. In the last few years, there has been much increased interest in bioenergy fuels such as biofuels, which can be used for transport; that is, biofuels, as mentioned, are liquid fuels (biodiesel and bioethanol) used for transport. Transport is currently at the center stage in finding renewable, alternate fuels to eventually replace hydrocarbon fuels. Unlike biofuels, solid biomass fuel is used primarily for electricity generation and heat supply.

Even though we have stated that bioenergy is one of the promising energy sources in and for the future (and it is), it is ironic whenever the experts (or anyone else) state this point without qualification. The qualification? The reality? Remember that only two hundred years ago, our economy was based primarily on bioenergy from biomass or carbohydrates rather than hydrocarbons. USDOE (2003a) points out that in

DOI: 10.1201/9781003605430-10

the late 1800s, the most significant chemicals sold were alcohols made from wood and grain, the first plastics were produced from cotton, and about 65% of the nation's energy came from wood.

By the 1920s, the economy started shifting toward the use of fossil resources, and after World War II, this trend accelerated as technological breakthroughs were made. By the 1970s, fossil energy was established as the backbone of the U.S. economy, and all but a small portion of the carbohydrate economy remained. In the industrial sector, plants accounted for about 16% of input in 1989, compared with 35% in 1925. In 2023, biomass accounted for about 5% of U.S. primary energy consumption.

Processing costs and the availability of inexpensive fossil energy resources continue to be driving factors in the dominance of hydrocarbon resources. In many cases, producing goods from petroleum or natural gas is still more economical than from plant matter. This trend is about to shift dramatically as we reach peak oil and as the world continues to demand unprecedented amounts of petroleum supplies from an ever-dwindling supply.

Technological advances in biological sciences and engineering, political change, and concern for the environment are aiding in this trend shift, which has begun to swing the economy back toward carbohydrates on several fronts. Technological advances are also reducing the cost of producing industrial products and fuels from biomass, making them more competitive with those produced from petroleum-based hydrocarbons. Developments in pyrolysis, ultracentrifuges, membranes, and the use of enzymes and microbes as biological factories are enabling the extraction of valuable components from plants at a much lower cost. As a result, the industry is investing in developing new bioproducts, steadily gaining a share of current markets (USDOE, 2003a).

New technology enables the chemical and food processing industries to develop new processes, allowing more cost-effective production of all kinds of industrial products from biomass. One example is a plastic polymer derived from corn that is now being produced at a 300 million pound-per-year plant in Nebraska, a joint venture between Cargill, the most prominent grain merchant, and Dow Chemical, the largest chemical producer (Fahey, 2001).

Other chemical companies are exploring low-cost biomass processes to make chemicals and plastics, which are now made from more expensive petrochemical processes (USDOE, 2003b). In this regard, new innovative methods such as biorefinery may become the foundation of the new bioindustry. The Biorefinery is similar in concept to a petroleum refinery, except that it is based on conversion of biomass feedstocks rather than crude oil. Biorefineries, in theory, would use multiple forms of biomass to produce a flexible mix of products, including fuels, power, heat, chemicals, and materials (see Figure 8.1). In a biorefinery, biomass would be converted into high-value chemical products and fuels (both gas and liquid). Byproducts, residues, and some of the fuels produced would be used for on-site power generation or cogeneration facilities producing heat and power.

The biorefinery concept has already proven successful in the U.S. agricultural and forest products industries, where such facilities now produce food, feed fiber, chemicals, heat, and electricity for plant operations.

Biorefineries offer the most potential for realizing the ultimate opportunities of the bioenergy industry.

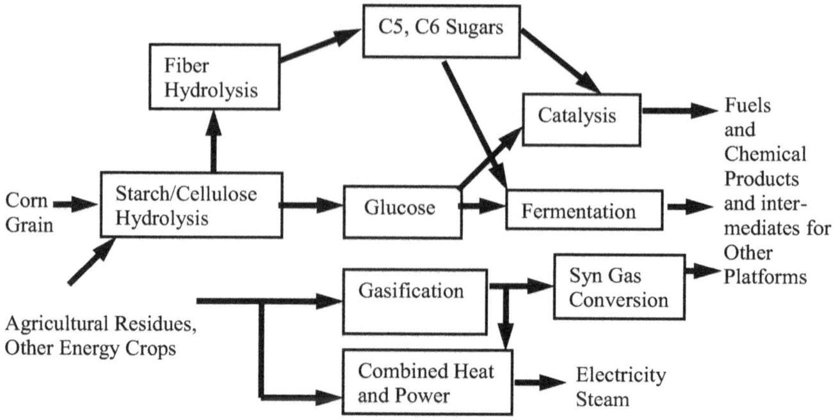

FIGURE 8.1 The biorefinery.

DID YOU KNOW?

Hanson (2004) estimated that in the United States alone, more than 100 billion gallons of light fuel oil per year can be produced from waste and biomass. The source waste and biomass included in this estimation are municipal solid waste, municipal sewage sludge, Hazmat waste, agricultural crop waste, feedlot manure, plastics, tires, heavy oil or tar sands, forestry waste, restaurant grease, and biomass crops (switchgrass) grown on idle land and cropland.

8.3 BIOMASS

Biomass (all Earth's living matter) consists of energy from plants and plant-derived organic-based materials; it is stored energy from the sun. Biomass can be biochemically processed to extract sugars, thermochemically processed to produce biofuels or biomaterial, or combusted to create heat to electricity. Biomass is also an input in other end-use markets, such as forestry products (pulpwood) and other industrial applications. This complicates the economics of biomass feedstock and requires that we differentiate between what is technically and economically feasible, considering relative prices and intermarket competition.

Biomass has been used since people began burning wood to cook food and keep warm. Trees have been the principal fuel of almost every society for over five thousand years, from the Bronze Age until the middle of the nineteenth century (Perlin, 2005). Wood is still the largest biomass energy resource today, but other biomass sources can also be used. These include food crops, grassy and woody plants, residues from agriculture or forestry, and the organic component of municipal and industrial wastes. Even the fumes from landfills (methane, a natural gas) can be used as a biomass energy source. It excludes organic material transformed by geological processes into substances such as coal or petroleum. The biomass industry is one of the fastest-growing industries in the United States.

DID YOU KNOW?

Biomass is converted to energy through various processes, including:

- Direct combustion (burning) to produce heat is the most common method for converting biomass to valuable energy.
- Thermochemical conversion to produce solid, gaseous, and liquid fuels using pyrolysis and gasification.
- Chemical conversion to produce liquid fuels
- Biological conversion to produce liquid and gaseous fuels

8.3.1 Feedstock Types

Various biomass feedstocks can produce transportation fuels, biobased products, and power. Feedstock refers to crops or products, like waste vegetable oil, which can be used or converted into biofuels and bioenergy. Concerning the advantages or disadvantages of one type of feedstock compared to another, this is gauged in terms of how usable material they yield, where they can grow, and how energy and water-intensive they are. Feedstock types are listed as first-generation or second-generation feedstocks. First-generation feedstocks are already widely grown and used for some form of bioenergy or biofuel production, which means there are food versus fuel conflicts. First-generation feedstocks include sugars (sugar beets, sugar cane, sugar palm, sweet sorghum, and Nipa palm), starches (cassava, corn, milo, sorghum, sweet potato, and wheat), waste feedstocks such as whey and citrus peels, and oils and fats (coconut oil, oil palm, rapeseed, soybeans sunflower seed, castor beans jatropha, jojoba, karanj, and waste vegetable oil and animal fat). Second-generation feedstock refers broadly to crops with high potential biofuel yields but that are not widely cultivated or not cultivated as an energy crop. It relates to cellulosic feedstocks or conventional crops such as miscanthus, prairie grass, switchgrass, will, and hybrid poplar trees. Algae and halophytes (saltwater plants) are other second-generation feedstocks.

Currently, most of the ethanol produced in the United States is made from corn or other starch-based crops. The present focus, however, is on developing cellulosic feedstocks—non-grain, nonfood-based feedstocks such as switchgrass, corn stover, and wood material—and on technologies to convert cellulosic material into transportation fuels and other products. Using cellulosic feedstocks can alleviate the potential concern of diverting food crops to produce fuel and has various environmental benefits (EERE, 2008).

Because such a wide variety of cellulosic feedstocks can be used for energy production, potential feedstocks are grouped into categories—or pathways. Figure 8.2 shows some of the specific feedstocks in each of these areas.

8.3.2 Composition of Biomass

The composition of the biomass feedstock determines the ease with which biomass can be converted to valuable products or intermediates. Biomass contains various components, some of which are readily accessible and others that are much more

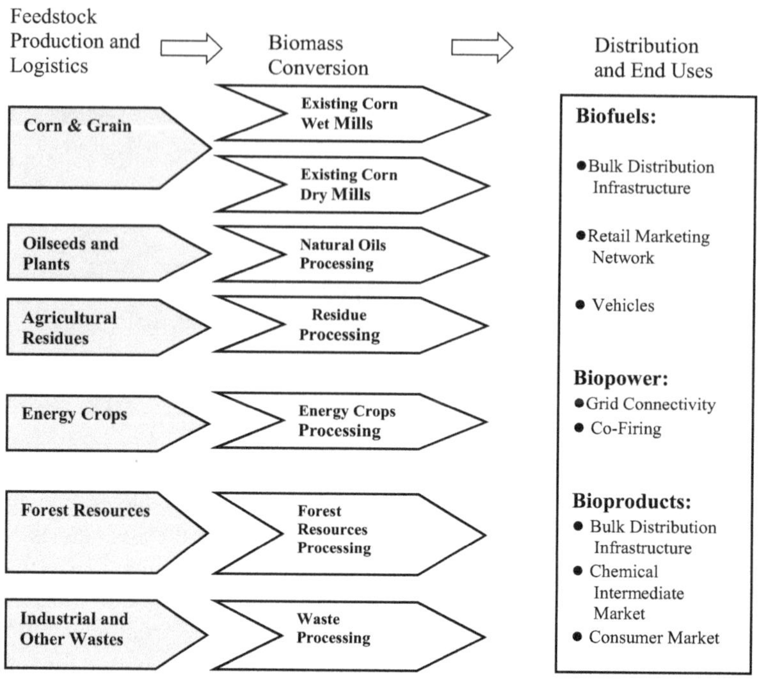

FIGURE 8.2 Resource-based biorefinery pathways.

difficult and costly to extract. The composition and subsequent conversion issues for current and potential biomass feedstock compounds are described below:

- **Starch (glucose)**—It is readily recovered and converted from grain (corn, wheat, and rice) into products. Starch from corn grain provides the primary feedstock for today's existing and emerging sugar-based bioproducts, such as polylactide and the entire fuel ethanol industry. Corn grain is the primary feedstock for starch manufacturing today's biobased products. Core wet mills use a multi-step process to separate starch from corn grain's germ, gluten (protein), and fiber components. The starch streams generated by wet milling are highly pure, and the acid of enzymatic hydrolysis is used to break the glycosidic linkages of starch to yield glucose. Glucose is then converted into a multitude of valuable products.
- **Lignocellulosic biomass**—It is the non-grain portion of biomass (e.g., cobs, stalks), often referred to as agricultural stover or resides, and energy crops such as switchgrass also contain valuable components. Still, they are not as readily accessible as starch. These lignocellulosic biomass resources (cellulosic) comprise cellulose, hemicellulose, and lignin. Lignocellulosic material contains 30%–50% cellulose, 20%–30% hemicellulose, and 20%–30% lignin. Some exceptions are cotton (98% cellulose) and flax (80% cellulose). Lignocellulosic biomass is perceived as a valuable and untapped resource for the future bioindustry. However, cost-effectively recovering the components represents a significant technical challenge.

- **Cellulose**—It is one of nature's polymers and is composed of glucose, a six-carbon sugar. The glucose molecules are joined by glycosidic linkages, allowing the glucose chains to assume an extended ribbon conformation. Hydrogen bonding between chains leads to the formation of flat sheets that lay on top of one another in a staggered fashion, like how staggered bricks add strength and stability to a wall. As a result, cellulose is very chemically stable and insoluble and serves as a structural component in plant walls.
- **Hemicellulose**—It is a polymer containing primarily 5-carbon sugars such as xylose and arabinose with some glucose and mannose dispersed throughout. It forms a short-chain polymer that interacts with cellulose and lignin to form a matrix in the plant wall, strengthening it. Hemicellulose is more readily hydrolyzed than cellulose. Much of the hemicellulose in lignocellulosic material is solubilized and hydrolyzed to pentose and hexose sugars.
- **Lignin**—It helps bind the cellulosic/hemicellulose matrix while adding flexibility to the mix. The molecular structure of lignin polymers is random and disorganized and consists primarily of carbon ring structures (benzene rings with methoxyl, hydroxyl, and propyl groups) interconnected by polysaccharides (sugar polymers). The ring structures of lignin have great potential as valuable chemical intermediates. However, separation and recovery of the lignin is difficult.
- **Oils and protein**—The seeds of certain plants offer two families of compounds with great potential for bioproducts: oils and protein. Oils and protein are found in the seeds of certain plants (soybeans and castor beans) and can be extracted in various ways. Plants raised for this purpose include soy, corn, sunflower, safflower, and rapeseed. A large portion of the oil and protein recovered from oilseeds and corn is processed for human or animal consumption. Still, they can also serve as raw materials for lubricants, hydraulic fluids, polymers, and other products.
- **Vegetable oils**—They are composed primarily of triglycerides, also called triacylglycerols. Triglycerides contain a glycerol molecule as the backbone with three fatty acids attached to glycerol's hydroxyl groups.
- **Proteins**—They are natural polymers with amino acids as the monomer unit. They are incredibly complex materials, and their functional properties depend on molecular structure. There are 20 amino acids, each differentiated by their side chain or R-group, and they can be classified as nonpolar and hydrophobic, polar uncharged, and ionizable. The interactions among the side chains, the amide protons, and the carbonyl oxygen help create the protein's 3D shape.

8.4 UNDERSTANDING BIOMASS: PLANT BASICS[1]

Optimizing plant biomass for more efficient processing requires a better understanding of plants and plant cell wall structure and function. The plant kingdom ranks second in importance only to the animal kingdom (at least from the human point of view). The importance of plants and plant communities to humans, bioenergy production, and their environment cannot be overstated. Some of the essential things plants provide are listed below:

- **Aesthetics**—Plants add to the beauty of the places we live.
- **Medicine**—80% of all medicinal drugs originate in wild plants.
- **Food**—90% of the world's food comes from only 20 plant species.
- **Industrial products**—Plants are essential for their goods (e.g., plant fibers provide clothing), and wood is used to build homes.
- **Recreation**—Plants form the basis for many critical recreational activities, including fishing, nature observation, hiking, and hunting.
- **Air quality**—The oxygen in the air we breathe comes from the photosynthesis of plants.
- **Water quality**—Plants aid in maintaining healthy watersheds, streams, and lakes by holding soil in place, controlling stream flows, and filtering sediments from water.
- **Erosion control**—Plant cover helps to prevent wind or water erosion of the top layer of soil that we depend on.
- **Climate**—The amount and type of plant cover impacts regional climates.
- **Fish and wildlife habitat**—Plants provide the necessary habitat for wildlife and fish populations.
- **Ecosystem**—Every plant species serves an important role or purpose in their community.
- **Feedstock for bioenergy production**—Some necessary fuel chemicals come from plants, such as ethanol from corn and soy diesel from soybeans.

Though both are essential kingdoms of living things, plants and animals differ in many important aspects. For example, plants contain chlorophyll, which can make their food. Animals cannot produce food and depend on plants and other animals. Plants give off oxygen and take in carbon dioxide. Animals give off carbon dioxide, which plants need to make food and take in oxygen that they need. Plants are rooted in one place and do not move on their own. Most animals can move freely. Plants have a fundamental ability to sense. Animals have a much more highly developed sensory and nervous system.

Before discussing the essential specifics of plants, defining a few key plant terms is necessary.

8.4.1 Plant Terminology

- **Apical meristem**—It consists of meristematic cells at a root or shoot's tip (apex).
- **Cambium**—It is the lateral meristem in plants.
- **Chloroplasts**—They are disk-like organelles with a double membrane found in eukaryotic plant cells.
- **Companion cells**—They are specialized cells in the phloem that load sugars into the sieve elements.

- **Cotyledons**—They are leaf-like structure (sometimes called seed leaf) present in flowering plants' seeds.
- **Dicot**—It is one of the two primary flowering plants, characterized by two cotyledons.
- **Diploid**—It means having two of each kind of chromosome (2n).
- **Guard cells**—They are specialized epidermal cells that flank stomata whose opening and closing regulate gas exchange and water loss.
- **Haploid**—It means having only a single set of chromosomes (n).
- **Meristem**—It is a group of plant cells that can divide indefinitely and provide new cells for the plant.
- **Monocots**—It is one of two central flowering plants characterized by a single cotyledon.
- **Periderm**—It is a layer of plant tissue derived from the cork cambium, then secondary tissue, replacing the epidermis.
- **Phloem**—It is a complex vascular tissue that transports carbohydrates throughout the plant.
- **Sieve cells**—They are conducting cells in the phloem of vascular plants.
- **Stomata**—They are pores on the underside of leaves that can be opened or closed to control gas exchange and water loss.
- **Thallus**—It is the main plant body, not differentiated into a stem or leaves.
- **Tropism**—It is the plant behavior, controlling the direction of plant growth.
- **Vascular tissue**—They are tissues found in the bodies of vascular plants that transport water, nutrients, and carbohydrates. The two primary kinds are xylem and phloem.
- **Xylem**—Xylem is the vascular tissue of plants that transports water and dissolved minerals from the roots upward to other parts of the plant. It often also provides mechanical support against gravity.

Although not typically acknowledged, plants are as intricate and complicated as animals. Plants evolved from photosynthetic protists and are characterized by photosynthetic nutrition, cell walls made from cellulose and other polysaccharides, lack of mobility, and a characteristic life cycle involving an alternation of generations. The phyla/division of plants and examples are listed in Table 8.1.

TABLE 8.1
The Main Phyla/Division of Plants

Phylum/Division	Examples
Bryophyta	Mosses, liverworts, and hornworts
Coniferophyta	Conifers such as redwoods, pines, and firs
Cycadophyta	Cycads and sago palms
Gnetophyta	Shrub trees and vines
Ginkophyta	*Ginkgo* is the only genus
Lycophyta	Lycopods (look like mosses)
Pterophyta	Ferns and tree-ferns
Anthophyta	Flowering plants, including oak, corn, maize, and herbs

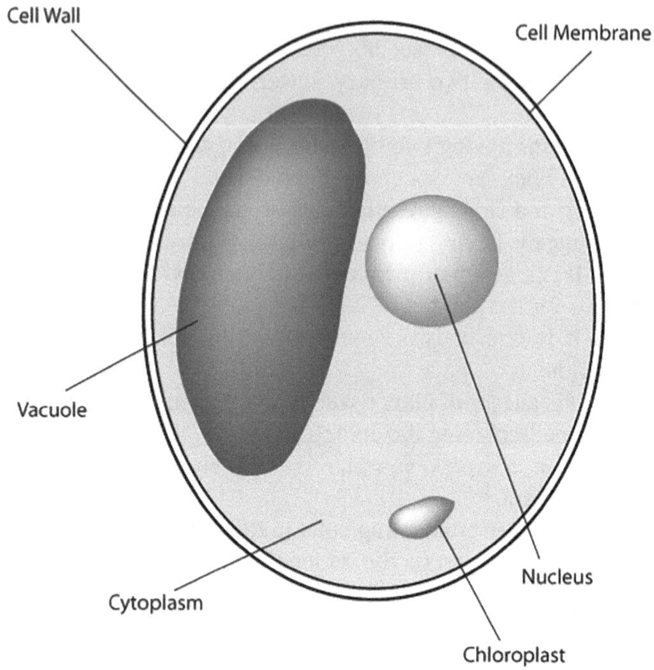

FIGURE 8.3 Plant cell.

8.4.2 The Plant Cell

A summary of plant cells is provided here (see Figure 8.3):

- *Plants have all animal cell organelles* (i.e., nucleus, ribosomes, mitochondria, endoplasmic reticulum, and Golgi apparatus).
- *Plants have chloroplasts.* Chloroplasts are unique organelles that contain chlorophyll and allow plants to conduct photosynthesis.
- *Plant cells can sometimes have large vacuoles for storage.*
- *Plant cells are surrounded by a rigid cell wall made of cellulose and the cell membrane surrounding* animal cells. Those walls provide support.

8.4.3 Vascular Plants

Vascular plants, called **Tracheophytes**, have vascular tissue to transport necessary liquids and minerals over long distances. Vascular tissues are composed of specialized cells that create "tubes" through which materials can flow throughout the plant body. These vessels are continuous throughout the plant, allowing for the efficient and controlled distribution of water and nutrients. In addition to this transport function, vascular tissues also support the plant. The two types of vascular tissue are the xylem and phloem.

- **The xylem** consists of a tube or a tunnel (pipeline) in which water and minerals are transported throughout the plant to leaves for photosynthesis. In addition to distributing nutrients, xylem (wood) provides structural support. After a time, the xylem at the center of older trees ceases to function in transport and only takes on a supportive role.
- **Phloem** tissue consists of cells called *sieve tubes* and *companion cells.* Phloem tissue moves dissolved sugars (carbohydrates), amino acids, and other products of photosynthesis from the leaves to different regions of the plant.

The two most important Tracheophytes are gymnosperms (gymno = naked; sperma = seed) and angiosperms (angio = vessel, receptacle, container).

- **Gymnosperms**—The plants we recognize as gymnosperms represent the sporophyte generation (i.e., the spore-producing phase in the life cycle of a plant that exhibits alternation of generation). Gymnosperms were the first tracheophytes to use seeds for reproduction. The seeds that develop in protective structures are called cones. A gymnosperm contains some female cones and some that are male. Female cones produce spores that, after fertilization, become eggs enclosed in seeds that fall to the ground. Male cones produce pollen, which the wind takes and fertilizes female eggs. Unlike flowering plants, the gymnosperm does not form actual flowers or fruits. Coniferous trees such as firs and pines are good examples of gymnosperms.
- **Angiosperms**, the flowering plants, are the most highly evolved and dominant plants. They have stems, roots, and leaves. Unlike gymnosperms, such as conifers and cycads, angiosperm seeds are found in flowers. Angiosperm eggs are fertilized and develop into seeds in an ovary, usually in a flower.

There are two types of angiosperms: monocots and dicots.

- **Monocots**—These angiosperms start with one seed leaf (cotyledon); thus, their name is derived from the presence of a single cotyledon during embryonic development. Monocots include grass, grains, and other narrow-leaved angiosperms. The central veins of their leaves are usually parallel and unbranched, the flower parts occur in multiples of three, and a fibrous root system is present. Monocots include orchids, lilies, irises, palms, grass, wheat, corn, and oats.
- **Dicots**—Angiosperms in this group grow two seed leaves (two cotyledons). Most plants are dicots, including maples, oaks, elms, sunflowers, and roses. Their leaves usually have a single central vein or three or more branched veins that spread out from the base of the leaf.

8.4.4 Leaves

The principal function of leaves is to absorb sunlight to manufacture plant sugars in photosynthesis. The leaves' broad, flattened surfaces gather energy from sunlight, while apertures on their undersides bring in carbon dioxide and release oxygen.

Leaves develop as a flattened surface to present a large area for efficient absorption of light energy. On its two exteriors, the leaf has layers of epidermal cells that secrete a waxy, impermeable cuticle (chitin) to protect against water loss (dehydration) and fungal or bacterial attack. Gases diffuse in or out of the leaf through **stomata**, which are small openings on the underside of the leaf. The opening or closing of the stomata occurs through the swelling or relaxing of **guard cells**. If the plant wants to limit the diffusion of gases and the transpiration of water, the guard cells swell together and close the stomata. The leaf thickness is kept to a minimum so that gases entering the leaf can diffuse quickly throughout the cells.

8.4.5 Chlorophyll/Chloroplast

The green pigment in leaves is **chlorophyll**. Chlorophyll absorbs red and blue light from the sunlight that falls on leaves. Therefore, the light reflected by the leaves is diminished in red and blue and appears green. The molecules of chlorophyll are large. They are not soluble in the aqueous solution that fills plant cells. Instead, they are attached to the membranes of disc-like structures, called **chloroplasts**, inside the cells. Chloroplasts are the site of photosynthesis, the process in which light energy is converted to chemical energy. In chloroplasts, the light absorbed by chlorophyll supplies the energy plants used to transform carbon dioxide and water into oxygen and carbohydrates.

Chlorophyll is not a stable compound; bright sunlight causes it to decompose. Plants continuously synthesize chlorophyll in their leaves to maintain the amount of chlorophyll in them. The synthesis of chlorophyll in plants requires sunlight and warm temperatures. Therefore, during summer, chlorophyll continuously breaks down and regenerates in the leaves of trees.

8.4.6 Photosynthesis

Because our quality of life, and indeed our very existence, depends on photosynthesis, it is essential to understand it. In photosynthesis, plants (and other photosynthetic autotrophs) use the energy from sunlight to create the carbohydrates necessary for cell respiration. More specifically, plants take water and carbon dioxide and transform them into glucose and oxygen:

$$6CO_2 + 6H_2O + \text{Light Energy} \rightarrow C_6H_{12}O_6 + 6O_2$$

This general equation of photosynthesis represents the combined effects of two different stages. The first stage is called the light reaction and the second is called the dark reaction. The **light reaction** is the photosynthesis process in which solar energy is harvested and transferred into the chemical bonds of ATP; it can only occur in light. The dark reaction is the process in which food (sugar) molecules are formed from carbon dioxide in the atmosphere using ATP; it can occur in the dark if ATP is present.

DID YOU KNOW?

Charles Darwin was the first to discuss how plants respond to light. He found that the new shoot of grasses bends toward the light because the cells on the dark side grow faster than those on the lighted side.

8.4.7 Roots

Roots absorb nutrients and water, anchor the plant in the soil, support the stem, and store food. They are usually below ground and lack nodes, shoots, and leaves. There are two significant types of root systems in plants. Taproot systems have a stout main root with a limited number of side-branching roots. Taproot system plants include nut trees, carrots, radishes, parsnips, and dandelions. Taproots make transplanting difficult. The second type of root system, fibrous, has many-branched roots. Examples of fibrous root plants are grass, marigolds, and beans. Radiating from the roots is a system of root hair that vastly increases the absorptive surface area of the roots. Roots also anchor the plant in the soil.

8.4.8 Growth in Vascular Plants

Vascular plants undergo two kinds of growth (primarily restricted to meristems): primary and secondary. **Primary growth** occurs close to the tips of roots and stems. Apical meristems initiate it and are mainly involved in extending the plant body. The tissues that arise during primary growth are called primary tissues, and the plant body composed of these tissues is called the primary plant body. Most primitive vascular plants are entirely made up of primary tissues. **Secondary growth** occurs in some plants, thickening the stems and roots. Secondary growth results from the activity of lateral meristems. Lateral meristems are called **cambia** (cambium), and there are two types:

1. **Vascular cambium**—Vascular cambium gives rise to secondary vascular tissues (secondary xylem and phloem). It gives rise to the xylem on the inside and the phloem on the outside.
2. **Cork cambium**—Cork cambium forms the **periderm** (bark). The periderm replaces the epidermis in woody plants.

8.4.9 Plant Hormones

Plant hormones control plant growth. They influence cell differentiation, elongation, and division. Some plant hormones also affect the timing of reproduction and germination.

- **Auxins**—Auxins affect cell elongation (tropism), apical dominance, and fruit drop or retention. They are also responsible for root development, secondary vascular cambium growth, lateral branching inhibition, and fruit development. Auxin engages in the absorption of vital minerals and fall color.

As a leaf reaches its maximum growth, auxin production declines. In deciduous plants, this triggers a series of metabolic steps that cause the reabsorption of valuable materials (such as chlorophyll) and their transport into the branch or stem for storage during the winter months. Once chlorophyll is gone, the other pigments typical of fall color become visible.

- **Kinins**—Kinins promote cell division and tissue growth in leaf, stem, and root. They are also involved in developing chloroplasts, fruits, and flowers. In addition, they have been shown to delay senescence (aging), especially in leaves, which is one reason that florists use cytokinins on freshly cut flowers—when treated with cytokinins, they remain green, protein synthesis continues, and carbohydrates do not break down.
- **Gibberellins**—Gibberellins are produced in the root-growing tips and act as a messenger to stimulate growth, mainly stem elongation. They can also end the dormancy period of seeds and buds by encouraging germination. Additionally, they play a role in root growth and differentiation.
- **Ethylene**—Ethylene controls the ripening of fruits. It may ensure that flows are carpellate (female), while gibberellin confers maleness on flowers. It also contributes to the senescence of plants by promoting leaf loss and other changes.
- **Inhibitors**—Inhibitors restrain growth and maintain the period of dormancy in seeds and buds.

8.4.10 Tropisms: Plant Behavior

Tropism is an organism's movement (and growth in plants) in response to an external stimulus. For example, tropisms, controlled by hormones, are a unique characteristic of sessile organisms such as plants that enable them to adapt to different features of their environment—gravity, light, water, and touch—so they can flourish. There are three main tropisms:

- **Phototropism**—The tendency of plants to grow or bend (move) in response to light. Phototropism results from the cells' rapid elongation on the plant's dark side, which causes the plant to bend in the opposite direction. For example, the stems and leaves of a geranium plant growing on the windowsill always turn toward the light.
- **Gravitropism**—It refers to a plant's tendency to grow toward or against gravity. A plant with positive gravitropism (plant roots) will grow downward toward the earth's center. That is, gravity causes the roots of plants to grow down so that the plant is anchored in the ground and has enough water to grow and thrive. Plants with negative gravitropism (plant stems) will grow upward, away from the earth. Most plants are negatively gravitropic. Gravitropism is also controlled by auxin. In a horizontal root or stem, auxin is concentrated in the lower half, pulled by gravity. In a positive gravitropic plant, this auxin concentration will inhibit cell growth on the lower side, causing the stem to bend downward. This auxin concentration in a harmful gravitropic plant will inspire cell growth on that lower side, causing the stem to bend upward.

- **Thigmotropism**—Some people notice that their houseplants respond to thigmotropism (i.e., growing or bending in response to touch), growing better when they touch and pay attention to them. Touch causes parts of the plant to thicken or coil as they feel or are touched by environmental entities. For instance, tree trunks grow thicker when exposed to strong winds, and vines tend to grow straight until they encounter a substrate to wrap around.

8.4.11 Photoperiodism

Photoperiodism is the response of an organism (e.g., plants) to naturally occurring changes in light for 24 hours. The site of perception of photoperiod in plants is leaves. For instance, sunflowers are known for their photoperiodism, or their ability to open and close in response to the sun's changing position throughout the day.

All flowering plants have been placed in one of three categories for photoperiodism: short-day plants, long-day plants, and day-neutral plants.

- **Short-day plants**—flowering promoted by day lengths shorter than a certain critical day length—include poinsettias, chrysanthemums, goldenrods, and asters.
- **Long-day plant—**flowering promoted by day lengths longer than a certain critical day length—includes spinach, lettuce, and most grains.
- **Day-neutral plants**—flowering response insensitive to day length—include tomatoes, sunflowers, dandelions, rice, and corn.

8.4.12 Plant Reproduction

Plants can reproduce both sexually and asexually. Each type of reproduction has its benefits and disadvantages. A comparison of sexual and asexual plant reproduction is provided in the following:

- Sexual Reproduction:
  - Sexual reproduction occurs when a sperm nucleus from the pollen grain fuses with an egg cell from the ovary of the pistil (pistil is defined as the female reproductive structures in flowers, consisting of the stigma, style, and ovary).
  - Each brings a complete set of genes and produces genetically unique organisms.
  - The resulting plant embryo develops inside the seed and grows when the seed is germinated.
- Asexual reproduction:
  - Asexual reproduction occurs when a vegetative part of a plant, root, stem, or leaf gives rise to a new offspring plant whose genetic content is identical to the "parent plant." An example would be a plant reproducing by root suckers, which are shoots that come from the root system. The breadfruit tree is an example.

- Asexual reproduction is also called vegetative propagation. Plant growers need to get many identical plants from one quickly.
- By sexual reproduction, plants can quickly spread and colonize an area (e.g., crabgrass).

8.4.13 Biomass Plant Cell Walls

Figure 8.3 shows the primary organelles within a standard plant cell, including the cell wall. It should be pointed out, however, that plants can have two types of cell walls: primary and secondary (see Figure 8.4). Primary cell walls contain cellulose consisting of hydrogen-bonded chains of thousands of glucose molecules, hemicellulose, and other materials woven into a network. Certain types of cells, such as those in vascular tissues, develop secondary walls inside the primary wall after the cell has stopped growing. These cell wall structures also contain lignin, which provides rigidity and resistance to compression. The area formed by two adjacent plant cells, the middle lamella, typically is enriched with pectin.

Cellulose in higher plants is organized into microfibrils, each measuring about 3–6 nm in diameter and containing up to 36 glucan chains having thousands of glucose residues. Like steel girders stabilizing a skyscraper's structure, the primary cell wall's mechanical strength is due to the microfibril scaffold (i.e., crystalline cellulose core) (USDOE, 2007).

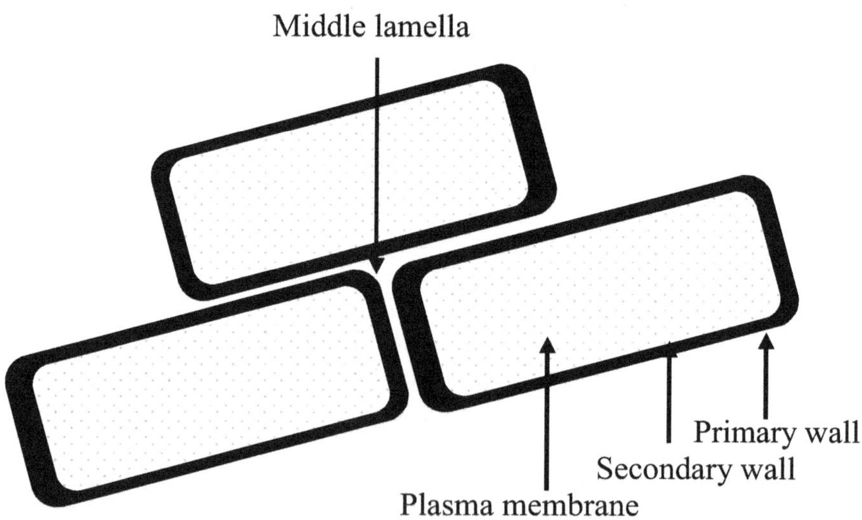

FIGURE 8.4 Plant cell walls.

DID YOU KNOW?

Cellulose microfibrils are linear chains of glucose molecules that hydrogen binds to form the microfibrils.

8.5 FIRST-GENERATION FEEDSTOCKS[2]

The type of primary end-use product is easily categorized for first-generation feedstocks. Corn and sugarcane are the most used feedstocks for ethanol production. Soybean and other vegetable oils and animal fats are used for biodiesel production (Bioproducts). Manure and landfill organic waste are used to produce methane and generate electricity. Corn is used for ethanol production and is currently the leading feedstock in the United States.

Several factors favor a positive outlook for further near-term growth in corn-ethanol production. Continued high oil prices will provide economic support for expanding all alternative fuel programs, including corn ethanol. Technology improvements that increase feedstock productivity, fuel conversion yields, and positive spillovers from second-generation technologies (biomass gasification in ethanol refiners) will also help lower corn ethanol production costs.

Among the factors likely to limit the future growth of corn-ethanol production are increased feedstock and other production costs, increased competition from unconventional liquid fossil fuels (such as oil sands, coal, heavy oil, and shale), the emergence of cellulosic ethanol as a low-cost competitor, and new policies to reduce greenhouse gas (GHG) emissions, which could favor advanced biofuels over corn ethanol.

Biodiesel is another biofuel experiencing expansion. While its production costs are higher than ethanol, biodiesel has some environmental advantages, including biodegradability and lower sulfur and carbon dioxide emissions when burned. Biodiesel production in the United States increased rapidly from less than 2 million gallons in 2000 to about 500 million in 2007. Policy incentives in the Energy Independence and Security Act of 2007 are expected to sustain demand for 1 billion gallons per year of this fuel after 2011.

Various oil-based feedstocks are converted to biodiesel using a process known as *transesterification*. This is the process of exchanging the organic group R″ of an ester with the organic group R′ of an alcohol. These reactions are often catalyzed by the addition of an acid or base as shown below:

$$\text{Transesterification: Alcohol + Ester} \rightarrow \text{Different Alcohol + Different Ester}$$

Oil-based feedstocks include vegetable oils (primarily soy), recycled oils, yellow grease, and animal fats like beef tallow. It takes 3.4 kg of oil/fat to produce 1 gallon of biodiesel (Baize, 2006). Biodiesel production costs are higher than ethanol's, with feedstocks accounting for 80% or more of total costs.

The biodiesel industry consists of many small plants that are highly dispersed geographically. Decisions about plant location are primarily determined by local availability and access to the feedstock. Recent expansion in biodiesel production is affecting the soybean market. Achieving a nationwide target of 2% biodiesel blend in diesel transportation fuel, for example, would require 2.8 million metric tons (MT) of vegetable oil, or about 30% of current U.S. soybean oil production (USDA-World Agricultural Outlook Board, 2008).

One of the critical biodiesel byproducts is glycerin. There is concern about producing glycerin mainly because there are no markets for the product. However, recent technological developments include an alternative chemical process that produces biodiesel without glycerin. In addition, new methods are being assessed that further transform glycerin into propylene glycol, which is used to manufacture antifreeze (Biodiesel, 2007).

8.6 SECOND-GENERATION FEEDSTOCKS: SHORT-TERM AVAILABILITY

Agricultural residues, a second-generation biomass feedstock, offer a potentially significant and readily available biomass resource, but sustainability and conservation constraints could place much of it out of reach. Given current U.S. cropland use, corn and wheat offer the most potentially recoverable residues. However, these residues are essential in recycling nutrients in the soil and maintaining long-term fertility and land productivity. Removing too much residue could aggravate soil erosion and deplete the soil of vital nutrients and organic matter.

Safe removal rate methodologies based on soil erosion have been developed. Methodologies to determine removal rates while safeguarding soil fertility and meeting conservation objectives still need to be developed. Studies have shown that under current tillage practices, the national average safe removal rate based on soil erosion for corn stover is less than 30%. Actual rates vary widely depending on local conditions. In other words, much of the crop residues generated may be out of reach for biomass use if soil conservation goals are to be achieved (Graham et al., 2007).

The estimated delivery costs for agricultural resides vary widely depending on crop type, load resource density, storage and handling requirements, distance, and transportation costs. Moreover, existing estimates are derived from engineering models, which may not account for economic conditions.

Agricultural residue feedstocks (such as corn stover) have a significant advantage in that they can be readily integrated into the expanding corn ethanol industry. However, dedicated energy crops (such as switchgrass) may have more benign environmental impacts.

Another significant biomass source is *forest biomass,* which can be immediately available if the bioenergy market develops. Logging residues are associated with timber industry activities and constitute significant biomass resources in many States, particularly in the Northeast, North Central, Pacific Northwest, and Southeast. In the Western States, the predominance of public lands and environmental pressure reduces the supply potential for logging residues. Still, there is a vast potential for biomass from thinning undertaken to reduce the risk of forest fires. However, the few analyses that have examined the recoverability of logging residues cite the need to account for factors such as the scale and location of biorefineries and biopower plants and regional resource density.

The potential for forest residues may be significant. Still, the quantities available for biomass conversion may be low due to the economics of harvesting, handling, and transporting the residues from forest areas to locations where they could be used.

How these residues compete with fossil fuels in Biopower and co-firing industries is unclear. In addition, these products have competing uses in the pulp and paper industry and different bioenergy end-uses. Economic studies on logging residues suggest a current lack of competitiveness with fossil fuels (coal, gas). However, logging residues could become more cost-competitive with further improvements in harvesting transportation technologies and with policies that require a fuller accounting of the social and environmental benefits of converting forest residues to biopower or biofuels.

Another source of forest residues that could be recovered in significant quantities is biomass from fuel treatments and thinning. Fuel treatment residues are the byproduct of efforts to reduce the risk of loss from fire, insects, and disease, and therefore present challenges different from logging residues.

The overall value of forest health benefits, such as clean air and water, is believed to exceed the cost of treatment. However, treated forests are often distant from end-use markets, resulting in high transportation costs to use the harvested material. Road or trail access, steep terrain, and other factors commonly limit thinning operations in Western forests.

Transportation costs can be a significant factor in the cost of recovering biomass. As much as half the cost of the material delivered to a manufacturing facility may be attributed to transportation. The offset to high-cost transportation of forest thinning is onsite biomass densification. This could entail pelletization, fast pyrolysis (to produce bio-oil), or baling.

The economics of transporting thinned woody residues versus onsite densification depends on the distance to end-use markets. Densification may be more economical if power generation facilities are far away. In addition to co-firing or co-generation facilities, improvements in thermochemical conversion efficiency and the establishment of small-scale conversion facilities using gasification and pyrolysis may favor using forest residues for biofuel production (Polagye et al., 2007).

A third major category of immediately available second-generation biomass is wood residues from *secondary mill products* and *urban wood waste*. Urban wood waste provides cheap feedstock to supplement other biomass resources (Wiltsee, 1998). Urban wood waste encompasses the biomass portion of commercial, industrial, and municipal solid waste (MSW). At the same time, secondary mill residues include sawdust, shavings, wood trims, and other byproducts generated from processing lumber, engineered wood products, or wood particles. Both urban wood waste and secondary mill residues have several primary uses and disposal methods. Urban wood waste that is not used in captive markets (such as the pulpwood industry) could be used as biomass to generate electricity or produce cellulosic ethanol when it becomes commercially viable.

The amount of urban wood waste produced in the United States is significant, and their use as biomass could be economically viable, particularly in large urban centers (Wiltsee, 1998). Several national availability estimates exist for various types of urban wood waste, but estimates vary depending on methodology, product coverage, and assumptions about alternative uses (Wiltsee, 1998; McKeever, 2004). One of the challenges facing the potential availability of urban woody waste is to sort out the available operation (not currently used) and determine alternative uses, including those used by captive markets (not likely to be diverted to bioenergy). One assessment of urban wood waste finds that 36 % of total biomass generated is currently solid to noncaptive markets, and 50 % of unused residues are unavailable due to contamination, quality, or recoverability.

Attributes of the Perfect Energy Crop

The perfect energy crop has

- High biomass
- Improved composition and structure
- Disease and pest resistance
- Optimized architecture
- Salt, pH, and aluminum tolerance
- Rapid and cost-effective propagation
- Stand establishment (cold germination and cold growth)
- Perennial
- Deep roots

One source of recurring and potentially available carbon feedstock is MSW. In 2005, the EPA estimated that 245.7 million tons of MSW was generated in the United States, of which 79 million tons were recycled, 33.4 million tons were diverted to energy recovery, and 133.3 million tons were disposed of in landfills. As such, landfilled material represents a significant source of renewable carbon that could be used for fuel/energy production or in support of biofuel production.

8.7 SECOND-GENERATION FEEDSTOCKS: LONG-TERM AVAILABILITY

In the long run, large-scale biofuel production will require other resources, including dedicated energy crops. Dedicated feedstocks are perennial grasses and trees grown as crops specifically to provide the necessary raw materials to bioenergy producers. A steady supply of low-cost, uniform, and consistent-quality biomass feedstock will be critical for the economic viability of cellulosic ethanol production. During the late 1980s, the Department of Energy sponsored research on perennial herbaceous (grassy) biomass crops, particularly switchgrass, considered a model energy crop because of its many perceived advantages: Native to North America, high biomass yield per acre, comprehensive regional coverage, and adaptability to marginal land conditions. An extensive research program on switchgrass in the 1990s generated a wealth of information on high-yielding varieties, regional adaptability, and management practices. Preliminary field trials show that the economic viability of switchgrass cultivation depends critically on the initial establishment success. During this phase, seed dormancy and seedling sensitivity to soil and weed conditions require that growers closely follow recommended practices. Viable yields require fertilization rates at about half the average for corn.

Switchgrass (*Panicum virgatum*) is a hardy, deep-rooted, perennial rhizomatous warm-season grass native to North America; it is believed to be most suitable for cultivation in marginal lands, low-moisture lands and lands with lower opportunity costs, such as pastures, including lands under the Conservation Reserve Program (CRP) where the Federal Government pays landowners annual rent for keeping land out of production (McLaughlin & Kzos, 2005). Additionally, much highly erodible land in the Corn Belt is unsuitable for straw or stover removal but potentially viable for dedicated energy crops such as switchgrass. Factors favoring the adoption of switchgrass include a selection of suitable lands, environmental benefits (carbon balances, improved soil nutrients, and quality), and using existing hay production techniques to grow the crop. Where switchgrass is grown on CRP lands, payments help to offset production costs. Factors discouraging switchgrass adoption include no possibility for crop rotation, farmers' risk aversion for producing a new crop because of lack of information, skills, and know-how, potential conflict with on-farm and off-farm scheduling activities, and compatibility with long-term land tenure. Overall, production budget and delivery cost assessments suggest switchgrass is a high-cost crop (undercurrent technology and price conditions) and may not compete with established crops, except in areas with low opportunity costs (e.g., pastureland and marginal lands).

DID YOU KNOW?

According to Parrish et al. (2008), no single method for establishing switchgrass can be suggested for all situations. No-till and convention tillage can be used to establish the crop. When seeded as part of a diverse mixture, planting guidelines should be followed for warm-season grass mixtures for conservation plantings.

Substantial variability is apparent in the economics of switchgrass production and assessments of production budgets and delivered costs. Factors at play include methods of storage and handling, transport distances, yields, and types of land used (cropland versus grassland). When the delivered costs of switchgrass are translated into break-even prices (compared with conventional crops), it becomes apparent that cellulosic ethanol or biopower plants would have to offer relatively high prices for switchgrass to induce farmers to grow it (Rinehart, 2006). However, the economics of switchgrass could improve if growers benefited from CRP payments and other payments tied to environmental services (such as carbon credits). In the long run, the viability of an energy crop like switchgrass hinges on continued reductions in cellulosic ethanol conversion costs and sustained improvements in switchgrass yield and productivity through breeding, biotechnology, and agronomic research.

DID YOU KNOW?

The main advantage of using switchgrass over corn as an ethanol feedstock is that its cost of production is generally about ½ that of grain corn and more biomass energy per hectare can be captured in the field (Samson et al., 2008).

Although a vital biofuel crop, switchgrass does have limitations. Switchgrass is not optimally grown everywhere. For example, in the upper Midwest, reed canary grass is more suitable under wet soil, while semi-tropical grass species are better adapted to the Gulf Coast region. State and local efforts are testing alternatives to switchgrass, such as, as previously mentioned, a reed canary grass (*Phalaris arundinacea,* tall, perennial grass that forms extensive single-species stands along the margins of lakes and streams). Also, *Miscanthus* (often confused with "Elephant Grass") is rapid growth, low mineral content, high biomass yield plant and other species are underway.

Regarding long-run sustainability, the ecology of perennial grassy crops favors a diversity of crops or even a mix of species within the same area. Both ecological and economic sustainability favor the development of a range of herbaceous species for optimal use of local soil and climatic conditions. A mix of energy crops in the same region would help reduce the risk of epidemic pests and disease outbreaks and optimize the supply of biomass to an ethanol or biopower plant since different grasses mature and can be harvested at different times. Moreover, the development of future energy corps must be evaluated from the standpoint of their water use efficiency, impact on soil nutrient cycling, effect on crop rotations, and environmental benefit (improved energy use efficiency and reduced GHG emissions, nutrient runoff, pesticide runoff, and land-use impacts). In the long run, developing a broad range of grassy corps for energy use is compatible with sustainability and economic viability criteria.

8.8 SHORT-ROTATION WOODY CROPS (SRWC)

Short-rotation woody crops (SRWC) represent another important category of future dedicated energy crops. Hybrid poplar, willow, American Sycamore, sweetgum, and loblolly have been extensively researched for their high biomass yield potential. Breeding programs and management practices continue to be developed for these species. SRWC is based on a high-density plantation system and more frequent harvesting (every 3–4 years for willow and 7 years for hybrid poplar).

- **Hybrid poplar**—This species is very site-specific and has a limited growing niche. It requires an abundant and continuous supply of moisture during the growing season. Soil should be moist but not continually saturated and have good internal drainage. It prefers damp, well-drained, fine sandy-loam soil near streams, where coarse sand is first deposited as flooding occurs. Typically, cottonwood trees are found at elevations 20 ft above the average level of adjacent streams. Hybrid poplars are among the fastest-growing trees in North America—in just six growing seasons, hybrid poplars reach 60 ft or more in height. They are well suited for producing bioenergy, fiber, and other biased products.

- **Willow species (Salix)**—In folklore and myth, a willow tree is believed to be quite sinister, capable of uprooting itself and stalking travelers. The reality is that willow is used for biofiltration, constructed wetlands, ecological wastewater treatment systems, hedges, land reclamation, landscaping, phytoremediation, stream bank stabilization (bioengineering), slope stabilization, soil erosion control, shelterbelt & windbreak, soil reclamation, tree bog compost toiler, and wildlife habitat. Willow is grown for biomass or biofuel in energy forestry systems due to its high energy-in-energy-out ratio, considerable carbon mitigation potential, and fast growth (Aylott, 2008). Willow is also grown and harvested to make charcoal.
- **American sycamore**—It prefers alluvial soils along streams in bottomlands. Sycamore growth and yield are less than cottonwood and willow.
- **Sweetgum** (*Liquidambar styraciflua*)—It is a species tolerant of various soils but grows best on rich, moist, alluvial clay and loam soils of river bottoms.
- **Loblolly pine** (*Pinus taeda*)—It is quite adaptable to various sites. It performs well on both poorly drained bottomland flats and modestly arid uplands. Biomass for energy is currently obtained from pre-commercial thinning and logging residue in loblolly pine stands. Utilization will undoubtedly increase, and loblolly pine energy plantations may become a reality.

In many parts of the country, plantations of willow, poplar, pine, and cottonwood have been established and are being commercially harvested. Willow plants are being planted in New York, particularly following the enactment of the Renewable Portfolio Standard (RPS) and other State incentives. Over 30,000 hectares of poplar are grown in Minnesota, and several thousand hectares are also grown as part of a DOE-funded project to provide biomass for a power utility company in southern Minnesota. The Pacific Northwest has a large plantation of hybrid poplars, estimated at 60,000 hectares in 2007. Most of these plantations are currently used for pulp wood, with little volume for bioenergy. Since SRWC can be used either for biomass or as feedstock for pulp and other products, pulp demand will influence the cost of using it for bioenergy production.

An essential consideration for energy crops (e.g., switchgrass, poplar, willow) is the potential for increasing yield and developing other desirable characteristics. Most energy crops are unimproved or have recently been bred for biomass yield. In contrast, corn and other commercial food crops have undergone substantial improvements in yield, disease resistance, and other agronomic traits. A more complex understanding of biological systems and applying the latest biotechnological advances would accelerate the development of new biomass crops with desirable attributes. These attributes include increased yields and processability, optimal growth in specific microclimates, better pest resistance, efficient nutrient use, and greater tolerance to moisture deficits and other sources of stress. Agronomic and breeding improvements of these new crops could significantly boost future energy crop production.

SIDEBAR 8.1 GASOLINE GALLON EQUIVALENT (GGE)

Before beginning our discussion about alternative renewable fuels used for transportation, it is essential to ensure that the reader understands the difference between conventional gasoline and diesel fuel energy output and non-conventional renewable products. Typically, this comparison is made utilizing a standard engineering parameter known as the gasoline gallon equivalent (GGE), which is the ratio of the number of British thermal units (Btu) available in one U.S. gallon (1 gal) of gasoline to the number of British thermal units available in 1 gal of the alternative substance in question. In 1994, NIST (1994) defined "gasoline gallon equivalent" (GGE) as 5,660 pounds of natural gas. The GGE parameter allows consumers to compare the energy content of competing fuels against a commonly known fuel—gasoline. Table 8.2 lists the listed fuels' GGE and Btu/unit value comparisons.

TABLE 8.2
Gasoline Gallon Equivalents (GGE)

Fuel	GGE	Btu/unit
Gasoline (Base)	1 US gallon	114,000 Btu/gal
Gasoline (conventional, summer)	0.996 US gallon	114,500 Btu/gal
Gasoline (convention, winter)	1.013 US gallon	112,500 Btu/gal
Gasoline (reformulated gasoline, ethanol	1.019 US gallon	111,836 Btu/gal
Gasoline (reformulated gasoline, ETBE)	1.019 US gallon	111,811 Btu/gal
Gasoline (reformulated gasoline, MTBE)	1.020 US gallon	111,745 Btu/gal
Gasoline (10% MBTE)	1.02 US gallon	112,000 Btu/gal
Gasoline (regular unleaded)	1 US gallon	114,100 Btu/gal
Diesel #2	0.88 US gallon	129,500 Btu/gal
Biodiesel (B100)	0.96 US gallon	118,300 Btu/gal
Bio Diesel (B20)	0.90 US gallon	127,250 Btu/gal
Liquid natural gas (LNG)	1.52 US gallons	75,000 Btu/gal
Compressed natural gas	1.26 ft^3 (3.587 m^3)	900 Btu/ ft^3
Hydrogen at 101.325 kPa	357.37 ft^3	319 Btu/ ft^3
Hydrogen by weight	0.997 kg (2.198 lb)	119.9 MJ/kg (51,500 Btu/lb)
Liquefied petroleum gas (LPG)	1.35 US gallons	84,300 Btu/gal
Methanol fuel (M100)	2.01 US gallons	56,800 Btu/gal
Ethanol fuel (E100)	1.500 US gallons	76,100 Btu/gal
Ethanol (E85)	1.39 US gallons	81,800 Btu/gal
Jet fuel (naphtha)	0.97 US gallon	118,700 Btu/gal

Source: USEPA (2007).

8.9 BIOETHANOL PRODUCTION: DRY CORN MILL PROCESS

Ethanol production from dry corn milling follows a seven-step process shown in Figure 8.5 (USDOE, 2010). First, corn feedstock is cleaned and milled (ground into corn meal); then milled corn is converted to slurry; slurry is liquefied (slurried with water—mash), and enzymes yield glucose; yeast is then added to the mash to convert starch into simple sugar (dextrose) (saccharification); after liquefaction, the mash is cooked in a saccharification tank to *reduce* bacterial levels and paled in a fermenter where fermentation takes place (sugar is converted to ethanol by yeast); the resulting "beer" containing 2%–12% ethanol is then distilled into ethanol at 95% alcohol and 5% water; the remaining solids (stillage) are collected during distillation, dried, and sold as an animal feed called dried distillers' grains (DDG) (see Figure 8.5). Water removal from ethanol beyond the last 5 % is called dehydration or drying.

DID YOU KNOW?

Fermentation is a series of chemical reactions that convert sugars to ethanol. The fermentation reaction is caused by yeast or bacteria, which feed on the sugars. Ethanol and carbon dioxide are produced as the sugar is consumed. The simplified fermentation reaction equation for the 6-carbon sugar, glucose, is:

$$\underset{\text{glucose}}{C_6H_{12}O_6} \rightarrow \underset{\text{ethanol}}{2\,CH_3CH_2OH} + \underset{\text{carbon dioxide}}{2\,CO_2}$$

8.9.1 Pros and Cons of Bioethanol

Ethanol can be used for transportation (propulsion) as an extender and octane enhancer in conventional gasoline. Its primary advantage is that its feedstock (plants) is renewable. It can also be used as a primary fuel (in E85), thus reducing dependence on petroleum products. Ethanol reduces environmental pollution from carbon monoxide (CO) gas and global carbon dioxide (CO_2). In some locations, E85 and gasohol are cheaper per gallon than conventional gasoline. As the production and use of ethanol as a fuel increases, farmers will benefit from the increased demand for their products. From a practical safety and health point of view, ethanol can be used to prevent gas-line freeze in freezing weather. Because it is not as flammable as gasoline, it is less likely to cause accidental explosions.

Ethanol used for transportation purposes does have a few negatives. Probably, the most apparent shortcoming is its non-availability at the local gas station. Gas stations selling E85 as part of their normal products are not as abundant as

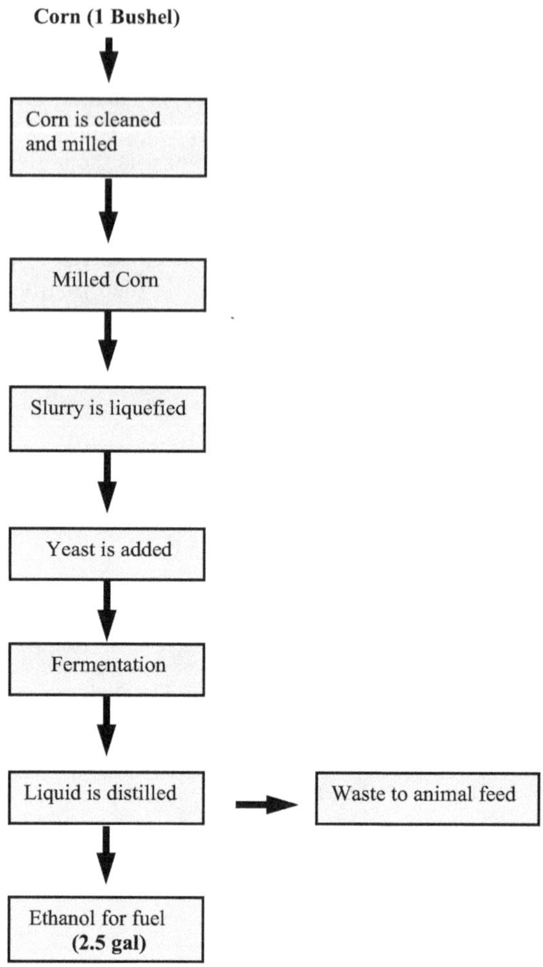

FIGURE 8.5 Flow diagram for production of ethanol from corn.

conventional gas stations. In some locations, E85 and gasohol are more expensive per gallon than traditional gasoline. E85 can damage vehicles that are not designed to burn it as fuel. Also, ethanol contains less energy than gasoline. This means your car won't go as far on a gallon of E85, and your fuel economy will decrease by 20%–30%. Ethanol is produced from plant matter that could otherwise be consumed as human and animal food products. Critics argue that food products used for fuel instead of feeding people and animals will indirectly contribute to world hunger. Moreover, it will lead to higher food prices. However, while it is true that higher corn prices increase animal feed and ingredient costs for farmers and food manufacturers, these are passed through to retail at a price rate of less than 10% of the corn price change.

SIDEBAR 8.2[3] CORN: ETHANOL PRODUCTION OR FOOD SUPPLY?

In 2007, record U.S. trade driven by economic growth in developing countries and favorable exchange rates, combined with tight global grain supplies, resulted in record or near-record prices for corn, soybeans, and other food and feed grains. For corn, these factors, along with increased demand for ethanol, helped push prices from under $2 per bushel in 2005 to $3.40 per bushel in 2007. By the end of the 2006/2007 crop year, over 2 billion bushels of corn (19% of the harvested crop) were used to produce ethanol, a 30% increase from the previous year. Higher corn prices motivated farmers to increase corn acreage at the expense of other crops, such as soybeans and cotton, raising their prices as well.

The pressing and pertinent question becomes: How do these higher commodity costs affect retail food prices? Retail food prices are generally much less volatile than farm-level prices and tend to rise by a fraction of the change in farm prices. The magnitude of the response depends on the retail costs beyond the raw food ingredients and the nature of competition in retail food markets. Ethanol's impact on retail food prices depends on how long the increased demand for corn drives up farm corn prices and how much higher corn prices are passed through to retail.

Retail food prices adjust as the cost of input into retail food production changes and the competitive environment in each market evolves. Intense competition among three to five retail stores in most U.S. markets has moderated food price inflation. Overall, retail food prices have been relatively stable over the past 20 years, with prices increasing by an average of 3.0%/year from 1987 through 2007, just below the overall inflation rate. Since then, food price inflation has averaged just 2.5%/year.

Field corn is the predominant corn type grown in the United States, primarily used for animal feed. Less than 10% of the U.S. field corn crop is used for direct domestic human consumption in corn-based foods such as corn meal, corn starch, and corn flakes. In contrast, the remainder is used for animal feed, exports, ethanol production, seed, and industrial uses. Sweet white and yellow corn is usually consumed as immature whole-kernel corn by humans and is also an ingredient in other corn-based foods. Still, only about 1% of U.S. corn production is produced by humans.

Because U.S. ethanol production uses field corn, the most direct impact of increased ethanol production should be on field corn prices and the price of food products based on field corn. However, even for those products heavily based on field corn, the effect of rising corn prices is dampened by other market factors. For example, an 18-ounce box of cornflakes contains about 12.9 ounces of milled field corn. When field corn is priced at $2.28 per bushel (the 20-year average), the actual value of corn in the box of cornflakes is about 3.3 cents (1 bushel = 56 pounds). (The remainder is packaging, processing, advertising, transportation, and other costs.) At $3.40 per bushel, the average price in 2007 was about 4.9 cents. The 49% increase in corn prices would be expected to raise the price of a box of corn flakes by about 1.6 cents, or 0.5%, assuming no other cost increases.

In 1985, Coca-Cola shifted from sugar to corn syrup in most U.S.-produced soda, and many other beverage makers followed suit. About 4.1% of U.S.-produced corn is made into high-fructose corn syrup. A 2-l bottle of soda contains about 15 ounces of corn in the form of high-fructose corn syrup. At $3.40 per bushel, the actual value of corn is 5.7 cents, compared with 3.8 cents when corn is priced at $2.28 per bushel. Assuming no other cost increases, the higher corn price in 2007 would raise soda prices by 1.9 cents per 2-l bottle, or 1%. These are notable changes in price measures and inflation, but relatively minor changes in the average household budget.

In addition to the impact on cornflakes and soda beverage prices, livestock prices are also impacted. This stands to reason when you consider that livestock rations traditionally contain a large amount of corn; thus, a more significant impact would be expected in meat and poultry prices due to higher feed costs than in other food products. Currently, 55% of corn produced in the United States is used as animal feed for livestock and poultry. However, estimating the corn used as feed to produce retail meat is a complicated calculation. Livestock producers have many options when deciding how much corn to include in a feed ration. For example, grass-fed cattle consume no corn at one extreme, while other cattle may have a diet consisting primarily of corn. For hog and poultry producers, ration variations may be less extreme but can still vary quite a bit. To estimate the impact of higher corn prices on retail meat prices, making a series of assumptions about feeding practices and grain conversion rates from animal to final retail meat products is necessary. To avoid downplaying potential impacts, this analysis uses upper-bound conversion estimates of 7 pounds of corn to produce 1 pound of beef, 6.5 pounds of corn to produce 1 pound of pork, and 2.6 pounds of corn to produce 1 pound of chicken.

Using these ratios and data from the Bureau of Labor Statistics, a simple pass-through model estimates the expected increase in meat prices given the higher corn prices. The logic of this model is illustrated by an example using chicken prices. Over the past 20 years, the average price of a bushel of corn in the United States has been $2.28, implying that a pound of chicken at the retail level uses 8 cents worth of corn or about 4% of the $2.05 average retail price for chicken breasts. Using the average price of corn for 2007 ($3.40 per bushel) and assuming producers do not change their animal-feeding practices, retail chicken prices would rise 5.2 cents or 2.5%. Using the same corn data, retail beef prices would go up 14 cents per pound, or 8.7%, while pork prices would rise 13 cents per pound, or 4.1%.

These estimates for meat, poultry, and corn-related foods, however, assume that the magnitude of the corn price change does not affect the rate at which cost increases are passed through to retail prices. Corn price fluctuations have little impact on retail food prices until they rise high enough for a long enough time to elicit a significant price adjustment by food producers and notably higher retail food prices.

On the other hand, these estimates may be overstating the effect of corn price increases on retail food prices since they do not account for potential substitution by producers from more expensive to less costly inputs. Such substitution would dampen the effect of higher corn prices on retail meat prices. Even assuming the upper-bound effects outlined above, the impact of rising corn commodity prices on overall food prices is limited. Given that less than a third of retail food contains corn as a major ingredient, these increasing prices for corn-related products would raise overall U.S. retail food prices by less than one percentage point per year above the average inflation rate.

Continuing elevated prices for corn will depend on the extent to which corn remains the most efficient feedstock for ethanol production, and ethanol remains a viable source of alternative energy. These conditions may change over time as other crops and biomass are used to produce ethanol, and other alternative energy sources develop.

DID YOU KNOW?

While higher commodity costs may have a relatively modest impact on U.S. retail food prices, there may be a more significant effect on retail food prices in lower-income developing countries. As a relatively low-priced food, grains have historically accounted for a large share of the diet in less developed countries. Even with incomes rising, consumers in such countries consume a less processed diet than is typical in the United States and other industrialized countries, so food prices are more closely tied to swings in both domestic and global commodity prices.

Even if these conditions do not change soon, market adjustments may dampen long-term impacts. In 1996, when field corn prices reached an all-time high of $3.55 per bushel due to drought-related tighter supplies in the United States and strong demand for corn from China and other parts of Asia, the effect on food prices was short-lived. At that time, retail prices rose for some foods, including pork and poultry, but these effects did not extend beyond the middle of 1997. Mostly, food markets adjusted to higher corn prices, and corn producers increased supply, bringing down prices.

Food producers, manufacturers, and retailers may adjust to the changing market conditions by adopting more efficient production methods and improved technologies to counter higher costs. For example, soft drink manufacturers may consider substituting sugar for corn syrup as a sweetener if corn prices remain high. At the same time, livestock and poultry producers may develop alternative feed rations that minimize corn needed for animal feed. Adjustments by producers, manufacturers, and retailers, along with continued retail solid competition, imply that U.S. retail food prices will remain relatively stable.

8.10 BIODIESEL

The diesel engine is the workhorse of heavy transportation and industrial processes; it is widely used to power trains, tractors, ships, pumps, and generators. Powering the diesel engine is conventional diesel fuel or biodiesel fuel. Biodiesel is a relatively dense liquid fuel comprising fatty acid alkyl esters, fatty acid methyl esters (FAME), or long-chain mono alkyl esters. It is produced from renewable sources such as new and used vegetable oils and animal fats and is a clearer-burning replacement for petroleum-based diesel fuel. It is non-toxic and biodegradable. This fuel is designed to be used in compression-ignition engines similar or identical to those that burn petroleum diesel. Biodiesel has physical properties (see Table 8.3) identical to petroleum diesel.

In the United States, most biodiesel is made from soybean oil or recycled cooking oils. Depending on costs and availability, animal fats, other vegetable oils, and other recycled oils can produce biodiesel. In the future, blends of all kinds of fats and oils may produce biodiesel.

Before providing a basic description of the vegetable and recycled grease to biodiesel process, defining and reviewing a few of the critical technological terms associated with the conversion process is essential.

- **Acid esterification**—Oil feedstocks containing more than 4% free fatty acids undergo an acid esterification process to increase biodiesel yield. These feedstocks are filtered and preprocessed to remove water and contaminants and then fed to the acid esterification process. The sulfuric acid catalyst is dissolved in methanol and mixed with the pretreated oil. The mixture is heated and stirred, and the free fatty acids are converted to biodiesel. Once the reaction is complete, it is dewatered and then fed to the transesterification process.
- **Transesterification**—Oil feedstocks containing less than 4% free fatty acids are filtered and preprocessed to remove water and contaminants and then fed directly to the transesterification process and any products of the acid esterification process. The catalyst, potassium hydroxide, is dissolved

TABLE 8.3
Biodiesel Physical Characteristics

Specific gravity	0.87–0.89
Kinematic viscosity @ 40°C	3.7–5.8
Cetane number	46–70
Higher heating value (Btu/lb)	16,928–17,996
Sulfur, wt%	0.0–0.0024
Cloud point °C	–11 to 16
Pour point °C	–15 to 13
Iodine numbers	60–135
Lower heating value (lbs/lb)	15,700–16,735

in methanol and mixed with the pretreated oil. If an acid esterification process is used, an extra base catalyst must be added to neutralize the acid added in that step. Once the reaction is complete, the significant co-products, biodiesel and glycerin, are separated into two layers.

- **Methanol recovery**—The methanol is typically removed after the biodiesel and glycerin have been separated to prevent the reaction from reversing. The methanol is cleaned and recycled back to the beginning of the process.
- **Biodiesel refining**—Once separated from glycerin, the biodiesel undergoes a clean-up or purification process to remove excess alcohol, residual catalysts, and soaps. This consists of one or more washings with clear water. It is then dried and sent to storage. Sometimes, the biodiesel undergoes an additional distillation step to produce a colorless, odorless, zero-sulfur biodiesel.
- **Glycerin refining**—The glycerin byproduct contains unreacted catalysts and soaps that are neutralized with acid. Water and alcohol are removed to produce 50%–80% crude glycerin. The remaining contaminants include unreacted fats and oils. In large biodiesel plants, glycerin can be further purified to 99% or higher purity for sale to the pharmaceutical and cosmetic industries.

The most popular biodiesel production process is transesterification (production of the ester) of vegetable oils or animal fats, using alcohol in the presence of a chemical catalyst. About 3.4 kg of oil/fat is required for each gallon of biodiesel produced (Baize, 2006). The transesterification of degummed soybean oil produces ester and glycerin. The reaction requires heat and a strong base catalyst such as sodium hydroxide or potassium hydroxide.

Some feedstocks must be pretreated before they can go through the transesterification process. Feedstocks with less than 4% free fatty acids, which include vegetable oils and some food-grade animal fats, do not require pretreatment. Feedstocks with more than 4% free fatty acids, including inedible animal fats and recycled greases, must be pretreated in acid esterification. In this step, the feedstock is reacted with an alcohol (like methanol) in the presence of a strong acid catalyst (sulfuric acid), converting the free fatty acids into biodiesel. The remaining triglycerides are converted to biodiesel in the transesterification reaction.

Figure 8.6 illustrates the basic technology for processing vegetable oils (such as soybeans) and recycled greases (used cooking oil and animal fat). When the feedstock is vegetable oil, the extracted oil is processed to remove all traces of water, dirt, and other contaminants. Free fatty acids are also removed. A combination of methyl alcohol and a catalyst, usually sodium hydroxide or potassium hydroxide, breaks the oil molecules apart in the esterification process. The resulting esters are then refined into usable biodiesel.

When the feedstock is used, cooking oil and animal fats are refined to produce biodiesel. The process is like biodiesel derived from vegetable oil, except an additional step is involved (Figure 8.6). Methyl alcohol and sulfur are used in a process called dilute acid esterification to obtain a substance resembling fresh vegetable oil, which is then processed in the same way as vegetable oil to get the final product.

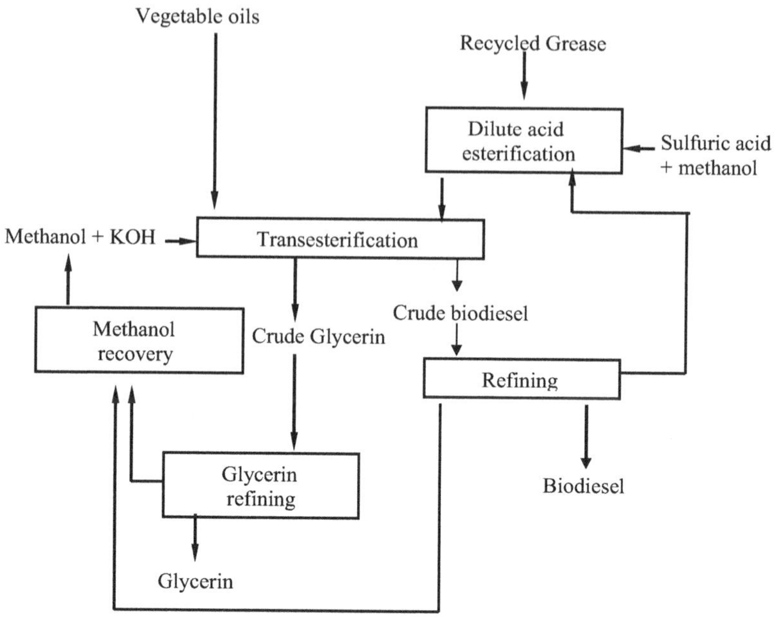

FIGURE 8.6 Biodiesel: process description.

DID YOU KNOW?

Biodiesel is routinely blended with petroleum diesel. The percentage of biodiesel in the blend is written following an uppercase letter B to denote the proportion. For example, a mixture of 20% biodiesel and 80% petroleum diesel is called B20, a mixture of equal parts biodiesel and petroleum diesel is B50, and neat (or pure) biodiesel is B100. Most conventional diesel engines can burn blends from pure petroleum diesel up to B20 without modification. With minor modifications, most diesel engines built since 1994 can use blends from B20 to B100. The diesel vehicle owner's manual and vehicle warranty should be checked before using any alternative fuel.

8.10.1 Algae to Biodiesel

Our discussion of biofuels has focused on terrestrial sources of biomass fuels. This section concerns photosynthetic organisms that grow in aquatic environments: macroalgae, microalgae, and emergent. Macroalgae, more commonly known as "seaweed," are fast-growing marine and freshwater plants that can grow considerably (up to 60 m in length). Emergents are plants that grow partially submerged in bogs and marshes. Microalgae are, as the name suggests, microscopic photosynthetic organisms.

Regarding the mass production of oil for alternative renewable energy purposes, the focus of this discussion is mainly on microalgae, organisms capable of photosynthesis that are less than 0.4 mm in diameter, including the diatoms and cyanobacteria, as opposed to macroalgae (as mentioned, seaweed). Microalgae are the preferred choice because they are less complex in structure, have a fast growth rate, and typically (for some species) contain a high concentration of oil. However, it is essential to note that recent research also focuses on using seaweed for algae fuel (biofuels), probably due to their high availability (Time Online, 2005).

DID YOU KNOW?

Algae fuel, also called algae fuel, algae oleum, or second-generation biofuel, is a biofuel derived from algae, that is, the production of natural oils for biodiesel.

The following algal species are currently being studied for their suitability as a mass-oil-producing crop in various locations worldwide (USDOE, 2010; Taipei Times, 2008):

Ulva
Botryococcus braunii
Chlorella
Dunaliella tertiolecta
Gracilaria
Pleurochrysis carterae
Sargassum (has 10 times the output volume of Gracilaria)

The algae-to-biodiesel alternative renewable fuel source program started during the Carter Administration in response to the 1970s Arab fuel embargo. At the same time, the Carter Administration consolidated all federal energy activities under the auspice of the newly established U.S. Department of Energy (DOE). Among its various programs established to develop all forms of alternative energy (related to solar power), DOE initiated research on using plant life as a source of transportation fuels (today called the Biofuels Program).

Before discussing the algae-to-biofuels process, the reader must understand basic algal concepts. In many ways, the study of microalgae is a relatively limited field. Algae are not nearly as well understood as other organisms that have found a role in today's biotechnology industry. The study of microalgae represents an area of higher risk and high gains; it also presents an opportunity for the curious and ambitious to break new ground in science by researching this growing area of interest.

DID YOU KNOW?

In the 1980s, DOE gradually shifted its focus to technologies that could have large-scale impacts on national fossil energy consumption. Much of DOE's publications from this period reflect a philosophy of energy research that might, somewhat pejoratively, be called "the quads mentality." A quad is a short-hand name for the unit of energy often used by DOE to describe the amounts of energy that a given technology might be able to displace. Quad is short for "quadrillion Btus"—a unit of energy representing 10^{15} (1,000,000,000,000,000) Btus of energy. This perspective led DOE to focus on the concept of immense algae farms.

8.10.1.1 Algae

The protists that perform photosynthesis are called **algae**. Algae can be both a nuisance and an ally; they possess Jekyll and Hyde-like characteristics (good versus bad characteristics). Many ponds, lakes, rivers, streams, and bays (e.g., Chesapeake Bay) in the United States (and elsewhere) are undergoing **eutrophication** (basically the killing off, although in many cases very slowly, of calm water environments, especially ponds, marshes, and lakes), the enrichment of an environment with inorganic substances (phosphorous and nitrogen). When eutrophication occurs, when filamentous algae like *Cladophora* break loose in a pond, lake, stream, or river and wash ashore, algae make a stinking, noxious presence known. More importantly, than the offensive odor dying algae give off, their deaths start the biodegradation process whereby instead of adding oxygen to their watery environment, they begin to use it up. As mentioned, algae have a good side too. For instance, algae are allies in many wastewater treatment operations. In addition, they can be valuable in long-term oxidation ponds, where they aid in the purification process by producing oxygen.

Before discussing the specifics and different types of algae, it is essential to be familiar with algal terminology.

8.10.1.1.1 Algal Terminology

- **Algae**—They are large and diverse assemblages of eukaryotic organisms that lack roots, stems, and leaves but have chlorophyll and other pigments for oxygen-producing photosynthesis.
- **Algology or phycology**—It is the study of algae.
- **Antheridium**—It is a unique male reproductive structure where sperm is produced.
- **Aplanospores**—They are nonmotile spores produced by sporangia.
- **Benthic**—They are the algae attached and living on the bottom of a body of water.
- **Binary fission**—It involves nuclear division followed by division of the cytoplasm.
- **Chloroplasts**—These are packets that contain ***chlorophyll*** and other pigments.
- **Chrysolaminarin**—It is the carbohydrates reserve in organisms of division **Chrysophyta**.
- **Diatoms**—They are photosynthetic, circular, or elliptical chrysophyte cells.
- **Dinoflagellates**—They are unicellular, photosynthetic protistan algae.

- **Dry mass factor**—It is the percentage of dry biomass with the fresh biomass; for example, if the dry mass factor is 5%, one would need 20 kg of wet algae (algae in the media) to get 1 kg of dry algae cells.
- **Epitheca**—It is the more significant part of the frustule (diatoms).
- **Euglenoids**—They contain chlorophylls **a** and **b** in their chloroplasts; the representative genus is *Euglena*.
- **Fragmentation**—It is a type of asexual algal reproduction in which the thallus breaks up, and each fragmented part grows to form a new thallus.
- **Frustule**—It is the distinctive two-piece wall of silica in diatoms.
- **Hypotheca**—It is the small part of the frustule (diatoms).
- **Lipid content**—It is the percentage of oil to the dry biomass needed to get it, i.e., if the algae lipid content is 40%, one would need 2.5 kg of dry algae to get 1 kg of oil.
- **Neustonic**—They are the algae that live at the water–atmosphere interface.
- **Oogonia**—They are vegetative cells functioning as female sexual structures in the algal reproductive system.
- **Pellicle**—It is a *Euglena* structure that allows for turning and flexing of the cell.
- **Phytoplankton**—It is made up of algae and small plants.
- **Plankton**—They are free-floating, mostly microscopic aquatic organisms.
- **Planktonic**— They are algae suspended in water instead of attached and living on the bottom (benthic).
- **Protothecosis**—It is a disease in humans and animals caused by the green algae *Prototheca piriformis*.
- **Thallus**—It is the vegetative body of algae.

Algae are autotrophic, contain green pigment chlorophyll, and are a form of aquatic plant. Algae differ from bacteria and fungi in their ability to carry out photosynthesis, which requires sunlight, carbon dioxide, and raw mineral nutrients. Photosynthesis takes place in chloroplasts. The chloroplasts are usually distinct and visible. They vary in size, shape, distribution, and numbers. In some algal types, chloroplasts may occupy most of the cell space. They usually grow near the water's surface because light cannot penetrate very far through water. Although in mass (multicellular forms like marine kelp), the unaided eye quickly sees them, many of them are microscopic. Algal cells may be nonmotile, motile by one or more flagella, or exhibit gliding **motility** as in diatoms. They occur most commonly in water (fresh and polluted water, as well as in salt water), in which they may be suspended (Planktonic) phytoplankton or attached and living on the bottom (benthic). A few algae live at the water–atmosphere interface and are termed Neustonic. They are important primary producers in fresh and saltwater environments (at the start of the food chain for other organisms). During their growth phase, they are essential oxygen-generating organisms and constitute a significant portion of the plankton in water.

According to the five-kingdom system of Whittaker, the algae belong to seven divisions distributed between two different kingdoms. Although seven divisions of algae occur, only five divisions are discussed in this section:

- **Chlorophyta**—Green algae
- **Euglenophyta**—Euglenoids
- **Chrysophyta**—Golden-brown algae, diatoms
- **Phaeophyta**—Brown algae
- **Pyrrophyta**—Dinoflagellates

The primary classification of algae is based on cellular properties. Several characteristics are used to classify algae, including (1) cellular organization and cell wall structure; (2) the nature of chlorophyll(s) present; (3) the type of motility, if any; (4) the carbon polymers that are produced and stored; and (5) the reproductive structures and methods.

Algae show considerable diversity in the chemistry and structure of their cells. Some algal cell walls are thin, rigid structures usually composed of cellulose modified by adding other polysaccharides. In other algae, the cell wall is strengthened by the deposition of calcium carbonate. Other forms have chitin present in the cell wall. Complicating the classification of algal organisms are the Euglenoids, which lack cell walls. In diatoms, the cell wall is composed of silica. As the fossil records indicate, diatoms' frustules (shells) have extreme resistance to decay and remain intact for long periods.

The principal feature used to distinguish algae from other microorganisms (for example, fungi) is the presence of chlorophyll and other photosynthetic pigments in algae. All algae contain chlorophyll *a*. Some, however, include different types of chlorophyll. The presence of these additional chlorophylls is characteristic of a particular algal group. In addition to chlorophyll, other pigments encountered in algae include fucoxanthin (brown), xanthophylls (yellow), carotenes (orange), phycocyanin (blue), and phycoerythrin (red).

Many algae have flagella (a threadlike appendage). As mentioned, the flagella are locomotor organelles that may be single-polar or multiple-polar types. The *Euglena* is a simple flagellate form with a single polar flagellum. Chlorophyta has either two or four polar flagella. Dinoflagellates have two flagella of different lengths. Sometimes, algae are nonmotile until they form motile gametes (a haploid cell or nucleus) during sexual reproduction. Diatoms do not have flagella but have gliding motility.

Algae can be either autotrophic or heterotrophic. Most are photoautotrophic; they require only carbon dioxide and light as their principal energy source and carbon. In the presence of light, algae carry out oxygen-evolving photosynthesis; in the absence of light, algae use oxygen. Chlorophyll and other pigments absorb light energy for photosynthetic cell maintenance and reproduction. One key characteristic of classifying algal groups is the nature of the reserve polymer synthesized due to utilizing carbon dioxide in water.

Algae may reproduce either asexually or sexually. Three types of asexual reproduction occur: binary fission, spores, and fragmentation. In some unicellular algae, binary fission occurs where the division of the cytoplasm forms new individuals like the parent cell following nuclear division. Some algae reproduce through spores. These spores are unicellular and germinate without fusing with other cells. The thallus breaks up in fragmentation, and each fragment grows to form a new thallus.

Sexual reproduction can involve the union of cells where eggs are formed within vegetative cells called Oogonia (which function as female structures), and sperm is produced in a male reproductive organ called antheridia. Algal reproduction can also occur by reducing chromosome numbers and the union of nuclei.

Algae are photosynthetic organisms that are far from monolithic. Biologists have categorized microalgae into various classes, mainly by pigmentation, life cycle, and basic cellular structure.

8.10.1.1.2 Characteristics of Algal Divisions

- **Chlorophyta** (Green Algae)—Most algae in ponds belong to this group; they can also be found in salt water and soil. Several thousand species of green algae are known today. Many are unicellular; others are multicellular filaments or aggregated colonies. The green algae have chlorophylls a and b, along with specific carotenoids, and they store carbohydrates in starch. Few green algae are found at depths greater than 7–10 m, mainly because sunlight does not penetrate that depth. Some species have a holdfast structure that anchors them to the bottom of the pond and other submerged inanimate objects. Green algae reproduce by both sexual and asexual means. Multicellular green algae have some division of labor, producing various reproductive cells and structures.
- **Euglenophyta** (Euglenoids)—A small group of unicellular microorganisms with animal and plant properties. Euglenoids lack a cell wall, possess a gullet, ingest food, assimilate organic substances, and, in some species, lack chloroplasts. They occur in fresh, brackish, salt waters and moist soils. A typical *Euglena* cell is elongated and bound by a plasma membrane; the absence of a cell wall makes it very flexible in movement. Inside the plasma membrane is a pellicle structure that gives the organisms a definite form and allows the cell to turn and flex. Photosynthetic euglenoids contain chlorophylls a and b and always have a red eyespot (**stigma**) sensitive to light (photoreceptive). Some Euglenoids move about using flagellum; others move about using contracting and expanding motions. The characteristic food supply for Euglenoids is lipopolysaccharide. Reproduction in Euglenoids is by simple cell division. Reproduction in euglenoids is by simple cell division.
- **Note**: Some autotrophic species of *Euglena* become heterotrophic when light levels are low.
- **Chrysophyta** (Golden Brown Algae)—The large Chrysophycophyta group has several thousand diversified members. They differ from green algae and euglenoids in that (1) chlorophylls a and c are present; (2) fucoxanthin, a brownish pigment, is present; and (3) they store food in the form of oils and leucosin, a polysaccharide. The combination of yellow pigments, fucoxanthin, and chlorophylls causes most of these algae to appear golden brown. Chrysophycophyta is also diversified in cell wall chemistry and flagellation. The division is divided into three major classes: golden-brown, yellow-brown algae, and diatom.

Some Chrysophyta lack cell walls, while others have intricately patterned coverings external to the plasma membrane, such as walls, plates, and scales. The diatoms are the only group with rigid cell walls of pectin, cellulose, or silicon, constructed in two halves (the epitheca and the hypotheca) called frustules. Two anteriorly attached flagella are typical among Chrysophyta, while others do not have flagella.

Most Chrysophyta are unicellular or colonial. Asexual cell division is the usual method of reproduction in diatoms; other forms of Chrysophyta can reproduce sexually. Diatoms have direct significance for humans. Because they make up most of the phytoplankton of the more excellent ocean parts, they are the ultimate food source for fish.

Water and wastewater operators understand the importance of their ability to function as indicators of industrial water pollution. Their specific tolerances to environmental parameters such as pH, nutrients, nitrogen, salt concentration, and temperature have been compiled as water quality indicators.

Interesting Point: Diatoms secrete a silicon dioxide shell (frustule) that forms the fossil deposits known as diatomaceous earth. This earth is used in filters and as abrasives in polishing compounds.

- **Phaeophyta** (Brown Algae)—Except for a few freshwater species, all algal species of this division exist in marine environments as seaweed. They are a specialized group of sessile multicellular organisms (attached and not free-moving). These algae contain essentially the same pigments seen in the golden-brown algae. Still, they appear brown because of the predominance of and the masking effect of a more significant amount of fucoxanthin. Brown algal cells store food as carbohydrate laminarin and some lipids. Brown algae reproduce asexually. Brown algae are used in foods, animal feeds, and fertilizers and as a source of alginate, a chemical emulsifier added to ice cream, salad dressing, and candy.
- **Pyrrophyta** (Dinoflagellates)—The principal members of this division are the dinoflagellates. The dinoflagellates comprise diverse biflagellated and nonflagellated unicellular, eukaryotic organisms. The dinoflagellates occupy a variety of aquatic environments, the majority of which live in marine habitats. Most of these organisms have a heavy cell wall composed of cellulose-containing plates. They store food such as starch, fats, and oils. These algae have chlorophylls *a* and *c* and several xanthophylls. The most common form of reproduction in dinoflagellates is by cell division, but sexual reproduction has also been observed.

DID YOU KNOW?

Cell division in dinoflagellates differs from that in most protistans. Chromosomes attach to the nuclear envelope and are pulled apart as the envelope stretches. In contrast, the nuclear envelope dissolves during cell division in most other eukaryotes.

Algal biomass contains three main components:

- Carbohydrates
- Protein
- Natural oils

  Biodiesel production applies exclusively to the natural oil fraction, which is this section's leading product of interest.

The bulk of natural oil made by oilseed crops is in the form of triacylglycerols (TAGs). TAGs consist of three long chains of fatty acids attached to a glycerol backbone. The algae species of concern can produce up to 60% of their body weight through TAGs. [Note: Recall that the species of concern, which are the oil producers, are *Ulva*, *Botryococcus braunii*, *Chlorella*, *Dunaliella tertiolecta*, *Gracilaria*, *Pleurochrysis carterae*, and *Sargassum*]. Thus, algae represent an alternative source of biodiesel that does not compete with the existing oilseed market.

8.10.1.2 Algae-to-Biodiesel Production

Christi (2007) points out that algae can produce up to 300 times more oil per acre than conventional crops, such as rapeseed, palms, soybean, or jatropha. Moreover, algae have a harvesting cycle of 1–10 days, permitting several harvests in a concise time frame, a strategy different from that of yearly crops. Algae can also be grown on land unsuitable for other established crops, such as arid land, excessively saline soil, and drought-stricken land. This minimizes the issue of taking away pieces of land from cultivating food crops (Schenk et al., 2008). Algae can grow 20–30 times faster than food crops (McDill, 2010).

Algae can be produced and harvested for biofuel using various technologies. These include photobioreactors (plastic tubes full of nutrients exposed to sunlight), closed-loop (not exposed to open air) systems, and open ponds. For illustration (even though there are many objectors and dissenters to open pond systems), we chose to discuss the open pond types configured in algae farms because the open pond raceway system is a relatively low-cost and easily understood process.

Open-Pond Algae Farm—Algae farms consist of open, shallow ponds in which some source of waste carbon dioxide (CO_2) could be bubbled into the ponds and captured by the algae (see Figure 8.7).

As shown in Figure 8.7, the ponds in an algae farm are "raceway" designs, in which the algae, water, and nutrients circulate a racetrack. Paddlewheels provide the flow. The algae are thus kept suspended in water. Algae are circulated back up to the surface on a regular frequency. The ponds are kept shallow because of the need to keep the algae exposed to sunlight and the limited depth to which sunlight can penetrate the pond water. The ponds operate continuously; water and nutrients are constantly fed to the pond while algae-containing water is removed at the other end. Some harvesting systems are required to recover algae containing substantial amounts of natural oil.

Figure 8.8 illustrates the concept of an "algae farm." The size of these ponds is measured in terms of surface area (as opposed to volume) because surface area is so critical to capturing sunlight. Their productivity is measured by biomass produced

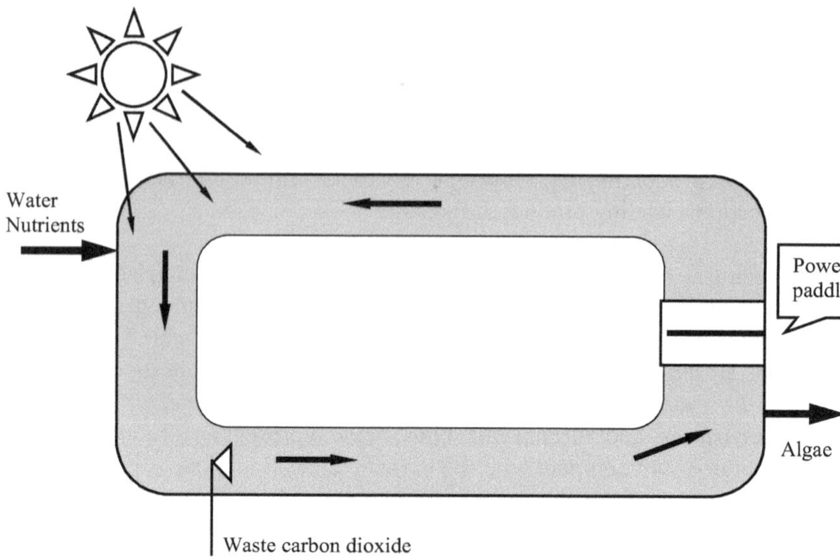

FIGURE 8.7 Raceway design algae pool.

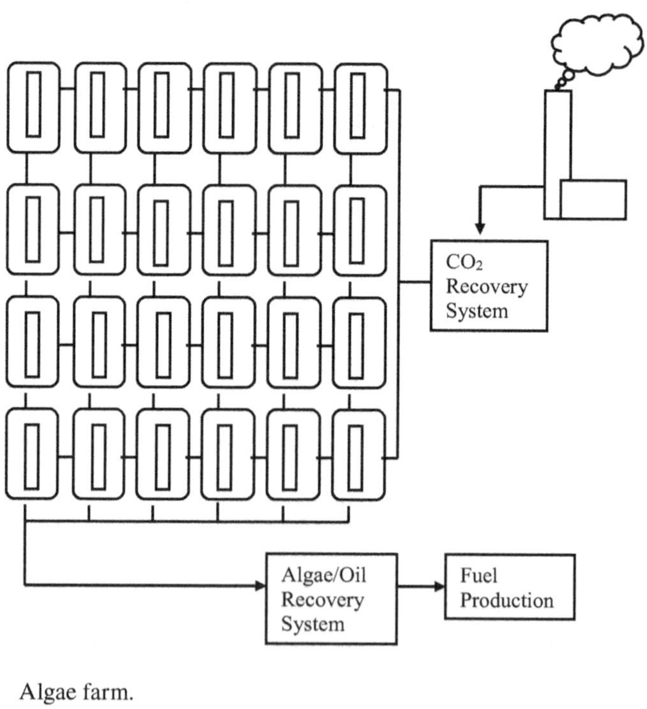

FIGURE 8.8 Algae farm.

per day per unit of available surface area. Such a system will require acres of land even at productivity levels that would stretch the limits of an aggressive research and development program. At such large sizes, it is more appropriate to think of these operations on the scale of a farm.

Waste CO_2 is readily available from several sources. Every operation involving fuel combustion for energy is a potential source. Generally, coal and other fossil fuel-fired power plants are targeted as the primary sources of CO_2. Typical coal-fired power plants emit flue gas from their stacks containing up to 13% CO_2. This high concentration of CO_2 enhances the transfer and uptake of CO_2 in the ponds. Coupling a coal-fired power plant with an algae farm provides a win-win approach to recycling CO_2 from coal combustion into a usable liquid fuel.

8.10.2 Jatropha to Biodiesel

The uninformed or misinformed might cringe when they discover that a byproduct of the plant *Jatropha curcas*—yes, the same plant known in some places as "black vomit nut" and in others as "bellyache bush" or "tuba-tub"—is being used as a product with any credible value to anyone. The Jatropha family of shrubs is a non-edible plant that grows mainly in countries like the Philippines. It is resistant to drought and can easily be planted or propagated through seeds or cuttings. It starts producing seeds within 14 months but reaches its maximum productivity level after 4–5 years. The Jatropha plant can make an oil content of 30%–58%, depending on the soil quality where it is planted. The seeds yield an annual equivalent of 0.75–2 tons or 1,892 L of biodiesel per hectare. The plant remains useful for around 30–40 years and can be planted in harsh climates where it won't compete for resources needed to grow food. Along with having the advantage of being a renewable fuel source, the Jatropha plant also reduces GHG emissions and our dependence on oil imports.

8.10.3 Pros and Cons of Biodiesel

The most significant advantage of biodiesel over conventional petroleum diesel is that biodiesel comes from renewable resources. The supply can be grown repeatedly. Biodiesel combustion also produces fewer emissions (except for nitrous oxides) than combustion of an equal amount of petroleum diesel. The widespread use of biodiesel can also reduce the dependency on imported oil. From a safety standpoint, biodiesel is safer than petroleum diesel because it is less combustible. From an environmental perspective, biodiesel is not persistent within ecological media (air, water, soil) when accidentally spilled because it is biodegradable. Biodiesel can also be produced from waste products such as cooking oils and grease.

The most pressing disadvantages or shortcomings of biodiesel, at least at present, are its lack of availability or accessibility and high-cost relative to petroleum diesel. This trend in non-availability and non-accessibility is bound to change as cheaper, more accessible petroleum diesel becomes more expensive and difficult to find. Additional disadvantages are that biodiesel requires special handling, storage, and transportation management compared to petroleum diesel. Regarding environmental considerations, biodiesel produces more nitrous oxide emissions than an equal amount of petroleum diesel when combusted. Another potential problem with biodiesel production is its dependence on soybeans as their primary feedstock; some concern is that the widespread use of biodiesel as fuel will contribute to higher food prices and indirectly to world hunger. There is a slight reduction in performance and mileage per gallon with biodiesel compared to petroleum diesel. Biodiesel can also act as a solvent in some diesel engines, causing loosened deposits that may clog filters.

8.11 BIOGAS (METHANE CH_4)

Primarily known as a fuel for interior heating systems, methane or biogas can also replace natural gas—a fossil fuel for electricity generation, cooking, and heating—and as an alternative fuel to gasoline. Methane is a natural gas that breaks organic material without oxygen in termite mounds, wetlands, and some animals. Humans are also responsible for the release of methane through biomass burning, rice production, cattle, and release from gas exploration. Methane can also be obtained directly from the earth; however, other methods of production have been developed, most notably the fermentation or composting of plant and animal waste.

The reasons for considering Biogas (methane) as a possible biofuel include:

- It is viable because of its potential use as an alternative fuel source.
- It is a viable alternative fuel to use to improve air quality.
- It can be produced locally, reducing the need for imported natural gas.

Methane is produced under anaerobic (no oxygen) conditions, where organic material is biodegraded or broken down by microorganisms. Figure 8.9 shows the three primary sources of feedstock material for anaerobic digestion, described below.

8.11.1 Anaerobic Digestion

Anaerobic digestion is the traditional method of managing waste, sludge stabilization, and releasing energy. It involves using bacteria that thrive without oxygen and is slower than aerobic destabilizing sludge and releasing energy. It uses bacteria that thrive without oxygen and is slower than aerobic digestion. However, anaerobic digestion has the advantage that only a tiny percentage of the wastes are converted into new bacterial cells. Instead, most organics are converted into carbon dioxide and methane gas.

Cautionary Note: Air should not be allowed to enter an anaerobic digester because it could mix with the gas produced, creating an explosive mixture.

USEPA (1979; 2006) and Spellman (2009) point out that there are four critical biological and chemical stages of anaerobic digestion (see Figure 8.10):

1. **Hydrolysis**—The process by which the proteins, cellulose, lipids, and other complex organics are broken down into smaller molecules and become soluble by utilizing water to split the chemical bonds of the substances.

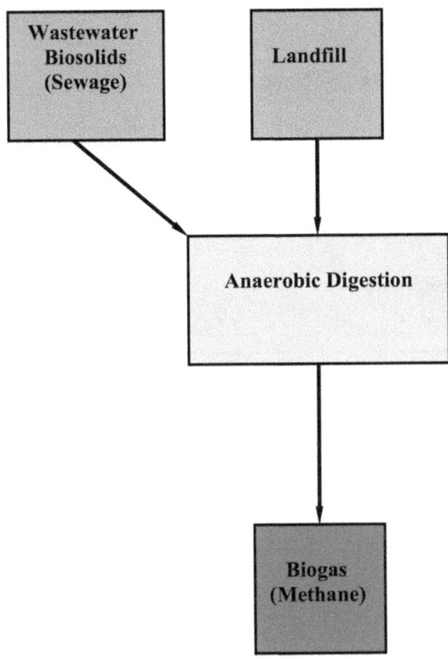

FIGURE 8.9 Production of biogas (methane CH_4).

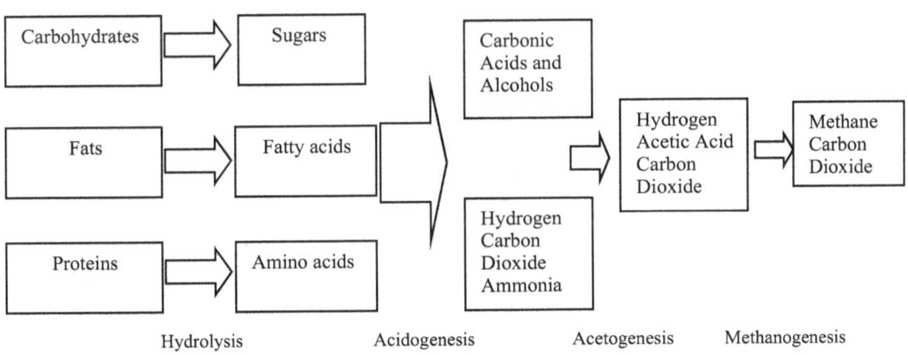

FIGURE 8.10 Key stages of anaerobic digestion.

2. **Acidogenesis**—The process by which the products of hydrolysis are converted into organic acids (where monomers are converted to fatty acids).
3. **Acetogenesis**—The process by which the fatty acids are converted to acetic acid, carbon dioxide, and hydrogen.
4. **Methanogenesis**—The process by which the organic acids produced during fermentation are converted to methane and carbon dioxide.

The efficiency of each phase is influenced by the temperature and the amount of time the process is allowed to react. For example, the organisms that perform hydrolysis

TABLE 8.4
Typical Contents of Biogas

Matter	Percentage (%)
Methane, CH_4	50–75
Carbon dioxide, CO_2	25–50
Nitrogen, N_2	0–10
Hydrogen, H_2	0–1
Hydrogen sulfide, H_2S	0–3
Oxygen, O_2	0–2

and volatile acid fermentation (often called the acidogenic bacteria) are fast-growing microorganisms that prefer a slightly acidic environment and higher temperatures than those that perform the methane formation step (the methanogenic bacteria).

A simplified generic chemical equation for the overall processes outlined above is as follows:

$$C_6H_{12}O_6 \rightarrow 3CO_2 + 3CH_4$$

Biogas is the ultimate waste product of the bacteria feeding off the input biodegradable feedstock and is mostly methane and carbon dioxide, with a small amount of hydrogen and trace hydrogen sulfide (see Table 8.4) (Spellman, 2009). Remember that the ultimate output from a wastewater digester is water; biogas (methane) is more of an off-gas that can be used as an energy source. Wastewater digestion and the production of biogas are discussed in the next section.

Equipment used in anaerobic digestion typically includes a sealed digestion tank with either a fixed or a floating cover or an inflatable gas bubble, heating and mixing equipment, gas storage tanks, solids and supernatant withdrawal equipment, and safety equipment (e.g., vacuum relief, pressure relief, flame traps, and explosion-proof electrical equipment). [**Caution**: Biosolids are inherently dangerous as possible sources of explosive gases and should never be entered without following OSHA's confined space entry permit requirements; only fully trained personnel should enter permit-required confined spaces].

Process residual (thickened or unthickened sludge) is pumped into the sealed digester. The organic matter is digested anaerobically through a two-stage process. Sugars, starches, and carbohydrates are converted to volatile acids, carbon dioxide, and hydrogen sulfide. The volatile acids are then converted to methane gas. This operation can occur in a single tank (single stage) or two tanks (two stages). In a single-stage system, supernatant and digested solids must be removed whenever flow is added. In a two-stage operation, solids and liquids from the first stage flow into the second stage each time fresh solids are added. The supernatant is withdrawn from the second stage to provide additional treatment space. Periodically, solids are withdrawn for dewatering or disposal. The methane gas produced in the process may be used for many plant activities.

DID YOU KNOW?

The primary purpose of a secondary digester is to allow for solids separation.

Various performance factors affect the operation of anaerobic digestion. For example, the percentage of volatile matter in raw sludge, digester temperature, mixing, volatile acids/alkalinity ratio, feed rate, percentage solids in raw sludge, and pH are all critical operational parameters the operator must monitor.

Digester operators must recognize normal/abnormal anaerobic digester performance parameters and know and understand standard operating procedures. These procedures include sludge additions, supernatant withdrawal, sludge withdrawal, pH control, temperature control, mixing, and safety requirements. Essential performance parameters are listed in Table 8.5 (Spellman 2009).

Sludge must be pumped (in small amounts) several times each day to achieve the desired organic loading and optimum performance.

Caution: In fixed-cover operations, additions must be balanced by withdrawals. Otherwise, structural damage occurs.

Supernatant withdrawal must be controlled for maximum sludge retention time. When sampling, sample all draw-off points and select the level with the best quality. Digested sludge is withdrawn only when necessary—always leave at least 25% seed. pH should be adjusted to maintain 6.8–7.2 by adjusting feed rate, sludge withdrawal, or alkalinity additions.

Note: The volatile acid/alkalinity relationship indicates the buffer capacity of an anaerobic digester. Decreases in alkalinity cause a corresponding increase in ratio.

If the digester is heated, the temperature must be controlled to an average temperature range of 90°F–95°F. Never adjust the temperature by more than 1°F per day. If the digester is equipped with mixers, mixing should be accomplished to ensure that organisms are exposed to food materials. Again, anaerobic digesters are inherently dangerous—several catastrophic failures have been recorded. Safety equipment such

TABLE 8.5
Anaerobic Digester—Sludge Parameters

Raw Sludge Solids	Impact
<4% Solids	Loss of alkalinity
	Decreased sludge retention time
	Increased heating requirements
	Decreased volatile acid:alk ratio
4%–8% Solids	Normal operation
>8% Solids	Poor mixing
	Organic overloading
	Decreased volatile acid:alk ratio

as pressure relief and vacuum relief valves, flame traps, condensate traps, and gas collection safety devices are installed to prevent such failures. These critical safety devices must be checked and maintained for proper operation.

Note: As mentioned, because of the inherent danger involved with working inside anaerobic digesters, they are automatically classified as permit-required confined spaces. Therefore, all internal entry operations must be done by OSHA's confined space entry standard. If safety questions concerning safe entry into confined spaces (of any type) arise, consult with a Certified Safety Professional (CSP), Certified Industrial Hygienist (CIH), or Professional Engineer (PE).

During operation, anaerobic digesters must be monitored and tested to ensure proper operation. Testing should be accomplished to determine supernatant pH, volatile acids, alkalinity, BOD or COD, total solids, and temperature. Sludge (in & out) should be routinely tested for % solids and % volatile matter. Typical operating parameters are listed in Table 8.6 (Spellman, 2009).

Process control calculations involved with anaerobic digester operation include determining the required seed volume, volatile acid-to-alkalinity ratio, sludge retention time, estimated gas production, volatile matter reduction, and percent moisture reduction in digester sludge. The following sections provide examples of how to make these calculations.

8.11.1.1 Required Seed Volume in Gallons

$$\text{Seed Volume (Gallons)} = (\text{Digester Volume})\ (\%\ \text{Seed}) \tag{8.1}$$

TABLE 8.6
Anaerobic Digester: Normal Operating Ranges

Parameter	Normal Range
Sludge Retention Time	
Heated	30–60 days
Unheated	180+ days
Volatile solids loading	0.04–0.1 lbs V.M/day/ft^3
Operating temperature	
Heated	90°F–95°F
Unheated	Varies with season
Mixing	
Heated – primary	Yes
Unheated – secondary	No
% Methane in gas	60%–72%
% Carbon dioxide in gas	28%–40%
pH	6.8–7.2
Volatile acids:Alkalinity ratio	≤0.1
Volatile solids reduction	40%–60%
Moisture reduction	40%–60%

Example 8.1

Problem:

The new digester requires a 25% seed to achieve regular operation within the allotted time. If the digester volume is 266,000 gallons, how many gallons of seed material will be required?

Solution:

$$\text{Seed Volume} = 266{,}000 \times 0.25 = 66{,}500 \text{ gals}$$

8.11.1.2 Volatile Acids to Alkalinity Ratio

The *volatile acids to alkalinity ratio* can be used to control the operation of an anaerobic digester:

$$\text{Ratio} = \frac{\text{Volatile Acids Concentration}}{\text{Alkalinity Concentration}} \tag{8.2}$$

Example 8.2

Problem:

The digester contains 240 mg/L of volatile acids and 1,860 mg/L alkalinity. What is the volatile acids/alkalinity ratio?

$$\text{Ratio} = \frac{240 \text{ mg/L}}{1{,}860 \text{ mg/L}} = 0.13$$

Note: Increases in the ratio typically indicate a potential change in the digester's operation condition, as shown in Table 8.7.

8.11.1.3 Sludge Retention Time

Sludge retention time is the length of time the sludge remains in the digester:

$$\text{SRT, Days} = \frac{\text{Digester Volume in Gallons}}{\text{Sludge Volume Added Per Day, gpd}} \tag{8.3}$$

TABLE 8.7
Opération Condition VA/Alkalinity ratio

Operating Condition	V.A./Alkalinity Ratio
Optimum	≤0.1
Acceptable range	0.1–0.3
% Carbon dioxide in gas increases	≥0.5
pH decreases	≥0.8

Example 8.3

Problem:
Sludge is added to a 525,000-gallon digester at 12,250 gallons per day.

Solution:

$$\text{SRT} = \frac{525{,}000\ \text{gallons}}{12{,}250\ \text{gpd}} = 42.9\ \text{days}$$

8.11.1.4 Estimated Gas Production in Cubic Feet/Day

The gas production rate is usually expressed as the volume of gas (ft^3) produced per pound of volatile matter destroyed. The total cubic feet of gas a digester will produce per day can be calculated by:

$$\text{Gas Prod.}\left(\text{ft}^3\right) = \text{Vol. Matter In, lbs/day} \times \%\ \text{Vol. Matter Red.} \times \text{Prod. Rate ft}^3/\text{lb} \tag{8.4}$$

Example 8.4

Problem:
The digester receives 11,450 lbs of volatile matter per day. Currently, it reduces volatile matter by 52%. The gas production rate is 11.2 cubic feet per pound of volatile matter destroyed.

Solution:

$$\text{Gas Prod.} = 11{,}450\ \text{lbs/day} \times 0.52 \times 11.2\ \text{ft}^3/\text{lb} = 66{,}685\ \text{ft}^3/\text{day}$$

8.11.1.5 Volatile Matter Reduction, Percent

Because of the changes occurring during sludge digestion, the calculation used to determine the percentage of volatile matter reduction is more complicated.

$$\%\ \text{Reduction} = \frac{\left(\%\ \text{Volatile Matter}_{\text{in}} - \%\ \text{Volatile Matter}_{\text{out}}\right) \times 100}{\left[\%\ \text{Volatile Matter}_{\text{in}} - \left(\%\ \text{Volatile Matter}_{\text{in}} \times \%\ \text{Volatile Matter}_{\text{out}}\right]\right.} \tag{8.5}$$

Example 8.5

Problem:
Using the data provided below, determine the % Volatile Matter Reduction for the digester.

Raw Sludge Volatile Matter 74%
Digested Sludge Volatile Matter 55%

$$\%\ \text{Volatile Matter Reduction} = \frac{(0.74 - 0.55) \times 100}{\left[0.74 - (0.74 \times 0.55)\right]} = 57\%$$

8.11.1.6 Percent Moisture Reduction in Digested Sludge

$$\% \text{ Moisture Reduction} = \frac{(\% \text{ Moisture}_{in} - \% \text{ Moisture}_{out}) \times 100}{\left[\% \text{ Moisture}_{in} - \left(\% \text{ Moisture}_{in} \times \% \text{ Moisture}_{out}\right)\right]} \tag{8.6}$$

Example 8.6

Problem:

Using the digester data provided below, determine the % Moisture Reduction and % Volatile Matter Reduction for the digester.

Solution:

Raw Sludge % Solids 6%
Digested Sludge % Solids 14%
Note: % Moisture = 100% – Percent Solids

$$\% \text{ Moisture Reduction} = \frac{(0.94 - 0.86) \times 100}{\left[0.94 - (0.94 \times 0.86)\right]} = 61\%$$

Animal waste accounts for 10% of methane emissions in the United States. Farm waste slurries can seriously pollute local watercourses when not correctly managed. Small anaerobic digesters have been installed on farms to treat excess animal slurries that can't be placed on the land. The biogas formed is commonly used for heat but can also fuel engines and other on-site energy needs such as electricity and heating. On-site biogas production and management also reduce offensive odors from overloaded or improperly managed manure storage facilities. These odors impair air quality and may be a nuisance to nearby communities. Anaerobic digestion of animal waste reduces these offensive odors because the volatile organic acids, the odor-causing compounds, are consumed by biogas-producing bacteria. In addition to biogas, ammonium is another vital byproduct of anaerobic digestion, the principal constituent of commercial fertilizer, and is readily available and utilized by crops. The bottom line on biogas production on the farm: Biogas recovery can improve profitability while improving environmental quality.

8.11.2 Landfill Biogas

Landfills can be a source of energy. Landfills produce methane as organic waste decomposes in the same anaerobic digestion process that converts wastewater and farm waste slurries into biogas. Most landfill gas results from the degradation of cellulose contained in municipal and industrial solid waste. Unlike animal manure digesters, which control the anaerobic digestion process, the digestion occurring in landfills is an uncontrolled process of biomass decay.

To be technically feasible, a landfill must be at least 40 feet deep and have at least one million tons of waste in place for landfill gas collection.

The efficiency of the process depends on the waste composition and moisture content of the landfill, cover material, temperature, and other factors. The biogas released from landfills, commonly called "landfill gas," is typically 50% methane, 45% carbon dioxide, and 5% other gases. The energy content of landfill gas is 400–550 Btu per cubic foot.

Figure 8.11 shows a landfill energy system. Such a system consists of a series of wells drilled into the landfill. A piping system connects the wells and collects the gas. Dryers remove moisture from the gas, and filters remove impurities. The gas typically fuels an engine-generator set or gas turbine to produce electricity. It can also fuel a boiler to produce heat or steam because waste-generated biogas is considered a "dirty gas" compared to natural gas. Further gas cleanup is required to improve the quality of biogas in pipelines, equivalent to natural gas. Reforming gas to hydrogen would make electricity production using fuel cell technology possible.

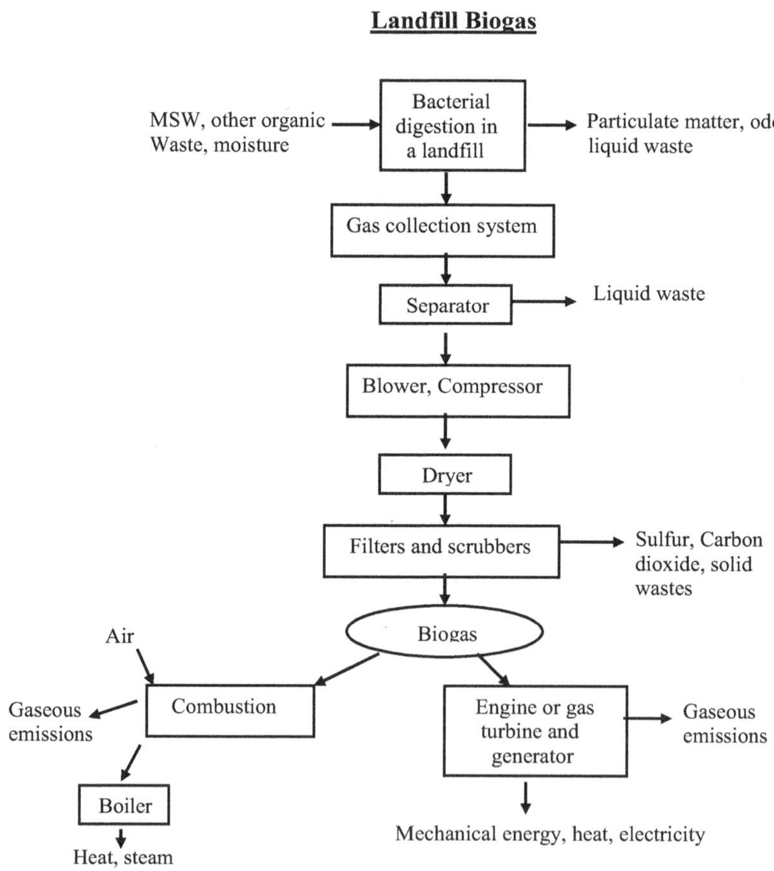

FIGURE 8.11 Landfill biogas system flow diagram.

8.12 BIOMASS FOR BIOPOWER

In addition to advanced fuels, biomass can also be used to produce biopower. This can be done in several ways, including direct combustion of biomass in dedicated power plants, co-firing biomass with coal, biomass gasification in a combined cycle plant to produce steam and electricity, or via anaerobic digestion (EPRI, 1997).

Combustion is the burning of biomass in the air. This involves converting chemical energy stored in biomass into heat, mechanical power, or electricity (McKendry, 2002). While it is possible to use all biomasses, combustion is preferable when biomass is more than 50% dry. High-moisture biomass is better suited for biological conversion processes. Net bioenergy conversion efficiencies for biomass combustion power plants range from 20% to 40%. Higher efficiencies are obtained with the combined heat and power (CHP) facilities and with large-size power-only systems (over 100 Mega-Watt-electrical—MWe) or when biomass is co-fired with coal in power plants (McKendry, 2002).

Co-firing biomass with coal is a straightforward and inexpensive way to diversify the fuel supply, reduce coal plant air emissions (NO_x, SO_2, CO_2), divert biomass from landfills, and stimulate the biomass power industry (Hughes, 2000). Moreover, biomass is the only renewable energy technology that can directly displace coal. Given the dominance of coal-based power plants in U.S. electricity production, co-firing with biomass fuel is the most economical way to reduce GHG emissions. Possible biomass fuel for co-firing includes wood waste, short-rotation woody crops, switchgrass, alfalfa stems, various types of manure, landfill gas, and wastewater treatment gas (Tillman, 2000). In addition, agricultural residues such as straw can also be used for co-firing.

A promising technology development currently at the demonstration stage is biomass integrated gasification/combined cycle (BIG/CC), where a gas turbine converts the gaseous fuel to electricity with a high conversion efficiency, reaching 40%–50% of the heating value of the incoming gas (McKendry, 2002). An essential advantage of gasification is working with a wider variety of feedstocks, such as high alkali fuels that are problematic with direct combustion. High alkali fuels such as switchgrass, straw, and other agricultural residues often cause corrosion. Still, gasification systems can easily remove the alkali species from the fuel gas before they are combusted. High silica and a problem with grass can also result in slagging in the reactor.

NREL (1993) points out that the slagging problem is not unique to one form of biomass but is familiar to many different types of biomass fuels. Slagging deposits can reduce heat transfer, minimize combustion efficiency, and damage combustion chambers when large particles break off. Research has focused on two alkali metals, potassium and sodium, and silica, all common elements in living plants. Generally, faster-growing plants (or faster-growing plant components such as seeds) tend to have higher concentrations of alkali metal and silica. Thus, materials such as straw, nut hulls, fruit pits, weeds, and grasses create more problems when burned than wood from a slow-growing tree.

Potassium and sodium metals, whether oxides, hydroxides, or metallo-organic compounds, tend to lower the melting point of ash mixtures containing other minerals such as silica (SiO_2). The ash's high alkali content (up to 35%) from burning annual crop residues lowers the fusion or "sticky temperature" of these ashes from 2,200°F for wood ash to as low as 1,300°F. This results in serious slagging on the boiler grate or the bed and fouling of convection heat transfer surfaces. Even small percentages (10%) of some of these high alkali residues burned with wood in conventional boilers will cause serious slagging and fouling in a day to two, necessitating a combustion system shutdown.

The coal industry has adopted a method to predict slagging and fouling from the combustion of biomass fuels, which has been adapted from the coal industry. The method involves calculating the weight in pounds of alkali ($K_2O + Na_2O$) per million Btu in the fuel as follows:

$$\frac{1 \times 10^6}{\text{Btu/lb}} \times \%\ \text{Ash} \times \%\ \text{Alkali of the Ash} = \frac{\text{lb Alkali}}{\text{MMBtu}}$$

This method combines all the pertinent data into one Index Number. A value below 0.4 lb/MMBtu (MMBtu = thousand thousand Btu) is considered a low-slagging risk. Values between 0.4 and 0.8 lb/MMBtu will probably slag with increasing certainty of slagging as 0.8 lb/MMBtu is approached. Above 0.8 lb/MMBtu, the fuel is virtually specific to slag and foul (see Table 8.8).

As mentioned, another process for biomass is the application of anaerobic digestion to produce biogas (methane) for electricity generation. Recall that anaerobic digestion involves bacteria-controlled breakdown of organic waste without oxygen.

TABLE 8.8
Alkali Content and Slagging Potential of Various Biofuels

Fuel	Total Alkali (lb/MMBtu)	Slagging Potential
	Wood	
Pine chips	0.07	Minimal
White oak	0.14	Minimal
Hybrid poplar	0.46	Probable
Urban wood waste	0.46	Probable
Clean Tree Trimmings	0.73	Probable
	Pits, Nuts, Shells	
Almond shells	0.97	Certain
Refuse derived fuel	1.60	Certain
	Grasses	
Switch grass	1.97	Certain
Wheat straw-average	2.00	Certain
Wheat straw-hi alkali	5.59	Certain
Rice straw	3.80	Certain

Source: Adaptation from Miles et al. (1993).

Primary agricultural feedstocks for anaerobic digestion include food processing waste and manure from livestock operations. The Energy Information Agency also projects a significant increase in the generation of electricity from municipal waste and landfill gas—to about 0.5% of U.S. electricity consumption (EIA, 2006)

8.13 BIOMASS FOR BIOPRODUCTS[4]

Bioproducts are industrial and consumer goods manufactured wholly or partly from renewable biomass (plant-based resources). Today's industrial bioproducts range from solvents and paints to pharmaceuticals, soaps, cosmetics, and building materials (see Table 8.9). Industrial bioproducts are integral to our way of life—a few sectors of the economy do not rely in some way on products made from biomass.

Corn, wood, soybeans, and plant oils are the primary resources that create this remarkable diversity of industrial and consumer goods. Biomass components are often combined with other materials, such as petrochemicals and minerals, to manufacture the final product. Soybean oil, for example, is blended with other components to produce paints, toiletries, solvents, inks, and pharmaceuticals. Some products, such as starch adhesives, are derived entirely from biomass.

The many derivatives of corn illustrate the diversity of products obtained from a single biomass resource. Besides being an essential source of food and feed, corn serves as a feedstock for ethanol and sorbitol (a sweetish, crystalline alcohol), industrial starches and sweeteners, citric and lactic acid, and many other products.

Biomass comprises carbohydrates and can produce products commonly manufactured from petroleum, natural gas, or hydrocarbons. Both resources contain the essential elements carbon and hydrogen, and they have sometimes captured a portion of the market share (see Table 8.10).

8.13.1 CLASSES OF BIOPRODUCTS

The thousands of different industrial bioproducts produced today can be categorized into five major areas:

TABLE 8.9
Common Products from Biomass

Biomass Resource	Uses
Corn	Solvents, pharmaceuticals, adhesives, starch, resins, Binders, polymers cleaners, ethanol
Vegetable oils	Surfactants in soaps and detergents, pharmaceuticals (inactive ingredients), inks, paints, resins, cosmetics, fatty acids, lubricants, biodiesel
Wood	Paper, building materials, cellulose for fibers and polymers, resins, binders, adhesives, coatings, paints, ins, fatty acids, road and roofing pitch

Source: USDOE (2003a).

TABLE 8.10
Products from Hydrocarbons versus Carbohydrates

Product	Total Production (millions of tons)	% Derived from Plants
Adhesives	5.0	40
Fatty acids	2.5	40
Surfactants	3.5	35
Acetic acid	2.3	17.5
Plasticizers	0.8	15
Activated carbon	1.5	12
Detergents	12.6	11
Pigments	15.5	6
Dyes	4.5	6
Wall paints	7.8	3.5
Inks	3.5	3.5
Plastics	30	1.8

Source: ILSR (2002).

- **Sugar and starch bioproducts** derived through fermentation and thermochemical processes include alcohols, acids, starch, xanthan gum, and other products derived from biomass sugars. Primary feedstocks include sugarcane, sugar beets, corn, wheat, rice, potatoes, barley, sorghum grain, and wood.
- **Oil- and lipid-based bioproducts** include fatty acids, alkyd resins, glycerin, and various vegetable oils derived from soybeans, rapeseed, or other oilseeds.
- **Gum and wood chemicals** include tall oil (liquid rosin), alkyd resins, rosins, pitch, fatty acids, turpentine, and other chemicals derived from trees.
- **Cellulose derivatives, fibers, and plastics** include products derived from cellulose, including cellulose acetate (cellophane) and triacetate, cellulose nitrate, alkali cellulose, and regenerated cellulose. The primary sources of cellulose are bleached wood pulp and cotton linters.
- **Industrial enzymes** are biocatalysts for various biochemical reactions in producing starch, sugar, alcohol, and oils. They are also used in laundry detergents, tanning of leathers, and textile sizing (Uhlig, 1998).

SIDEBAR 8.3 THE BRAZILIAN EXAMPLE[5]

Brazil has the world's second-largest ethanol program and is capitalizing on plentiful soybean supplies to expand into biodiesel. More than half of the nation's sugarcane crop is processed into ethanol, which accounts for about 20% of the country's fuel supply.

Initiated in the 1970s after the OPEC oil embargo, Brazil's policy program was designed to promote the nation's energy independence and to create an alternative and value-added market for sugar producers. The government has spent billions to support sugarcane producers, develop distilleries, build up a distribution infrastructure, and promote the production of pure-ethanol-burning and, later, flex-fuel vehicles [able to run on gasoline, ethanol-gasoline blends, or pure hydrous (or wet; 7-4% water content) ethanol]. Advocates contend that, while the costs were high, the program saved far more in foreign exchange from reduced petroleum imports.

In the mid-to-late 1990s, Brazil eliminated direct subsidies and price setting for ethanol. It pursued a less intrusive approach with two main elements—a blending requirement (now about 25%) and tax incentives favoring ethanol use and purchasing ethanol-using or flex-fuel vehicles. Today, more than 80% of Brazil's newly produced automobiles have flexible fuel capability, up from 30% in 2004. With ethanol widely available at almost all of Brazil's 32,000 gas stations, consumers choose primarily between 100% hydrous ethanol and a 2% ethanol-gasoline blend based on relative prices.

Approximately 20% of current fuel use (alcohol, gasoline, and diesel) in Brazil is ethanol, but it may be challenging to raise the share as Brazil's fuel demand grows. Brazil is a middle-income economy with a per capita energy consumption of only 15% of the United States and Canada. Current ethanol production levels in Brazil are not much higher than in the late 1990s. Domestic off- and on-shore petroleum resource production has grown more rapidly than ethanol, accounting for a larger share of expanding fuel use than ethanol in the last decade.

8.14 THE BOTTOM LINE

With near-record oil prices, the future of biofuel—made from plant material (biomass)—is of keen interest worldwide. Using biofuels in the power industry, private vehicles, and personal appliances has several benefits over using conventional fuels. For example, environmental benefits include using biomass energy to reduce GHG emissions significantly—burning biomass releases about the same amount of carbon dioxide as burning fossil fuels. However, fossil fuels release carbon dioxide captured by photosynthesis millions of years ago—an essentially "new" GHG. Biomass, on the other hand, releases carbon dioxide primarily balanced by the carbon dioxide captured in its growth (depending on how much energy was used to grow, harvest, and process the fuel).

Another benefit of using biomass for fuel is reducing dependence on foreign oil. Biofuels are the only renewable liquid transportation fuels available.

Moreover, biomass energy supports U.S. agricultural and forest-product industries. The primary biomass feedstocks for power are paper mill residue, lumber mill scrap, and municipal waste. For biomass fuels, the feedstocks are corn (for ethanol) and soybeans (for biodiesel), both surplus crops. Shortly—and with developed technology—agricultural residues such as corn stover (the stalks, leaves, and husks of the plant) and wheat straw will also be used. Long-term plans include growing and using dedicated energy crops, such as fast-growing trees and grasses that can grow sustainably on land that will not support intensive food crops.

The preceding lists many of the benefits of using biomass fuel; that is all good, but the reality is that the future role of biofuels depends on profitability and new technologies. Technological advances and efficiency gains—higher biomass yields per acre and more gallons of biofuel per ton of biomass—could steadily reduce biofuel production's economic cost and environmental impact. Biofuel production will likely be most profitable and environmentally benign in tropical areas where growing seasons are longer per acre, biofuel yields are higher, and fuel and other input costs are lower. For example, Brazil uses bagasse, a byproduct from sugar production to power ethanol distilleries, whereas the United States uses natural gas or coal.

Biofuels will likely be part of a portfolio of solutions to high oil prices, including conservation and other alternative fuels. The role of biofuels in global fuel supplies will likely remain modest because of its land intensity. In the United States, replacing all current gasoline consumption with ethanol would require more land in corn production than in all agricultural production. Technology will be central to boosting the role of biofuels. If the energy is widely available, cellulose materials could be economically harnessed worldwide, and biofuel yards per acre could more than double, reducing land requirements significantly (USDA, 2007).

AI and ML bioenergy applications are being researched, pilot-tested, or operated—in systems with novel or improved functions for basic and applied biological research, including enzymes, pathways, and whole genomes of microorganisms, plants, and microbial communities (microbiomes). Intricate connectivity and complexity are part and parcel of using AI/ML in bioenergy. This is not a turn-on-the-switch operation; it is complex but doable. It is achievable because AL/ML advancements that enable computers to learn automatically from experience have emerged in recent years as potentially powerful tools to address challenges (Carbonell et al., 2019; Volk et al., 2022).

AI has emerged as an innovative approach to computational modeling and optimization of the anaerobic digestion process.

NOTES

1 This section adapted from Spellman, F.R., 2009. *Biology for the Non-Biologist.* Lanham, MD: Government Institutes Press.
2 Much of this section is from USDA et al., 2008. *The Economics of Biomass Feedstocks in the United States.* Washington, DC: Biomass Research and Development Board.
3 Information from USDA Amber Waves, Feb. 2008 edition. *Ephraim Leibtag.* Accessed 03/08/24 @ http://www.ers.usda.gov/amberwaves/February08/Featrues/CornPrices.htm.
4 Information in this section from USDOE, 2003a. *Industrial Bioproducts: Today and Tomorrow.* Washington, DC: Office of the Biomass program.
5 Example taken from USDA, 2007. *Amber Waves: The Future of Biofuels*; a Global Perspective. November.

REFERENCES

Aylott, M.J., 2008. Yield and spatial supply of bioenergy poplar and willow short-rotation coppice in the UK. *New Phytologist* 178(2): 358–370.

Baize, J., 2006. *Bioenergy and Biofuels*. Agricultural Outlook Forum, February 17. Washington, DC: USDA.

Biodiesel, 2007. FUMPA biofuels. *Biodiesel Magazine*, February. Accessed 1/25/25 @ https://biodieselmagazine.com/articles.

Carbonell, P., et al., 2019. Opportunities at the intersection of synthetic biology, machine learning, and automation. *ACS Synthetic Biology* 8(7): 1474–1477.

Christi, Y., 2007. Biodiesel from microalgae. *Biotechnology Advances* 25: 294–306.

EERE, 2008. *Biomass Program*. Accessed 03/04/24 @ https://www1.eere.energy.gov/biomass/feedstocks_types.html.

EIA, 2006. Energy Information Administration 2006. *Annual Energy Outlook 2007 with Projections to 2030*, Washington, DC: DOE-EIA-0382. February.

EPRI, 1997. Electric Power Research Institute. *Renewable Energy Technology Characterizations*. Washington, DC: TR-109496, Department Energy, December.

Fahey, J., 2001. Shucking petroleum, *Forbes Magazine*, November 26, p. 206.

Friedman, T.L., 2010. The fat lady has sung. *New York Times*. Accessed 02/20/10.

Graham, R., Nelson, R., Sheehan J., Perlack, R., and Wright L., 2007. Current and potential U.S. corn stover supplies. *Agronomy Journal* 99: 1.11.

Hanson, B.J., 2004. *Energy Power Shift*. Maple, WI: Lakota Scientific Press.

Hughes, E., 2000. Biomass cofiring: Economics, policy and opportunities. *Biomass and Bioenergy* 19: 45–465.

ILSR, 2002. *Accelerating the Shift to a Carbohydrate Economy: The Federal Role: Executive Summary of the Minority Report of the Biomass Research & Development Technical Advisory Committee*. Washington, DC: Institute for Local Self-Reliance.

McDill, S., 2010. *Can Algae Save the World—Again?* Accessed 03/12/24 @ https://www.Reuters.Com/article/environmentalnews/id/USTRES5196HB200902107.

McKeever, D., 2004. Inventories of woody residues and solid wood waste in the United States. In *2002 Ninth International Conference, Inorganic-Bonded Composite Materials*. Vancouver, British Columbia, October 10–13.

McKendry, P., 2002. Energy production from biomass (part I): Overview of biomass. *Bioresource Technology* 83: 37–46.

McLaughlin, S.B., and Kzos, L.A., 2005. Development of Switchgrass as a bioenergy feedstock in the U.S. *Biomass & Bioenergy* 28: 515–535.

Miles, T.R., Miles, T.R., Jr., Baxter, L.L., Jenkins, B.M., and Oden, L.L., 1993. Alkali slagging problems with biomass fuels. In *First Biomass Conference of the American Energy, Environment, Agriculture, and Industry, Volume 1*. Burlington, Vermont.

NIST, 1994. *Handbook 44 Appendix D Definitions*. Washington DC: U.S. Department of Commerce.

NREL, 1993. *Alkali Content and Slagging Potential of Various Biofuels*. Accessed 03/17/24 @ https://cta.ornl.gov/bedb/biopower/alkali_conten_salgging_potential_of_variousbiofules.xls.

Parrish, D.J., Fike, J.H., Bransby, D.I., and Samson, R., 2008. Establishing and managing switchgrass as an energy crop. Forage and Grazinglands xi: 114–119.

Perlin, J., 2005. *A Forest Journey: The Story of Wood and Civilization*. Woodstock, VT: The Countryman Press.

Polagye, B., Hodgson, K., and Malte, P., 2007. An economic analysis of bioenergy options using thinning from overstocked forests. *Biomass ad Bioenergy* 31: 105–125.

Rinehart, L., 2006. *Switchgrass as a Bioenergy*. Butte, MT: National Center for Appropriate Technology.

Samson, R., et al., 2008. Developing energy crops for thermal applications: Optimizing fuel quality, energy security and GHG mitigation. In D. Pimental (Ed.), *Solar and Wind Are Renewable Energy Systems in Biofuels: Benefits and Risks*. Berlin: Springer Science.

Schenk, P., Thomas-Hall, S., Stephens, R., Marx, U., Mussgnug, J., Posten, C., Kruse, O., Hankamer, B, 2008. Second generation biofuels: High-efficiency microalgae for industrial production. *BioEnergy Research* 1(1): 20–45.

Spellman, F.R., 2009. *Handbook of Water and Wastewater Treatment Plant Operations*, 2nd edition. Boca Raton, FL: CRC Press.

Taipei Times, 2008. *Algae Eyed as Biofuel Alternative*. Accessed 7/27/24 @ https://www.taipietimes.com/news/taiwan/archives/2008/01-12-2003396760.

Tillman, D., 2000. Biomass cofiring: The technology, the experience, the combustion consequences. *Biomass and Bioenergy* 19: 365–384.

Time Online, 2005. *Seaweed Breathes New Life into the Fight against Global Warming*. Accessed 7/26/24 @ www.timeonline.Co.Uk/tot/news.world/article522203.ece.

Uhlig, H., 1998. *Industrial Enzymes and Their Applications*. New York: John Wiley & Sons, Inc.

USDA, 2007. *Amber Waves: Future of Biofuels*. Accessed 03/20/24 @ https://www.ers.usda.gov/AmberWaves/november07/features/biofuels.htm.

USDA-World Agricultural Outlook Board, 2008. *World Agricultural Supply and Demand Estimates*. Washington: DCL USDA.

USDOE, 2003a. *Industrial Bioproducts: Today and Tomorrow*. Washington, DC: U.S. Department of Energy.

USDOE, 2003b. *The Bioproducts Industry: Today and Tomorrow*. Washington, DC: U.S. Department of Energy.

USDOE, 2007. *Understanding Biomass: Plant Cell Walls*. Washington, DC: U.S. Department of Energy.

USDOE, 2010. *Fuel Ethanol Production*. Accessed 7/25/24 @ https://genomicScience.energy.gov/biofuels.ethanolproduction.shtml.

USEPA, 1979. *Process Design Manual Sludge Treatment and Disposal*. Washington, DC: United States Environmental Protection Agency, Office of Research and Development, EPA 625/625/1-79-011.

USEPA, 2006. *Biosolids Technology Fact Sheet: Multi-Stage Anaerobic Digestion*. Washington, DC: United States Environmental Protection Agency Office of Water, EPA832-F-06-031.

USEPA, 2007. *Fuel Economy Impact Analysis of RFG*. Accessed 03/11/24 @ http.www.epa.gov/oms/rfsecon.htm.

Volk, M.J., et al., 2022. Metabolic engineering: Methodologies and applications. *Chemical Reviews* 16: 27–33.

Wiltsee, G., 1998. *Urban Wood Waste Resource Assessment NREL/SR-570-25918*. Golden, CO: National Renewable Energy Laboratory.

9 Geothermal Energy and AI

9.1 INTRODUCTION[1]

The U.S. Geological Survey has calculated that the heat energy in the upper 10 km of the Earth's crust in the U.S. equals over 600,000 times the country's annual non-transportation energy consumption. Probably, no more than a tiny fraction of this energy could ever be extracted economically. However, just one-hundredth of the total is equal to half the country's current non-transportation energy needs for over a century, with only a fraction of the pollution from fossil-fueled energy sources (McLarty et al., 2000). Geothermal heat is the only renewable energy source created naturally by the Earth. Suppose we utilize waste biomass, solar (passive and thermal), wind (onshore), photovoltaic, geothermal, and other renewable resources available to us in the United States. In that case, we will exceed the demand (what we need) by at least five times as much energy as we need, all from clean, renewable sources. The growth and expansion of AI data centers require high-performance processors that are very energy-intensive to the point where they are becoming energy hogs and constantly pushing for more energy and energy sources.

Approximately 4,000 miles below the Earth's surface is the core, where temperatures can reach 9,000°F. This heat—geothermal energy (geo, meaning Earth, and thermos, meaning heat)—flows outward from the core, heating the surrounding area. This can form underground reservoirs of hot water and steam. These reservoirs can be tapped for various uses, such as generating electricity or heating buildings.

Note: Renewable energy consumption in the United States was approximately 9%. In 2023, geochemical energy accounted for 1% of energy used in the United States and about 0.4% of total electricity produced.

The geothermal energy potential in the uppermost 6 miles of the Earth's crust amounts to 50,000 times the energy of all oil and gas resources globally. In the United States, most geothermal reservoirs are in the western states, Alaska and Hawaii. However, geothermal heat pumps (GHPs), which take advantage of the shallow ground's stable temperature to heat and cool buildings, can be used almost anywhere.

Humans have used renewable resources since immemorial, from solar power to burning biomass (wood) in caves and elsewhere. For example, hot springs have been used for bathing since Paleolithic times or earlier (DOE, 2009). The early Romans used hot springs to feed public baths and underfloor heating. France's world's oldest geothermal district heating system has operated since the 14th century (Lund, 2007). The history of geothermal energy use in the United States is lengthy. In the following, a brief chronology of major geothermal events in the United States is provided (DOE, 2006).

DOI: 10.1201/9781003605430-11

9.1.1 Geothermal Timeline

9.1.1.1 8,000 B.C. (and Earlier)

Paleo-Indians used hot springs for cooking, refuge, and respite. Hot springs were neutral zones where members of warring nations bathed together in peace. Native Americans have a history with every major hot spring in the United States.

9.1.1.2 1807

As European settlers moved westward across the continent, they gravitated toward these springs of warmth and vitality. In 1807, the first European to visit the Yellowstone area, John Colter [(c. 1774–c.1813) widely considered to be the first mountain man], probably encountered hot springs, leading to the designation "Colter's Hell." Also, in 1897, settlers found the city of Hot Springs, Arkansas, where, in 1830, Asa Thompson charged one dollar each for three spring-fed baths in a wooden tub, and the first known commercial use of geothermal energy occurred.

9.1.1.3 1847

William Bell Elliot, a member of John C. Fremont's survey part, stumbles upon a steaming valley just north of what is now San Francisco, California. Elliot calls the area The Geysers—a misnomer—and thinks he has found the gates of Hell.

9.1.1.4 1852

The Geysers is developed into a spa called The Geysers Resort Hotel. Guests include J. Pierpont Morgan, Ulysses S. Grant, Theodore Roosevelt, and Mark Twain.

9.1.1.5 1862

At springs located southeast of The Geysers, businessman Sam Brannan pours an estimated half million dollars into an extravagant development dubbed "Calistoga," replete with a hotel, bathhouse, skating pavilion, and racetrack. Brannan's was one of many spas reminiscent of those of Europe.

9.1.1.6 1864

Homes and dwellings have been built near springs through the millennia to take advantage of the natural heat of these geothermal springs, but the construction of the Hot Lake Hotel near La Grande, Oregon, marks the first time that the energy from hot springs is used on a large scale.

9.1.1.7 1892

Boise, Idaho, provided the world's first district heating system, piping water from hot springs to town buildings. Within a few years, the system served 200 homes and 40 downtown businesses. Today, four district heating systems in Boise provide heat to over 5 million square feet of residential, business, and governmental space. There are now 17 district heating systems in the United States and dozens more around the World.

9.1.1.8 1900

Hot springs water is piped to homes in Klamath Falls, Oregon.

9.1.1.9 1921

John D. Grant drills a well at The Geysers to generate electricity. This effort was unsuccessful, but Grant met with success across the valley at another site one year later, and the United States' first geothermal power plant went into operation. Grant uses steam from the first well to build a second one, and several wells later, the operation produces 250kW, enough electricity to light the buildings and streets at the resort. The plant, however, is not competitive with other power sources, and it soon falls into disuse.

9.1.1.10 1927

Pioneer Development Company drills the first exploration at Imperial Valley, California.

9.1.1.11 1930

The first commercial greenhouse use of geothermal energy is undertaken in Boise, Idaho. The operation uses a 1,000-foot well drilled in 1926. In Klamath Falls, Charlie Lieb develops the first downhole heat exchanger (DHE) to heat his house. Today, more than 500 DHEs are in use around the country.

9.1.1.12 1940

Nevada's first residential space heating began in the Moan area in Reno.

9.1.1.13 1948

Geothermal technology moved east when Carl Nielsen developed the first ground-source heat pump for his residence. J.D. Krocker, an engineer in Portland, Oregon, pioneered the first commercial building use of a groundwater heat pump.

9.1.1.14 1960

The country's first large-scale geothermal electricity-generating plant begins operation. Pacific Gas and Electric operates the plant located at the Geysers. The first turbine produces 11 MW of net power and has operated successfully for over 30 years. Today, 69 generating facilities operate at 18 resource sites around the country.

9.1.1.15 1978

Geothermal Food Processors, Inc. opened the first geothermal food-processing (crop-drying) plant in Brady Hot Springs, Nevada. The Load Guaranty Program provided \$3.5 million for the facility.

9.1.1.16 1979

The first electrical development of a water-dominated geothermal resource occurs at the East Mesa field in the Imperial Valley in California. The Plant is named for B.C. McCabe, the geothermal pioneer who, with his Magma Power Company, did field development work at several sites, including The Geysers.

9.1.1.17 1980

TAD's Enterprises of Nevada pioneered using geothermal energy for the cooking, distilling, and drying processes associated with alcohol fuel production. UNOCAL built the country's first flash plant, generating 10 MW at Brawley, California.

9.1.1.18 1982

Economic electrical generation begins at California's Salton Sea geothermal field using crystallizer-clarifier technology. The technology resulted from a government/ industry effort to manage the high-salinity brines at the site.

9.1.1.19 1984

A 20-MW plant begins generating power at Utah's Roosevelt Hot Springs. Nevada's first geothermal electricity is generated with a 1.3-MW binary power plant that begins operation.

9.1.1.20 1987

Geothermal fluids were used in the first geothermal-enhanced heap leaching project for gold recovery near Round Mountain, Nevada.

9.1.1.21 1989

The world's first hybrid (organic Rankine/gas engine) pressure-geothermal power plant begins operation at Pleasant Bayou, Texas. It uses both the heat and the methane of a pressured resource.

9.1.1.22 1992

Electrical generation begins at the 25-MW geothermal plant in the Puna field of Hawaii.

9.1.1.23 1993

A 23-MW binary power plant is completed at Steamboat Springs, Nevada.

9.1.1.24 1995

Integrated Ingredients dedicate an annual food-dehydration facility at Empire, Nevada, that processes 15 million pounds of dried onions and garlic. A DOE low-temperature resource assessment of 10 western states identifies nearly 9,000 thermal wells and springs and 271 communities collocated with a geothermal resource greater than 50.

9.1.1.25 2002

GeoPowering the West organized geothermal development working groups in five states: Nevada, Idaho, New Mexico, Oregon, and Washington. Group members represent all stakeholder organizations. The working groups identify barriers to geothermal development in their state and bring together all interested parties to arrive at mutually beneficial solutions.

9.1.1.26 2003

The Utah Geothermal Working Group is formed.

9.2 GEOTHERMAL ENERGY: THE BASICS

As mentioned, geothermal energy processes evolve around the Earth's natural heat, which can be used for beneficial purposes when the heat is collected and transported to the surface. To properly understand geothermal energy as it is currently used, it is essential to define enthalpy, the heat content of a substance per unit mass. The term "enthalpy" is used because temperature alone is not sufficient to specify the applicable energy content of a steam/water mixture. A mass of steam at a given temperature and pressure can provide much more energy than the same mass of water under the same conditions. Enthalpy is a function of:

$$\text{Pressure} + \text{Volume} + \text{Temperature} = \text{Enthalpy}$$

Geothermal practitioners usually classify geothermal resources as "high enthalpy" (water and steam at temperatures above 180°C–200°C), "medium enthalpy" (about 100°C–180°C), and "low enthalpy" (<100°C). For this text, it is sufficient to think of temperature and enthalpy going hand in hand.

9.2.1 Earth's Layers

The Earth comprises three main compositional layers: crust, mantle, and core (see Figure 9.1). The crust has variable thickness and composition: The continental crust is 10–50 km thick, while the oceanic crust is 8–10 km thick. Silicon, oxygen, aluminum, and iron make up the Earth's crust. Like the shell of an egg, the Earth's crust is brittle and can break. The Earth's continental crust is 35 km thick. The oceanic crust is 7 km thick.

Based on seismic (earthquake waves) that pass through the Earth, we know that below the crust is the mantle, a dense, hot layer of semi-solid (plastic-like liquid) rock approximately 2,900 km thick. The mantle, which contains silicon, oxygen,

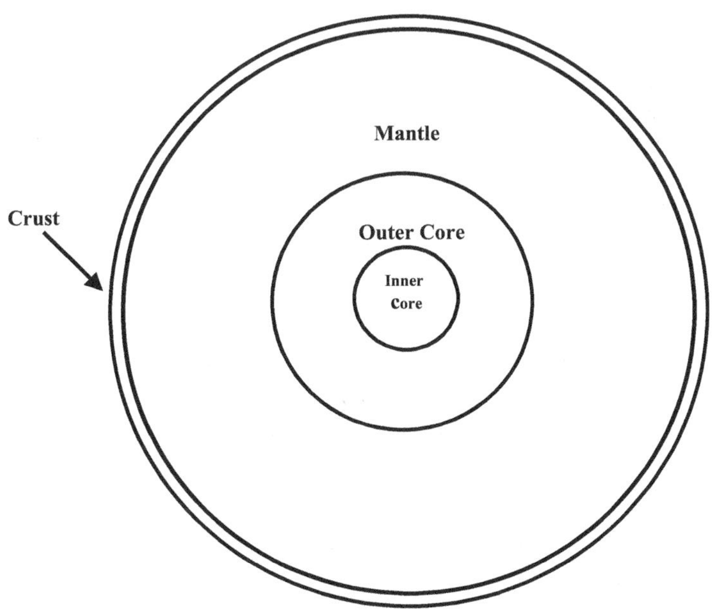

FIGURE 9.1 Earth's layers.

aluminum, iron, magnesium, and calcium more than the crust, is hotter and denser because temperature and pressure inside the Earth increase with depth. As a comparison, the mantle might be thought of as the white of a boiled egg. The Moho layer is the 30-km-thick transitional layer between the mantle and crust. The temperature at the top of the mantle is 870°C. The temperature at the bottom of the mantle is 2,200°C.

At the center of the Earth lies the core, which is nearly twice as dense as the mantle because its composition is metallic [Iron (Fe)-Nickel (NI) alloy] rather than stony. Unlike the yolk of an egg, however, the Earth's core comprises two distinct parts: a 2,200-km-thick liquid outer core and a 1,250-km-thick solid inner core. The liquid outer core spins as the planet rotates, creating the Earth's magnetic field. Several essential points related to the Earth's structure and geothermal properties include:

- Heat flows outward from the center because of radioactive decay.
- The crust (about 30 and 60 km thick) insulates us from the interior heat.
- A solid inner core followed by a liquid out of the core, with a semi-molten mantle.
- Temperature at the base of the crust is about 1,000°C, increasing slowly into the core.
- Hot spots are located 2–3 km from the surface.

9.2.2 Crustal Plates

Within the past 45 or 50 years, geologists have developed the theory of plate tectonics (tectonics: Greek, "builder"). The theory of plate tectonics deals with the formation, destruction, and large-scale motions of significant segments of the Earth's surface (crust), called *plates*. This theory relies heavily on the older concepts of continental drift (developed during the first half of the 20th century) and seafloor spreading (understood during the 1960s), which help to explain the cause of earthquakes and volcanic eruptions and the origin of fold mountain systems.

Regarding crustal plate activity and geothermal energy, and as mentioned, large quantities of heat (that are economically extractable) tend to be concentrated in places where hot or even molten (magma) exists at relatively shallow depths in the Earth's outmost layer (the crust). Such "hot" zones generally are near the boundaries of the dozen or so slabs of rigid rock (or plates) that form the Earth's lithosphere.

These crustal plates are composed of great slabs of rock (lithosphere), about 100 km thick, and cover many thousands of square miles (they are thin compared to their length and width); they float on the ductile asthenosphere, carrying both continents and oceans. Many geologists recognize at least eight main plates and numerous smaller ones. These **main** plates include

- African Plate covering Africa—Continental plate
- Antarctic Plate covering Australia—Continental plate
- Australian Plate covering Australia—Continental plate
- Eurasian Plate covering Asia and Europe—Continental plate

- Indian Plate covering the Indian subcontinent and a part of the Indian Ocean—Continental plate
- Pacific Plate covering the Pacific Ocean—Oceanic plate
- North American Plate covering North America and north-east Siberia—Continental plate
- South American Plate covering South America—Continental plate

The **minor** plates include

- Arabian Plate
- Caribbean Plate
- Juan de Fuca Plate
- Cocos Plate
- Nazea Plate
- Philippine Plate
- Scotia Plate

The plates ride in the asthenosphere, the upper mantle's ductile, soft, plastic-like zone. Crustal plates move around one another at one of three types of plate boundaries: convergent (collision boundaries), divergent (spreading boundaries), or transform boundaries. These boundaries between plates are typically associated with deep-sea trenches, significant faults, fold mountain ranges, and mid-oceanic ridges.

9.2.3 Convergent Boundaries

Convergent boundaries (or active margins) develop where two plates slide toward each other, commonly forming either a subduction zone (if one plate subducts or moves underneath the other) or a continental collision (if the two plates contain continental crust). One plate is deformed and slips below the other to relieve the stress created by the colliding plates.

9.2.4 Divergent Boundaries

Divergent boundaries occur when two plates slide apart. Oceanic ridges are examples of these divergent boundaries. They are where new oceanic, melted lithosphere materials well up, resulting in basaltic magmas that intrude and erupt at the oceanic ridge, creating new oceanic lithosphere and crust (new ocean floor). Along with volcanic activity, the mid-oceanic ridges are also areas of seismic activity.

9.2.5 Transform Plate Boundaries

Transform or shear/constructive boundaries do not separate or collide; instead, they slide past each other horizontally with a shearing motion. Most transform boundaries occur where oceanic ridges are offset on the sea floor. The San Andreas Fault in California is an example of a transform fault.

DID YOU KNOW?

- Plates are in constant motion (several cm/year).
- When collision or grinding occurs, it can create mountains, volcanoes, geysers, and earthquakes.
- Near the junction of these plates is where heat travels rapidly from the interior.

9.3 ENERGY CONVERSION

The conversion of heat to electricity is common to most power plants. This is the case even though the energy source may be coal, gas, nuclear, wind, solar, water, or geothermal. Powering the non-transportation section of our economy is essential, of course, so using any fuel source to convert to electrical power for industrial use is one of our premium objectives in our constant and insatiable appetite for energy. Because of our increasing need for more electricity generation, one might surmise that we should focus solely on electricity production to power our economy, industrial complex, and everything else—one genie in one bottle to accomplish everything. Liquid fuels are the fuels of choice now because they are accessible, available, and relatively inexpensive. Thus, though our ongoing research for other energy sources persists, we still do not have that absolute pressing need to develop a liquid fuel replacement—not yet we don't.

When our energy-needs-focus shifts due to necessity, absolute or otherwise, geothermal is available, and we need to continue our research in this critical area. Moreover, energy conversion occurs at present with all forms of energy in use, and with geothermal conversion, it is nothing new; only the procedure and methodology differ. In geothermal energy, "conversion" refers to the power-plant technology that converts the hot geothermal fluids into electric power.

Even though geothermal power plants have much in common with traditional power-generating stations, turbines, generators, heat exchanges, and other standard power-generating equipment, there are essential differences between geothermal and other power-generating technologies. For example, each geothermal site has its own unique set of characteristics and operating conditions. For instance, geothermal sites are essentially site-specific; these conditions must be considered. That is, the fluid produced from a geothermal well can be steam, brine, or a mixture of the two, and the temperature and pressure of the resource can vary substantially from site to site. The chemical composition of the resource can contain dissolved minerals, gases, and other hard-to-manage substances. Because these site-specific conditions can profoundly affect efficiency, productivity, and economic viability, engineers strive to fine-tune geothermal conversion technology, precisely matching plant design to the site-specific conditions.

9.4 GEOTHERMAL POWER PLANT TECHNOLOGIES

Geothermal power plants fall into one of three conversion categories: direct steam, flash, and binary plants. The type of conversion used depends on the state of the fluid (whether steam or water) and its temperature. Dry steam power plants were the first type of geothermal power generation plants built. They use the steam from the geothermal reservoir from wells and route it directly through turbine/generator units to produce electricity. Flash steam plants are the most common type of geothermal power generation plants in operation today. They use water temperatures greater than 360°F (182°C) pumped under high pressure to the generation equipment at the surface. Binary cycle geothermal power generation plants differ from Dry Steam and Flash Steam systems in that the water or steam from the geothermal reservoir never meets the turbine/generator units.

DID YOU KNOW?

The Earth's temperature increases, on average, by about 28°C for every kilometer, or 80°F for every mile, of depth below the surface for the first several kilometers down.

9.4.1 Dry Steam Power Plants

Location-specific, dry-steam power plants take advantage of subterranean rocks that are so hot that the water from a geothermal power plant vaporizes on its way up through the production well (see Figure 9.2). This geothermal power plant can use steam directly from the ground to drive the turbine. This is the oldest type of geothermal power plant. It was first used at Larderello in Italy in 1904 and is still very effective. These geothermal plants emit only excess steam and few gases (USDOE, 2008).

9.4.2 Flash Steam Power Plants

Figure 9.3 is a functional diagram of a *flash steam geothermal power plant.* Fluid is sprayed into a tank held at a much lower pressure than the fluid, causing some fluid to vaporize or "flash rapidly." The vapor then drives a turbine, which drives a generator. If any liquid remains in the tank, it is returned to the groundwater pump to be forced down into the Earth again and can be flashed again to extract more energy (USDOE, 2008).

9.4.3 Binary Cycle Power Plants

In a binary cycle geothermal power plant (see Figure 9.4), water is pumped into the Earth and back up hot, just as in the flash-steam system. Instead of going into a flash tank, the hot water enters a *heat exchanger* (see Figure 9.4), where most of its energy is transferred to another fluid called a *binary liquid.* This fluid can be water, but more often, it is a volatile liquid resembling refrigerant that boils quickly into vapor at a lower temperature than the water. The liquid-to-vapor conversion occurs in a particular low-temperature boiler. Then, the vapor leaves the turbine, is cooled back into liquid by a condenser, and is recirculated to the boiler. Because this is a closed-loop system, virtually nothing is emitted to the atmosphere. Moderate-temperature water

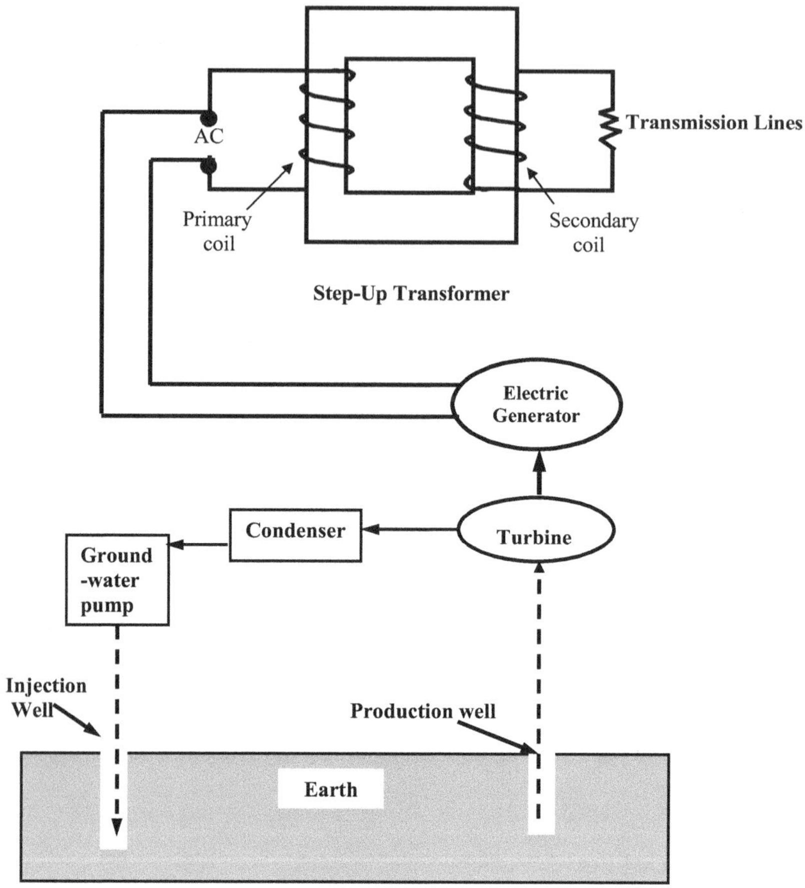

FIGURE 9.2 Simplified functional diagram of a dry steam geothermal power plant.

is by far the most common geothermal resource, and most geothermal power plants in the future will be binary-cycle plants (USDOE, 2008).

9.5 ENHANCED GEOTHERMAL SYSTEMS (EGS)

Enhanced geothermal systems (EGS), sometimes called engineered geothermal systems, can dramatically expand geothermal energy use. Present geothermal power generation comes from hydrothermal reservoirs and is limited in geographic application to specific ideal places in the western United States. EERE (2006) points out that this represents the "lower-hanging fruit" of geothermal energy potential.

EGS offers the chance to extend geothermal resources to large areas of the western United States and into new geographic regions in the United States. More than 100,000 MWe (megawatt electrical) of economically viable capacity may be available in the continental United States, representing a 40-fold increase over the present geothermal power generating capacity. This potential is about 10% of the overall

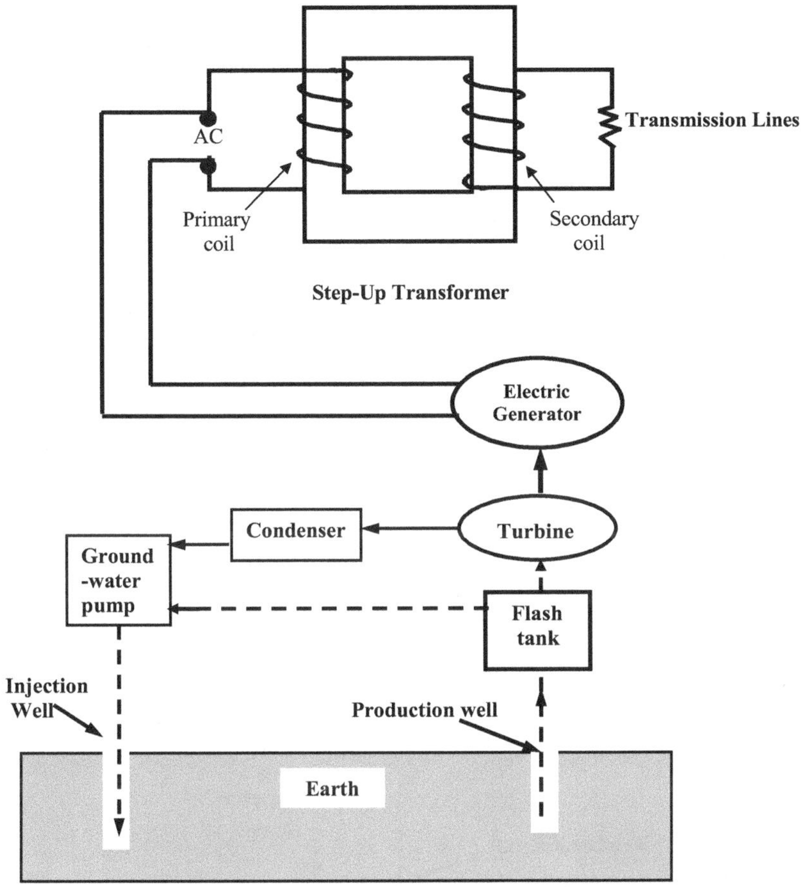

FIGURE 9.3 Simplified functional diagram of a flash steam geothermal power plant.

U.S. electrical capacity today, and as mentioned, means a domestic energy source that is clean, reliable, and proven.

The EGS concept extracts heat by creating a subsurface fracture reservoir where injection wells can add water. A production injection well is drilled into hot basement rock with limited permeability and fluid content. This type of geothermal resource is sometimes called "hot, dry rock" and represents an enormous potential energy resource. Creating an enhanced or engineered geothermal system requires improving the natural permeability of the rock. Rocks are permeable due to minute fractures and pore spaces between mineral grains. Injected water is at sufficient pressure to ensure fracturing is heated by contact with the rock and returns to the surface through production wells, as in naturally occurring hydrothermal systems. Additional production-injection wells are drilled to extract heat from large volumes of rock mass to meet power generation requirements. A previously unused but extensive energy resource is now available for clean geothermal power generation.

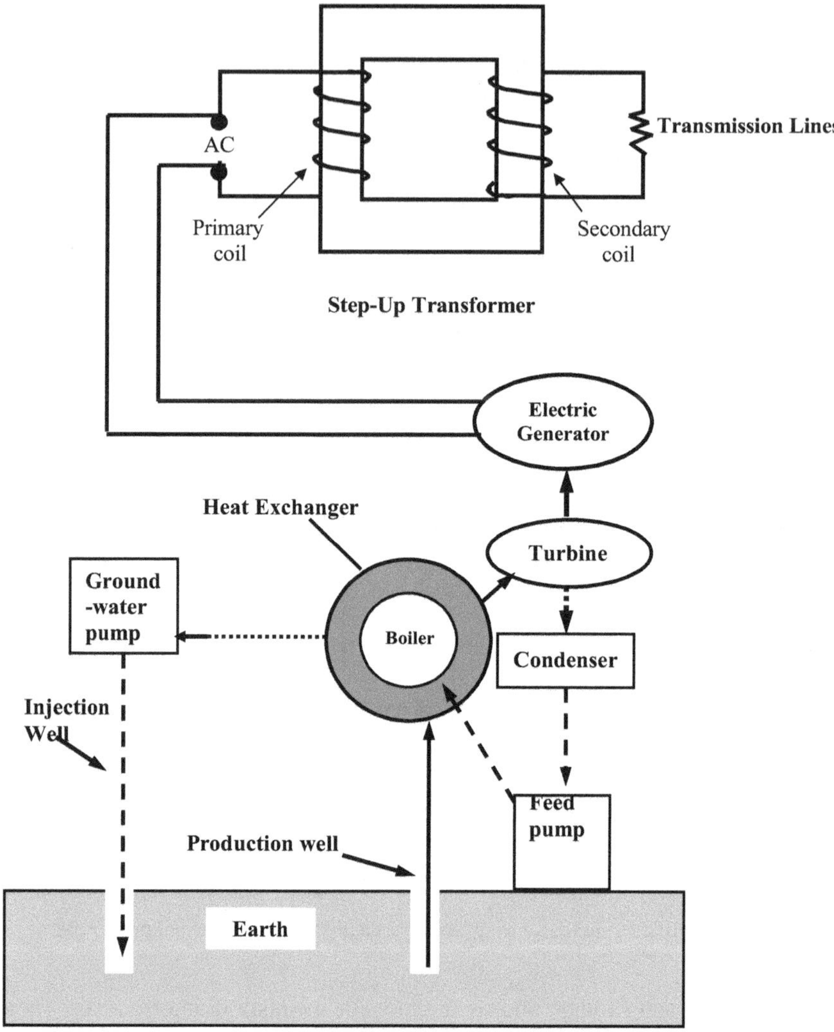

FIGURE 9.4 Simplified functional diagram of a binary cycle geothermal power plant.

9.6 GEOTHERMAL HEAT PUMPS

Geothermal heat pumps (sometimes referred to as GeoExchange, Earth-coupled, ground-source, or water-source heat pumps) have been in use since the late 1940s. GHPs use the constant temperature of the Earth as the exchange medium instead of the outside air temperature. This allows the system to reach high efficiencies (300%–600%) on the coldest winter nights, compared to 175%–250% for air-source heat pumps on cool days. GHPs are used for space heating, cooling, and water heating. Its great advantage is that it concentrates naturally existing heat rather than producing heat through fossil fuel combustion. The system includes three principal components (see Figure 9.5):

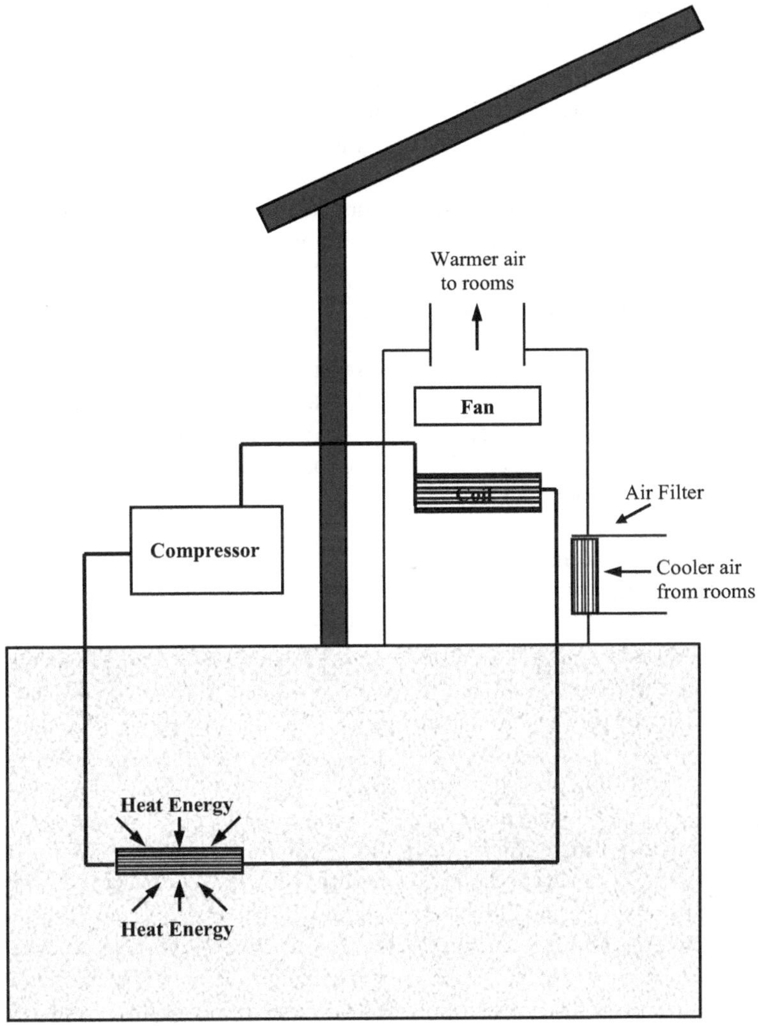

FIGURE 9.5 Geothermal heat pump system. (Adapted from Gibilisco, 2007, p. 54.)

- **Geothermal Earth connection subsystems**—Using the Earth as a heat source/sink, a series of pipes, commonly called a "loop," is buried near the building to be conditioned. The loop can be buried either vertically or horizontally. It circulates a fluid (water or a mixture of water and antifreeze) that absorbs heat from or relinquishes heat to the surrounding soil, depending on whether the ambient air is colder or warmer than the soil.
- **Geothermal heat pump subsystem**—For heating, a GHP removes heat from the fluid in the Earth connection, concentrates it, and then transfers it to the building. For cooling, the process is reversed.
- **Geothermal heat distribution subsystem**—Conventional ductwork distributes heated or cooled air from the GHP throughout the building.

In addition to space conditioning, GHPs can provide domestic hot water when the system is operating. Many residential systems now have desuperheaters that transfer excess heat from the GHP's compressor to the house's hot water tank. A desuperheater provides no hot water during the Spring and Fall when the GHP system is not operating; however, because the GHP is so much more efficient than other means of water heating, manufacturers are beginning to offer "full demand" systems that use a separate heat exchanger to meet all of a household's hot water needs. These units cost-effectively provide hot water as quickly as any competing system (EERE, 2009).

9.6.1 Types of Geothermal Heat Pumps

Energy Savers (2009) lists four basic ground-loop GHP system types. Three closed-loop systems are horizontal, vertical, and pond/lake. The fourth type is the open-loop option. The best depends on the site's climate, soil conditions, available land, and local installation costs. All these approaches can be used for residential and commercial building applications.

- **Closed-loop systems**—
  - **Horizontal**—This type of installation is generally most cost-effective for residential installations, particularly for new construction where sufficient land is available. It requires trenches at least four feet deep. The most common layouts use two pipes, one buried at six feet and the other at four feet, or two pipes placed side by side at five feet in the ground in a two-foot-wide trench. The Slinky™ method of looping pipe allows more pipe in a short trench, cutting installation costs and making horizontal installation possible in areas without conventional horizontal applications.
  - **Vertical**—Large commercial buildings and schools often use vertical systems because the land area required for horizontal loops would be prohibitive. Vertical loops are also used where the soil is too shallow for trenching, and they minimize the disturbance to existing landscaping. For a vertical system, holes (approximately 4 inches in diameter) are drilled about 20 ft apart and 100–400 ft deep. Two pipes connected at the bottom with a U-bend to form a loop into these holes. The vertical loops are connected with a horizontal manifold in trenches, linking to the building's heat pump.
  - **Pond/lake**—This may be the lowest-cost option if the site has an adequate water body. A supply line pipe runs underground from the building to the water and is coiled into circles at least eight feet under the surface to prevent freezing. The coils should only be placed in a water source that meets minimum volume, depth, and quality criteria.
- **Open-loop system**—It uses well or surface body water as the heat exchange fluid that circulates directly through the GHP system. Once it has circulated through the system, the water returns to the ground through the well, a recharge well, or surface discharge. This option is practical only when there is an adequate supply of relatively clean water, and all local codes and regulations regarding groundwater discharge are met.

DID YOU KNOW?

Geothermal system life is estimated at 25 years for the inside components and 50+ years for the ground loop. Approximately 50,000 GHPs are installed in the United States each year. Even with the large number of annual installations, it is essential to remember that heat pumps are relatively expensive. This is especially true of the deep-ground source type. A new system may take a long time to pay for itself.

9.7 THE BOTTOM LINE

The supply of geothermal energy is vast and can be considered renewable if each site is engineered correctly and operated to ensure that excessive water is not pumped into the Earth in one location in too short a time. Geothermal energy can be accessed by drilling water or steam wells in a process like drilling for oil. Geothermal energy is an enormous, underused heat and power resource that is clean (emits little or no greenhouse gases), reliable (average system availability of 95%), and homegrown (making us less dependent on foreign oil).

Geothermal resources range from shallow ground to hot water and rock several miles below the Earth's surface and even farther down to the boiling molten rock called magma. Mile-or-more-deep wells can be drilled into underground reservoirs to tap steam and boiling water to be brought to the surface for various applications. In the United States, most geothermal reservoirs are in the western states, Alaska and Hawaii. However, before geothermal electricity can be considered a vital element of the U.S. energy infrastructure, it must become cost-competitive with traditional forms of energy.

Artificial intelligence (AI) has been leveraged for various energy purposes, including geothermal, and has played a transformative role in the geothermal industry. It can improve the efficiency of geothermal technologies through preventive maintenance, cost reduction, output resource allocation, energy storage optimization, advanced control systems, and grid integration and stability.

NOTE

1 Based on information from DOE, 2001. *Renewable Energy: An Overview.* Washington, DC: U.S. Department of Energy.

REFERENCES

DOE, 2006. *Geothermal Technologies Program: A History of Geothermal Energy in the United States*. Accessed 06/18/24 @ https://www1.eere.energy.gov/geothermal/printableversison/history.html.

DOE, 2009. *Fossil Fuels*. U.S. Dept of Energy. Accessed 06/10/24 @ https://www.energy.gov/energysources/fossilfuels.htm.

EERE, 2006. *How an Enhanced Geothermal System Works*. Accessed 03/29/24 @ https://www1.eere.energy.gov/gelotherm/printable_versions/egs_animation.html.

EERE, 2009. *Geothermal Heat Pumps*. Accessed 03/27/24 @ https://www1.eere.energy.gov/geothermal/heatpumps.html?print.

Energy Savers, 2009. *Types of Geothermal Heat Pump Systems*. Washington, DC: US Department of Energy. Accessed 12/12/24 @ https://www.energysavers.gov/your_home/space_heatingcooling/index.

Gibilisco, S., 2007. *Alternative Energy Demystified*. New York: McGraw Hill.

Lund, J.W., 2007. Characteristics, development, and utilization of geothermal sources. *Geo-Heat Centre Quarterly Bulletin* 28(2): 1–9.

McLarty, L., Grabowski, P., Entingh, D., and Robertson-Tait, A. 2000. Enhanced Geothermal Systems R&D in the United States. *World Gas Conference* 2000, pp. 3793–3796. Tokyo, Japan.

USDOE, 2008. *Hydrothermal Power Systems*. Accessed 03/26/24 @ www1.eere.energy.gov.geothermal/printable_versions/powerplants.html.

10 Ocean Energy

One day, King Canute told his followers to carry him and his throne to the seashore at low tide. "Set me down at the water's edge right there," he said. "And I will command the sea to stay away. I will order the tide not to rise."

"Oh wow!" said his followers as they obeyed the King. "This ought to be something to

see!"

King Canute sat. He shouted, "Sea, stay away! Tide, do not rise!" But slowly, the tide came in, and the sea rose over his feet, past his knees, and up to his waist. The crowd of followers pulled back a little to keep their feet dry. They were puzzled. Why did the sea not obey the King?

The water rose higher and higher, up to the King's shoulders, over his chin. "Loyal glub subjects!" he gurgled. "Now you see that glub something no man, be he glub king or commoner, can glub do! Now pull me the glub out of here!"

—Based on the classic fairy tale

10.1 INTRODUCTION

The ocean is never still.

Waves transmit energy across the surface of the sea.

The bigger the wave, the more energy it contains.

The seas rise and fall.

As King Canute reportedly discovered, the rise and fall of the seas represent a vast and relentless natural phenomenon, certainly beyond the absolute control of all earthly subjects. The ocean is never still. Waves are caused by energy passing through the water, causing the water to move in a circular motion. We are tapping into this natural energy and developing energy-capturing techniques.

The ocean can produce two types of energy: *thermal energy* from the sun's heat and *mechanical energy* from tides and waves. Generating technologies for deriving electrical power from the ocean include tidal power, wave power, ocean thermal energy conversion, ocean currents, ocean winds, and salinity gradients. The three most well-developed technologies are tidal power, wave power, and Ocean Thermal Energy Conversion (OTEC). Tidal power requires large tidal differences, which, in the United States, occur only in Maine and Alaska. Ocean thermal energy conversion is limited to tropical regions, such as Hawaii, and a portion of the Atlantic coast. Wave energy has a more general application, with potential along the California boast. The western coastline has the highest wave potential in the United States; in California, the most significant potential is along the northern coast. Ocean thermal energy can be used for many applications, including electricity generation. Electricity conversion systems use warm surface water or boiling seawater to turn a turbine, activating a generator.

DOI: 10.1201/9781003605430-12

The conversion of electricity into both tidal and wave energy usually involves mechanical devices. It is essential to distinguish tidal energy from hydropower. Recall that hydropower is derived from the hydrological climate cycle and is powered by solar energy, usually harnessed via hydroelectric dams. In contrast, tidal energy results from the interaction of the moon's gravitational pull and, to a lesser extent, the sun on the seas. Processes that use tidal energy rely on the twice-daily tides and the resultant upstream flows and downstream ebbs in estuaries and the lower reaches of some rivers, as well as, in some cases, tidal movement out at sea. A dam typically converts tidal energy into electricity by forcing the water through turbines and activating a generator. Meanwhile, wave energy, a tremendous potential resource to be tapped, uses mechanical power to directly activate a generator and transfer it to a working fluid, water, or air, which drives a turbine/generator.

Before beginning a detailed discussion of the ocean's thermal and mechanical energy potential, it is essential to have a basic understanding of oceans, especially their margins; it is at the margins of coastal regions where most, if not all, ocean energy is harnessed using present technology. Accordingly, the following section provides foundational information to better understand the ocean energy concepts presented in this chapter.

10.2 OCEANS AND THEIR MARGINS[1]

Oceans are a principal component of the hydrosphere and the storehouse of the Earth's water. Oceans cover about 71% of the Earth's surface. The average depth of the Earth's oceans is about 3,800 m, with the most incredible ocean depth recorded at 11,036 m in the Mariana Trench. At present, the oceans contain a volume of about 1.35 billion cubic kilometers (96.5% of the Earth's total water supply), but the volume fluctuates with the growth and melting of glacial ice.

The composition of ocean water has remained constant throughout geological time. The major constituents dissolving in ocean water (from rivers and precipitation and the result of weathering and degassing of the mantle by volcanic activity) is composed of about 3.5% by weight of dissolved salts including chloride (55.07%), sodium (30.62%), sulfate (7.72%), magnesium (3.68%), calcium (1.17%), potassium (1.10%), bicarbonate (0.40%), bromine, (0.19%) and strontium (0.02%).

The most significant factor related to ocean water that everyone is familiar with is the salinity of the water. *Salinity*, a measure of the amount of dissolved ions in the oceans, ranges between 33 and 37 parts per thousand. Often, the concentration is the amount (by weight) of salt in water, as expressed in "parts per million" (ppm—analogous to a full whiskey shot glass of water taken from an Olympic-sized swimming pool). Water is saline if it has a concentration of more than 1,000 ppm of dissolved salts; ocean water contains about 35,000 ppm of salt (USGS, 2007). Chemical precipitation, absorption of clay minerals, and plants and animals prevent seawater from containing even higher salinity concentrations. However, salinity does vary in the oceans because surface water evaporates, rain and stream water are added, and ice forms or thaws.

Along with salinity, another essential property of seawater is temperature. The temperature of surface seawater varies with latitude, from near 0°C near the poles to 29°C near the equator. Some isolated areas can have temperatures up to 37°C. Temperature decreases with ocean depth.

10.3 THE OCEAN FLOOR

The bottoms of ocean basins (ocean floors) are marked by mountain ranges, plateaus, and other relief features like (although not as rugged as) those on land.

As shown in Figure 10.1, the ocean floor has been divided into four divisions: the continental shelf, continental slope, continental rise, and deep-sea floor or abyssal plain.

- **Continental shelf**—These are the continents' flooded, nearly flat, accurate margins. Varying in width to about 40 miles and a depth of approximately 650 ft, continental shelves slope gently outward from the shores of the continents (see Figure 10.1). Continental shelves occupy roughly 7.5% of the ocean floor.
- **Continental slope**—A relatively steep slope descending from the continental shelf (see Figure 10.1) rather abruptly to the deeper parts of the ocean. These slopes occupy about 8.5% of the ocean floor.
- **Continental rise**—This is a broad, gentle slope below the continental slope, with sediment accumulating along parts of it.
- **Abyssal plain**—This is a sediment-covered deep-sea plain about 12,000–18,000 ft below sea level. This plane makes up about 42% of the ocean floor.

The deep ocean floor does not consist exclusively of the abyssal plain. In places, there are areas of considerable relief. Among the more essential features are:

- **Seamounts**—These are isolated mountain-shaped elevations more than 3,000 ft high.
- **Mid-oceanic ridge**—These are submarine mountains extending more than 37,000 miles through the oceans, generally 10,000 ft above the abyssal plain.
- **Trench**—This is a deep, steep-sided trough in an abyssal plain.

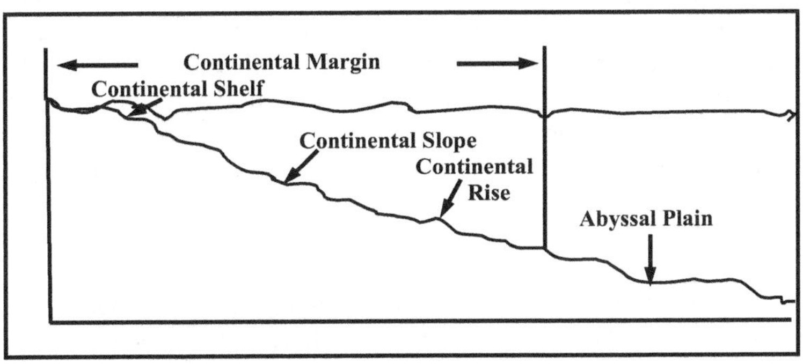

FIGURE 10.1 Cross section of ocean floor showing significant elements of topography.

- **Guyots**—They are flat-topped seamounts that were once volcanoes. They rise from the ocean bottom and usually are covered by 3,000–6,000 ft of water.

10.4 OCEAN TIDES, CURRENTS, AND WAVES

Water is the master sculptor of the Earth's surfaces. The ceaseless, restless motion of the sea is a highly effective geologic agent. Besides shaping inland surfaces, water sculpts the coast. Coasts include sea cliffs, shores, and beaches. Seawater set in motion erodes cliffs, transports eroded debris along shores, and dumps it on beaches. Therefore, most coasts retreat or advance. In addition to the unceasing causes of motion—wind, seawater density, and earth rotation—the chief agents in this process are tides, currents, and waves.

10.4.1 Tides

The periodic rise and fall of the sea (once every 12 hours and 26 minutes) produces the tides. Tides are due to the gravitational attraction of the moon and, to a lesser extent, the sun on the Earth. The moon has a more significant effect on tides and causes the Earth to bulge toward the moon. It is interesting to note that at the same time, the moon causes a bulge on Earth; a bulge occurs on the opposite side of the Earth due to inertial forces (further explanation is beyond the scope of this text). The effect of the tides is not too noticeable in the open sea; the difference between high and low tides amounts to about 2 ft. However, the tidal range may be considerably greater near shore. It may range from less than 2 ft to as much as 50 ft. The tidal range will vary according to the moon's phase and distance from the Earth. The type of shoreline and the physical configuration of the ocean floor will also affect the tidal range.

10.4.2 Currents

The oceans have localized movements of masses of seawater called ocean currents. These result from the drift of the upper 50–100 m of the sea due to drag by the wind. Thus, surface ocean currents generally follow the same patterns as atmospheric circulation except that atmospheric currents continue over the land surface while the land deflects ocean currents. Along with wind action, current may also be caused by tides, variation in salinity of the water, rotation of the earth, and turbid or muddy water concentrations. Temperature changes in water affect water density, which, in turn, causes currents—these currents cause seawater to circulate vertically.

10.4.3 Waves

Waves, varying wildly in size, are produced by wind friction on open water. Wave height and power depend upon wind strength and fetch—the amount of unobstructed ocean over which the wind has blown. In a wave, water travels in loops. Essentially, due to an up-and-down movement of the water, the diameter of the loops decreases in depth. The diameter of loops at the surface is equal to wave height (h). Breakers are formed when the wave comes into shallow water near the shore. The lower part of the wave is retarded by the ocean bottom, and the top, having more incredible momentum, is hurled forward, causing the wave to break. These breaking waves may damage coastal property as the race across coastal lowlands is driven by winds, gales, or hurricane velocities.

10.5 COASTAL EROSION, TRANSPORTATION, AND DEPOSITION

Like previously discussed geologic agents, the sea's geologic work involves erosion, transportation, and deposition. The sea accomplishes its work of coastal landform sculpting essentially using waves and wave-produced currents, which may have quite pronounced effects on the seacoast. The coast, accompanying coastal deposits, and landform development represent a balance between wave energy and sediment supply.

10.5.1 Wave Erosion

Waves attack shorelines and erode by a combination of several processes. The resistance of the rocks composing the shoreline and the intensity of wave action to which it is subjected are the factors that determine how rapidly the shore will be eroded. Wave erosion works chiefly by hydraulic action, corrosion, and attrition. As waves strike a sea cliff, *hydraulic action* crams air into rock crevices, putting tremendous pressure on the surrounding rock; as waves retreat, the explosively expanding air enlarges cracks and breaks off chunks of rock (*scree*). Chunks hurled by waves against the cliff break off more scree (via a sandpapering action), a process called *corrosion*. When the sea rubs and grinds rocks together, a scree is formed, and it is thrown into the cliffs, reducing broken rocks to pebbles and sand grains, a process called *attrition* (Lambert, 2007).

Marine erosion forms several features—different combinations of wave action, rock type, and rock beds produce these features. Below are some of the more typical erosion-formed features of shorelines:

- **Sea cliffs or wave cut cliffs**—These are formed by wave erosion of underlying rock followed by the caving-in of the overhanging rocks. As waves travel further inland, they leave a wave-cut beach or platform. Such cliffs are essentially vertical and are expected at certain localities along North America's New England and Pacific coasts.
- **Wave-cut bench—**This results from wave action not having enough time to lower the coastline to sea level. Because of the resistance to erosion, a relatively flat wave-cut bench develops. If subsequent uplift of the wave-cut bench occurs, it may be preserved above sea level as a wave-cut bench.
- **Headlands**—These are finger-like projections of resistant rock extending out into the water. Indentations between headlands are termed *coves*.
- **Sea caves, sea arches, and stacks**—These are formed by continued wave action on a sea cliff. Wave action hollows out cavities or caves in the sea

cliffs. Eventually, waves may cut entirely through a headland to form a sea arch; if the arch's roof collapses, the rock left separated from the headland is called a stack.

10.5.2 Marine Transportation

Waves and currents are essential transporting agents. Rip currents and undertows carry rock particles back to the sea, and long-shore currents pick up sediments (some of them in solution), moving them out from shore into deeper water. Materials carried in solution or suspension may drift seaward for great distances and eventually be deposited far from shore. During transportation, sediments undergo additional erosion and become reduced in size.

10.5.3 Marine Deposition

Marine deposition takes place whenever currents and waves suffer reduced velocity. Some rocks are thrown up on the shore by wave action. Most of the sediments thus deposited consist of rock fragments derived from the mechanical weathering of the continents, and they differ considerably from terrestrial or continental deposits. Due to the input of sediments from rivers, deltas may form, and due to beach drift, features such as spits and hooks, bay barriers, and tombolos may form. Depositional features along the coasts are discussed below:

- **Beaches**—These are transitory coastal deposits of debris that lie above the low-tide limit in the shore zone.
- **Barrier islands**—These are long, narrow accumulations of sand lying parallel to the shore and separated from the shore by a shallow lagoon.
- **Spits and hooks**—These are elongated, narrow barriers of sand and pebbles extending out into the water but attached by one end to the land.
- **Tombolos—**These are bars of sand or gravel connecting an island with the mainland or another island.
- **Wave-built terraces**—These are structures built up from sediments deposited in deep water beyond a wave-cut terrace.

10.6 WAVE ENERGY[2]

Waves are caused by the wind blowing over the surface of the ocean. In many areas of the world, the wind blows with enough consistency and force to provide continuous waves. Wave energy does not have the tremendous power of tidal fluctuations. However, the regular pounding of the waves should not be underestimated because there is immense energy in the ocean waves. The total waves breaking on the world's coastlines are between 2 and 3 million megawatts. In optimal wave areas, more than 65 MW of electricity could be produced along a single mile of shoreline, according to the U.S. Office of Energy Efficiency and Renewable Energy (EERE, 2004). In essence, because the wind is originally derived from the sun, we can consider the energy in ocean waves to be a stored, moderately high-density form of solar energy.

According to specific estimates, if fully developed, wave technologies could feasibly fulfill 10% of the global electricity supply (World Energy, 2010). The west coast of the United States and Europe and the coast of Japan and New Zealand are good sites for harnessing wave energy.

10.6.1 Wave Energy: Facts, Parameters, and Equations

- Three main processes create waves: Air flowing over the sea exerts tangential stress on the water surface, resulting in the formation and growth of waves; turbulent air flow close to the water surface creates rapidly varying shear stresses and pressure fluctuations (when these oscillations are in phase with existing waves, further wave development occurs); and when waves have reached a specific size, the wind can exert a more potent force on the up-wind face of the wave, resulting in additional wave growth.
- Waves located within or close to the areas where they are generated are called storm waves.
- *Swell waves* can develop at great distances from the point of origin.
- The distance over which wind energy is transferred into the ocean to form waves is called fetch.
- *Sea state* is the general condition of the free surface on a large body of water—concerning wind waves and swell—at a specific location and moment. The World Meteorological Organization sea state code essentially adopts the "wind se" definition of the Douglas Sea Scale.

WMO Sea State Code	Wave Height (m)	Characteristics
0	0	Calm (glassy)
1	0–0.1	Calm (rippled)
2	0.1–0.5	Smooth (wavelets)
3	0.5–1.25	Slight
4	1.25–2.5	Moderate
5	2.5–4	Rough
6	4–6	Very rough
7	6–9	High
8	9–14	Very high
9	Over 14	Phenomenal

- The shape of a typical wave is described as *sinusoidal* (that is, it has the form of a mathematical sine function; see Figure 10.2). The difference in height between the peaks troughs is known as the *height*, H, and the distance between successive peaks (or troughs) of the wave is known as the wavelength, λ. The time in seconds taken for consecutive peaks (or troughs) to pass a given fixed point is known as the *period*, T. The *frequency*, v, of the wave describes the number of peak-to-peak (or trough-to-trough) oscillations of the wave surface per second, as seen by a fixed observer, and is the reciprocal of the period. That is, $v = 1/T$.

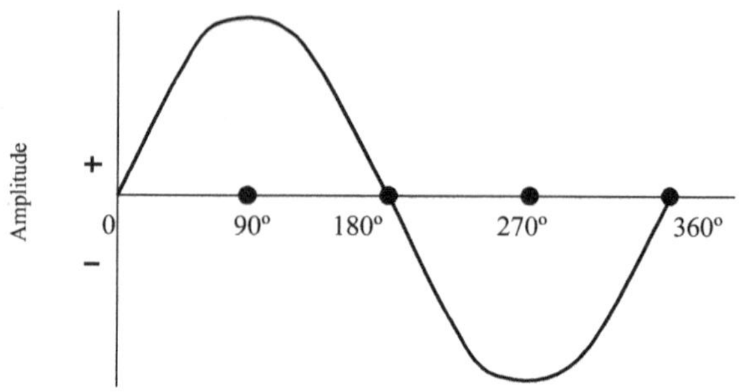

FIGURE 10.2 Sinusoidal waves show wavelength and amplitude.

If a wave travels at a velocity v past a given fixed point, it will travel a distance equal to its wavelength λ in a time equal to the wave period T (i.e., $v = \lambda/T$). The power, P (kW/m), of an idealized ocean wave is approximately equal to the square of the height, H (m), multiplied by the wave period, T (seconds). The exact expression is the following:

$$P = \frac{g^2 H^2 T}{32}$$

where P is in units of W/m and g is the acceleration due to gravity (9.81 m/s^2) (Phillips, 1977).

- **Deep water waves**—The velocity of a long ocean wave can be shown to be proportional to the period (if the depth of water is greater than about half of the wavelength λ) as follows:

$$v = \frac{gT}{2\pi}$$

A helpful approximation can be derived from this: velocity in m/s is about 1.5 times the wave period in seconds. The result leads to deep ocean waves traveling faster than the shorter waves. Moreover, if the above relationships hold, we can find the deep-water wavelength, λ, for any given wave period:

$$\lambda = \frac{gT}{2\pi}$$

- **Intermediate depth waves**—As the water becomes shallower, the properties of the waves become increasingly dominated by water depth. When waves reach shallow water, their properties are entirely governed by the water depth. Still, in intermediate depths (i.e., between $d = \lambda/4\pi$), the properties of the waves will be influenced by both water depth d and wave period T (Phillips, 1977; Goda, 2000).
- **Shallow water waves**—As waves approach the shore, the seabed starts to affect their speed, and it can be shown that if the water depth d is less than a quarter of the wavelength, the velocity is given by:

$$v = \sqrt{gd}$$

- As waves propagate, their energy is transported. The energy transport velocity is the group velocity. As a result, the wave energy flux, through a vertical plane of unit width perpendicular to the wave propagation direction, is equal to:

$$P = Ec_g$$

where c_g is the group velocity (m/s).

DID YOU KNOW?

The wave height increases with increases in:

- wind speed
- time duration of the wind blowing
- fetch
- water depth

10.7 WAVE ENERGY CONVERSION TECHNOLOGY

In the early 1970s, the harnessing of wave power focused on using floating devices such as Cockerell Rafts (a wave power hydraulic device), Salter Duck (a curved-cam-like device that can capture 90% of waves for energy conversion), Rectifier (concerts A-C to D-C electricity), and the Clam (a floating rigid toroid—i.e., doughnut-shaped—that converts wave energy to electrical energy). Wave energy converters can be classified by location: fixed to the seabed, generally in shallow water, floating offshore in deep water, or tethered in intermediate depths. These floating devices are not cost-effective and have challenging moving problems. So, the current practice is to move inshore, sacrificing some energy, but fixed devices, according to Tovey (2005), have several advantages, including:

- easier maintenance
- easier landing on the device
- no mooting problem
- easier power transmission
- enhanced productivity
- better design life

Wave energy devices can be classified using their reaction system, but discussing how they interact with the wave field is often more instructive. Each moving body may act as a displacer or a reactor in this context.

DID YOU KNOW?

Hydro (water) Kinetic Energy is the energy a body of water possesses because of its motion ($KE = 1.2\ mv^2$). Hydrostatic (at rest) Energy is the energy a body possesses because of its position or location at an elevation or height above a reference or datum ($PE = mgh$).

- **Displacer**—This is the body moved by the waves. It might be a buoyant vessel or a mass of water. If buoyant, the displacer may pierce the surface of the waves or be submerged.
- **Reactor**—This is the body that provides reaction to the displacer. As suggested above, it could be a body fixed to the seabed or the seabed itself. It could also be another structure or mass that is not fixed but moves to create reaction forces (e.g., by moving by a different amount or at other times). A degree of control over the forces acting on each body and acting between the bodies (particularly stiffness and damping characteristics) is often required to optimize the amount of energy captured.

Note: In some designs, the reactor is inside the displacer, while in others, it is an external body. Internal reactors are not subject to wave forces, but external ones may experience loads that cause them to move in ways like a displacer. This can be extended to the view that some devices do not have dedicated reactors, but rather a system of displacers whose relative emotion creates a reaction system (INEL, 2005).

Three well-known Wave Energy Conversion devices exist: Point absorbers, terminators, and attenuators (see Figure 10.3; INEL, 2005).

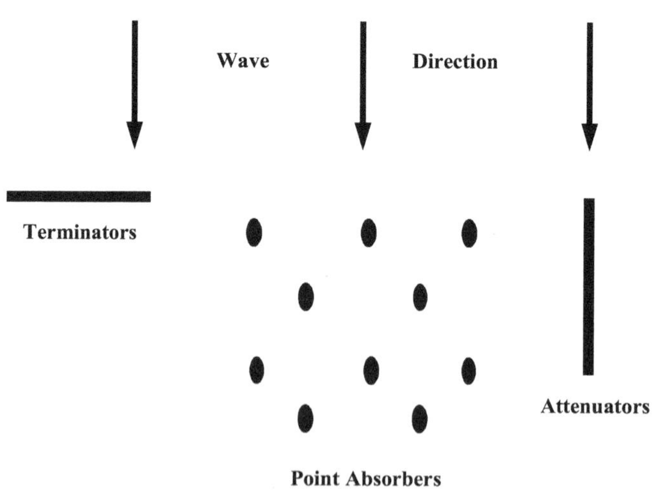

FIGURE 10.3 Types of wave energy converters. (Adaptation from Tovey, 2005.)

- **Point absorber**—A floating structure that absorbs energy in all directions by moving at or near the water surface (see Figure 10.4). It may be designed to resonate—moving with larger amplitudes than the waves. This feature helps maximize the amount of power available for capture. The power take-off system may take several forms, depending on the configuration of displacers/reactors.
- **Terminator**—This is also a floating structure that moves at or near the water surface but absorbs energy in only a single direction (see Figure 10.5). The device extends in the direction normal to the predominant wave direction, so as waves arrive, the device restrains them. Again, resonance may be employed, and the power take-off system may take various forms.

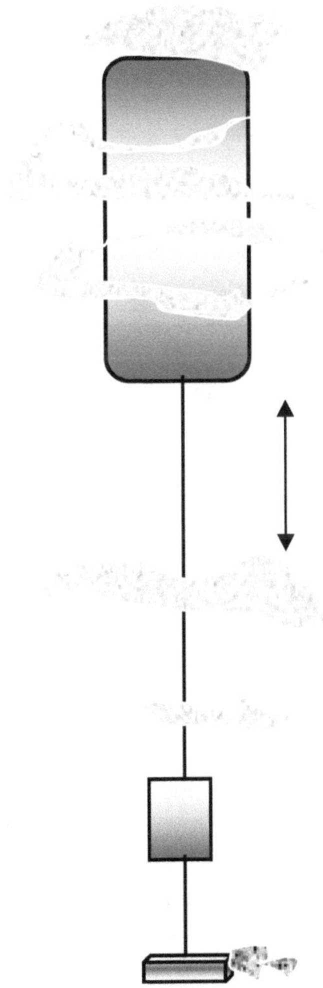

FIGURE 10.4 Point absorber.

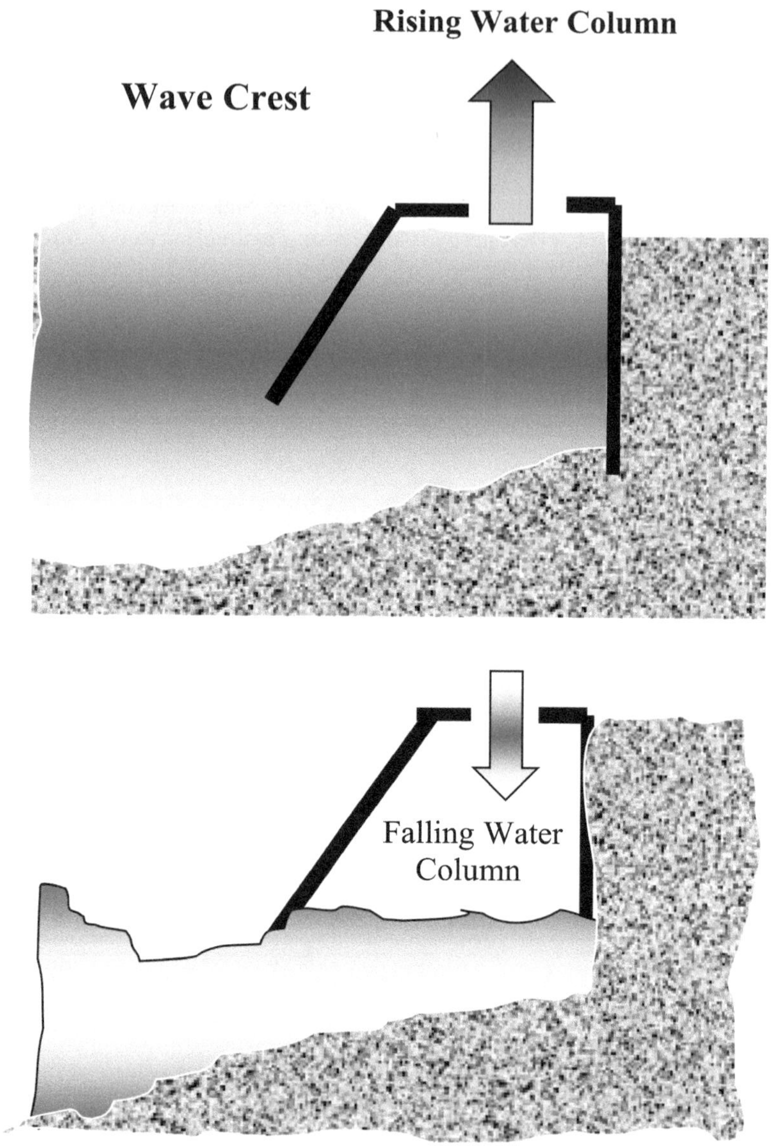

FIGURE 10.5 Terminator.

- **Attenuator**—This device is a long floating structure like the terminator but is orientated parallel to the waves (see Figure 10.6). It rides the waves like a ship, and the device's movement at its bow and along its length can be restrained to extract energy. A theoretical advantage of the attenuator over the terminator is that its area normal to the waves is small, and therefore, the forces it experiences are much lower.

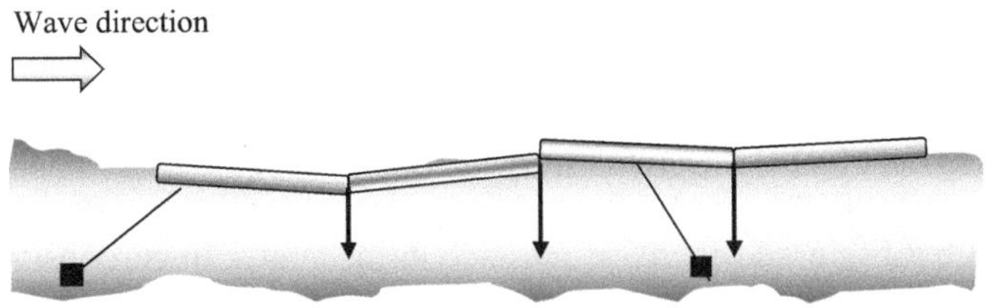

FIGURE 10.6 Attenuator.

10.8 TIDAL ENERGY

The tides rise and fall in eternal cycles. Tides are changes in the level of the oceans caused by the gravitational pull of the moon and sun and the Earth's rotation. The relative motions of these cause several different tidal cycles, including a semi-diurnal cycle—period 12 hours 25 minutes; a semi-monthly cycle—i.e., Spring or Neap Tides corresponding with the position of the moon; a semi-annual cycle—period about 178 days, which is associated with the inclination of the Moon's orbit. This causes the highest spring tides to occur in March and September and other long-term cycles, e.g., a nineteen-year cycle of the moon. Nearshore water levels can vary up to 40 ft, depending on the season and local factors. Only about 20 locations have suitable inlets and an extensive enough tidal range—about 10 ft—to produce energy economically (DOE, 2010).

DID YOU KNOW?

The Spring Tides have a range about twice that of neap tides, while the other cycles can cause further variations of up to 15%. The tidal range is amplified in estuaries, and in some situations, the shape of the estuary is such that near resonance occurs.

10.8.1 Tidal Energy Technologies

Some of the oldest ocean energy technologies use tidal power. Tidal power is more predictable than solar power and wind energy. All coastal areas consistently experience two high and two low tides over slightly more than 24 hours. To harness those tidal differences into electricity, the difference between high and low tides must be at least five meters or more than 16 ft. Only about 40 sites on the earth have tidal ranges of this magnitude. Currently, there are no tidal power plants in the United States. However, conditions suit tidal power generation in the country's Pacific Northwest and the Atlantic Northeast regions. Tidal energy technologies include the following:

- **Tidal barrages**—These are simple generation systems for tidal plants that involve a dam, known as a barrage, across an inlet. Sluice gates (commonly used to control water levels and flow rates) on the barrage allow the tidal basin to fill on the incoming high tides and empty through the turbine system on the outgoing tide, also known as the ebb tide. Two-way systems generate electricity on both the incoming and outgoing tides. A potential disadvantage of a barrage tidal power system is the effect a tidal station can have on plants and animals in estuaries. Tidal barrages can change the tidal level in the basin and increase the amount of matter in suspension in the water (turbidity). They can also affect navigation and recreation.

- **Tidal fences**—These look like giant turnstiles. A tidal fence has vertical axis turbines mounted in a fence. All the water that passes is forced through the turbines. Some of these currents run at 5–8 knots (5.6–9 miles per hour) and generate as much energy as winds of much higher velocity. Tidal fences can be used in areas such as channels between two landmasses. Tidal fences are cheaper to install than tidal barrages and have less environmental impact. Note that tidal barrages can disrupt the movement of large marine animals.
- **Tidal turbines**—Wind turbines in the water located anywhere there is a strong tidal flow (function best where coastal currents run between 3.6 and 4.9 knots—4–5.5 mph). Because water is about 800 times denser than air, tidal turbines must be much sturdier than wind turbines. Tidal turbines are heavier and more expensive to build but capture more energy.

10.9 OCEAN THERMAL ENERGY CONVERSION

As mentioned, our planet's most plentiful renewable energy source is solar radiation: 170,000 TW ($170{,}000 \times 10^{12}$ W) fall on Earth. Because of its dilute and erratic nature, however, it isn't easy to harness. To do so, that is, to capture this energy, we must employ the use of large collecting areas and large storage capacities; these requirements are satisfied on Earth only by the tropical oceans. We are all taught at an early age that oceans (and water in general) cover about 71% (or two-thirds) of the Earth's surface. In a fitting reference to the vast oceans covering most of Earth, Ambrose Bierce (1842–1914) commented: "A body of water occupying about two-thirds of the world made for man who has no gills." So, true, we have no gills; thus, those who look out upon those vast bodies of water that cover the surface might ask: What is their purpose? And, of course, this is a good question with several possible answers. Regarding renewable energy, we can look at those vast seas and wonder: How can we use this massive storehouse of energy for our own needs? Because it is so extensive and deep, it absorbs much of the heat and light that comes from the sun. One thing seems inevitable: Our origin, past, present, and future, lies within those massive wet confines we call oceans.

10.9.1 Ocean Energy Conversion Process[3]

The ocean is a gigantic solar collector. The energy from the sun heats the surface water. In tropical regions, the surface water can be 40 or more degrees warmer than the deep water. This temperature difference can be used to produce electricity. OTEC has the potential to produce more energy than tidal, wave, and world energy combined.

The OTEC systems can be open or closed. In a closed system, an evaporator turns warm surface water into a stream under pressure (see Figures 10.7 and 10.8). This steam spins a turbine generator to produce electricity. Water pumps bring cold, deep water through pipes to a condenser on the surface. The cold water condenses the steam, and the closed cycle begins again. In an open system, the steam is turned into fresh water, and new surface water is added. A transmission cable carries electricity to the shore.

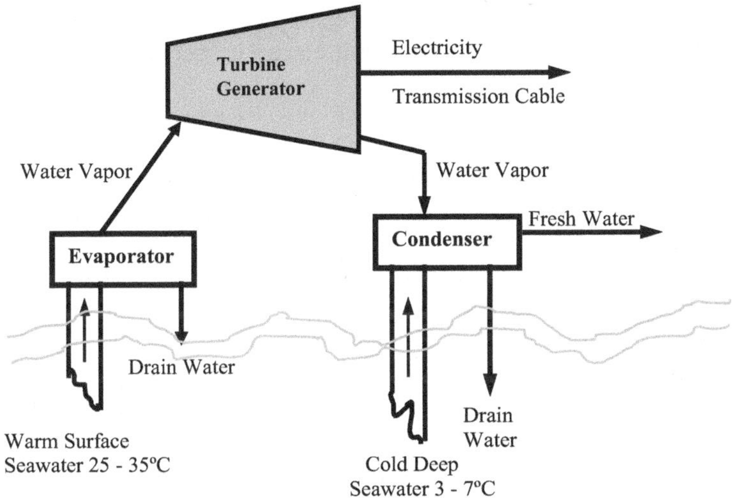

FIGURE 10.7 Schematic of ocean thermal energy conversion system.

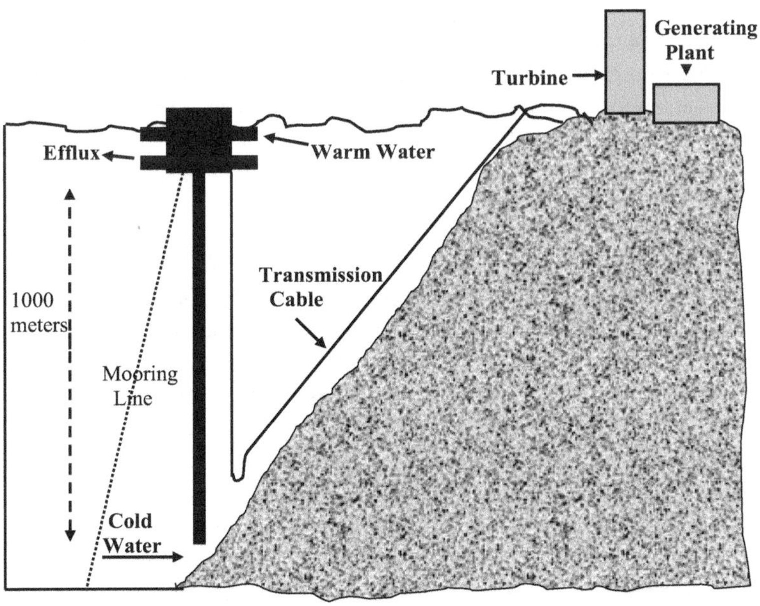

FIGURE 10.8 OTEC floating platform.

OTEC systems must operate with a temperature difference of about 25°C. This limits OTEC's use in tropical regions where the surface waters are hot and the water is icy. With its tropical climate, Hawaii has experimented with OTEC systems since the 1970s.

Because widespread use presents many challenges, there are no large or significant operations but several experimental OTEC plants. Pumping water is a giant engineering challenge, and because of this, OTEC systems are not very energy efficient. DOE (2010) estimates that it will probably be 10–20 years before the technology is available to produce and transmit electricity economically from OTEC systems.

- **Closed-cycle**

  These systems use fluids with a low boiling point, such as ammonia, to rotate a turbine to generate electricity. Warm surface seawater is pumped through a heat exchanger, where the lower-boiling-point fluid is vaporized. The expanding vapor turns the turbo-generator. Cold deep-seawater—pumped through a second heat exchanger—condenses the vapor back into a liquid, which is then recycled through the system.
- **Open-cycle**

  These systems use the tropical oceans' warm surface water to make electricity. When warm seawater is placed in a low-pressure container, it boils. The expanding steam drives a low-pressure turbine attached to an electrical generator. The steam, which has left its salt behind in the lower-pressure container, is almost pure fresh water. It is condensed back into a liquid by exposure to cold temperatures from deep-ocean water.
- **Hybrid**

  These systems combine the features of both closed-cycle and open-cycle systems. In a hybrid system, warm seawater enters a vacuum chamber and is flash-evaporated into steam, like the open-cycle evaporation process. The steam vaporizes a lower-boiling-point fluid (in a closed-cycle loop), which drives a turbine to produce electricity.

10.10 THE BOTTOM LINE

The three forms of ocean energy—tidal, wave, and OTEC systems—are all renewable. This is a significant advantage over fossil fuels and other energy forms that pollute the environment. Although all three forms of ocean energy promise future development, OTEC systems appear to be the most beneficial. OTEC systems provide both economic and non-economic benefits.

Economic benefits include:

- Helps produce fuels such as hydrogen, ammonia, and methanol.
- Provides moderate-temperature refrigeration.
- Produces baseload electrical energy.
- Produces desalinated water for industrial, agricultural, and residential uses.
- Serves as a resource for on-shore and near-shore mariculture operations.

- Provides air conditioning for buildings.
- Has significant potential to provide clean, cost-effective electricity for the future.

Non-economic benefits include:

- Enhances energy independence and energy security.
- Promotes competitiveness and international trade.
- Promotes international sociopolitical stability.
- Has the potential to mitigate greenhouse gas emissions resulting from burning fossil fuels.

Practitioners of ocean energy have found ways to enhance advancements, including using artificial intelligence (AI) and machine learning (ML), which are emerging as essential tools in optimizing the efficiency and sustainability of these renewable energy sources. AI and ML analyze real-time data, enabling the identification of optimal conditions for energy generation. AI and ML are also used for monitoring and maintenance, ensuring the smooth operation of ocean energy systems.

NOTES

1 Material in this section is adapted from F.R. Spellman, F.R., 2009. *Geology for Non-Geologists.* Lanham, MD: Government Institutes Press.
2 From DOE, 2010. *Ocean Energy.* Accessed 03/31/24 @ www.mms.gov.
3 Much of the information in this section is from DOE, 2010. *Ocean Energy.* Accessed 03/31/24 @ www.mms.gov.

REFERENCES

DOE, 2010. *Ocean Energy*. Washington, DC: Department of Interior. Accessed 12/12/24 @ www.mms.gov.
EERE, 2004. *Oceans*. Accessed 03/31/23 @ www.eere.gov/RE/ocean.html.
Goda, Y. 2000. *Random Seas and Designs of Maritime Structures*. World Scientific. Accessed 04/10/24 @ www.esru.strath.ac.uk/EandE/Web_sites/01-02/RE_info/wave%20power.htm.
INEL, 2005. *DOE HydroKinetic Workshop*. Accessed 04.06/24 @ https://www.hydropower.inel.Gov/hydrokinetic_wave.pdf.
Lambert, D., 2007. *The Field Guide to Geology*. New York: Checkmark Books.
Phillips, O.M., 1977. *The Dynamics of the Upper Ocean*, 2nd edition. Cambridge: Cambridge University Press.
Tovey, N.K., 2005. *ENV-2E02 Energy Resources 2005 Lecture*. Accessed 03/01/24 @ www2.env.ac.UK.gmmc/energy.
USGS, 2007. *The Water Cycle: Water Storage in Oceans*. Accessed 7/11/24 @ https://ga.water.usgs.Gov/edu/watercycleoceans.html.
World Energy, 2010. *Survey of Energy Sources: Wave Energy*. Accessed 03/31/24 @ https://www.worldenergy.org/wec-gies/publications.

11 Energy/Fuel Cells

11.1 INTRODUCTION

In casual conversations with energy experts, faculty, students, and a few associates, when energy cells (aka fuel cells) for vehicles are part of the conversation, most of them state that the day will come when we finally end the internal combustion engine's 100-year reign as the dominant power source for personal transportation. They say it will be a winning situation. All consumers will get an efficient power source, communities will get zero emissions, and automakers will get another significant business opportunity—a growth opportunity. Energy/fuel cells are not just fodder for casual conversation, science fiction, or pie-in-the-sky technology. They are used in some applications today, and many wonder why they are not more commonly and universally used.

Fuel cells are like batteries in that they are composed of positive and negative electrodes separated by an electrolyte or membrane. Both are highly efficient. However, fuel cells do not need to be recharged like batteries, so they may run for extended periods if fuel and air are provided.

Energy/fuel cells are the future. However, consider both sides and begin discussing hydrogen energy/fuel cells.

11.2 ENERGY/FUEL CELLS: A REALISTIC VIEW

11.2.1 Hydrogen[1]

Hydrogen, the chemical symbol H, is the simplest element on Earth. It contains only one electron and one proton. Hydrogen is a diatomic molecule—each molecule has two atoms of hydrogen (which is why pure hydrogen is commonly expressed as H_2). Although abundant on Earth as an element, hydrogen combines readily with other components and is almost always found as part of another substance, such as water hydrocarbons or alcohols. Hydrogen is also found in biomass, which includes all plants and animals.

- Hydrogen is an energy carrier, not an energy source. Hydrogen can store and deliver usable energy, but it doesn't typically exist by itself; it must be produced from compounds containing it.
- Hydrogen can be produced using diverse domestic resources, including nuclear, natural gas, coal, biomass, and other renewables, including solar, wind, hydroelectric, or geothermal energy. This diversity of domestic energy sources makes hydrogen a promising energy carrier essential to our nation's energy security. It is expected and desirable for hydrogen to be produced using a variety of resources and process technologies (or pathways).

DOI: 10.1201/9781003605430-13

- DOE focuses on hydrogen-production technologies that result in near-zero net greenhouse gas emissions and use renewable energy sources, nuclear energy, and coal (when combined with carbon sequestration). Energy efficiency is also essential to ensure sufficient clean energy for our overall energy needs.
- Hydrogen can be produced via various process technologies, including thermal (natural gas reforming, renewable liquid and bio-oil processing, and biomass and coal gasification), electrolytic (water splitting using a variety of energy resources), and photolytic (splitting water using sunlight via biological and electrochemical materials).
- Hydrogen can be produced in extensive, central facilities (50–300 miles from the point of use), smaller semi-central (located within 25–100 miles of use), and distributed (near or at the point of use). Learn more about distributed vs. centralized production.
- For hydrogen to be successful in the marketplace, it must be cost-competitive with the available alternatives. This competitive requirement in the light-duty vehicle transportation market means hydrogen must be available untaxed at $2–$3/gge (gasoline gallon equivalent). This price would result in hydrogen fuel cell vehicles having the exact cost to the consumer on a cost-per-mile-driven basis as a comparable conventional internal combustion engine or hybrid car.
- DOE is engaged in research and development of various hydrogen production technologies. Some are further along in development than others—some can be cost-competitive for the transition period (beginning in 2015), and others are considered long-term technologies (cost-competitive after 2030).

Infrastructure is required to move hydrogen from where it's produced to the dispenser at a refueling station or stationary power site. Infrastructure includes pipelines, trucks, railcars, ships, and barges that deliver fuel, as well as the facilities and equipment needed to load and unload them.

Delivery technology for hydrogen infrastructure is currently available commercially, and several U.S. companies deliver bulk hydrogen today. Some of the infrastructure is already in place because hydrogen has long been used in industrial applications. Still, supporting widespread consumer use of hydrogen as an energy carrier is insufficient. Because hydrogen has a relatively low volumetric energy density, its transportation, storage, and final delivery to the point of use comprise a significant cost and result in some energy inefficiencies associated with using it as an energy carrier.

DID YOU KNOW?

The hydrogen fuel cell paradox: Energy must first be used to produce hydrogen only to gain less energy back when hydrogen is used to generate electricity.

Options and trade-offs for hydrogen delivery from central, semi-central, and distributed production facilities to the point of use are complex. The choice of a hydrogen production strategy greatly affects the cost and method of delivery.

For example, more extensive, centralized facilities can produce hydrogen at relatively low costs due to economies of scale. Still, the delivery costs for centrally produced hydrogen are higher than those for semi-central or distributed production options (because the point of use is farther away). In comparison, distributed production facilities have relatively low delivery costs. Still, hydrogen production costs are likely higher—lower volume production, which means higher equipment costs per unit of hydrogen basis.

The key challenges to hydrogen delivery include reducing delivery costs, increasing energy efficiency, maintaining hydrogen purity, and minimizing hydrogen leakage. Further research is needed to analyze the trade-offs between the hydrogen production and delivery options taken together as a system. Building a national hydrogen delivery infrastructure is a big challenge. It will take time to develop and will likely include combinations of various technologies. Delivery infrastructure needs and resources will vary by the region and type of market (e.g., urban, interstate, or rural). Infrastructure options will evolve as hydrogen demand grows and delivery technologies develop and improve.

11.2.1.1 Hydrogen Storage

Storing enough hydrogen on board a vehicle to achieve a driving range of greater than 300 miles is a significant challenge. On a weight basis, hydrogen has nearly three times the energy content of gasoline (120 MJ/kg for hydrogen versus 44 MJ/kg for gas). However, on a volume basis, the situation is reversed (8 MJ/L for liquid hydrogen versus 32 MJ/L for gasoline). On-board hydrogen storage in the 5–13 kg H_2 range is required to encompass the entire platform of light-duty vehicles.

Hydrogen can be stored in various ways, but for it to be a competitive fuel for vehicles, the hydrogen vehicle must be able to travel a comparable distance to conventional hydrocarbon-fueled vehicles.

Hydrogen can be physically stored as either a gas or a liquid. Storage as a gas typically requires high-pressure tanks (5,000–10,000 psi tank pressure). Storage as a liquid requires cryogenic temperatures because the boiling point of hydrogen at one-atmosphere pressure is 252.8°C.

Hydrogen can also be stored on the surfaces of solids (by adsorption) or within solids (by absorption). In adsorption, hydrogen is attached to the material's surface as hydrogen molecules or hydrogen atoms. In absorption, hydrogen is dissociated into H-atoms and then the hydrogen atoms are incorporated into the solid lattice framework.

Hydrogen storage in solids may make it possible to store large quantities of hydrogen in smaller volumes at low pressures and temperatures close to room temperature. It is also possible to achieve volumetric storage densities greater than liquid hydrogen because the hydrogen molecule is dissociated into atomic hydrogen within the metal hydride lattice structure.

Finally, hydrogen can be stored through the reaction of hydrogen-containing materials with water (or other compounds such as alcohol). In this case, hydrogen is effectively stored in both the material and the water. The term "chemical hydrogen storage" or chemical hydrides describes this form of hydrogen storage. It is also possible to store hydrogen in the chemical structures of liquids and solids.

11.2.1.2 Hydrogen Fuel Cell

The fuel cell uses hydrogen's chemical energy to cleanly and efficiently produce electricity, with water and heat as byproducts. Fuel cells are unique in terms of the variety of their potential applications; they can provide energy for systems as large as a utility power station and as small as a laptop computer.

Fuel cells have several benefits over conventional combustion-based technologies currently used in many power plants and passenger vehicles. They produce much smaller quantities of greenhouse gases and none of the air pollutants that create smog and cause health problems. If pure hydrogen is used as a fuel, fuel cells emit only heat and water as byproducts.

DID YOU KNOW?

Hydrogen fuel cell vehicles (FCVs) emit approximately the same amount of water per mile as vehicles using gasoline-powered internal combustion engines (ICEs).

A *fuel cell* is a device that uses hydrogen (or hydrogen-rich fuel) and oxygen to create electricity by an electrochemical process. A single fuel cell comprises an electrolyte and two catalyst-coated electrodes (a porous anode and cathode). While there are different fuel cell types, all fuel cells work similarly:

- Hydrogen, or a Hydrogen-rich fuel, is fed to the anode, where a catalyst separates hydrogen's negatively charged electrons from positively charged ions (protons).
- At the cathode, oxygen combines with electrons and, in some cases, with species such as protons or water, resulting in water or hydroxide ions.
- For polymer electrolyte membranes and phosphoric acid fuel cells, protons move through the electrolyte to the cathode to combine oxygen and electrons, producing water and heat.
- For alkaline, molten carbonate, and solid oxide fuel cells, negative ions travel through the electrolyte to the anode, combining hydrogen to generate water and electrons.
- The electrons from the anode cannot pass through the electrolyte to the positively charged cathode; they must travel around it via an electrical circuit to reach the other side of the cell. This movement of electrons is an electrical current.

11.3 AI APPLICATIONS FOR FUEL CELLS

Promising materials must be found to manufacture a capable and efficient fuel cell. AI incorporation in developing and producing these materials is ongoing. Machine learning (ML) can enable performance predictions, service-life predictions, and fault diagnosis in fuel cells. AI can be used to increase hydrogen production, storage, and transport using algorithms and models. Moreover, AI can simulate, control, and optimize fuel cell operation.

11.4 THE BOTTOM LINE

The discussion about fuel cells is often expected to be chained to powering electric vehicles. Fuel cells can be used in various applications, providing power across multiple sectors, including transportation, industrial/commercial/residential buildings, and long-term energy storage for the grid in reversible systems. Another truth is that fuel cells have several benefits over conventional combustion-based technologies currently used in many power plants and vehicles. For example, fuel cells can operate at higher efficiencies than combustion engines and convert the chemical energy in the fuel directly to electrical energy with efficiencies that exceed 60%. Fuel cells have lower to zero emissions than combustion engines; regarding global climate change, hydrogen fuel cells emit only water, no carbon dioxide emissions, and no smog or health problems at the point of operation. The bottom-line truth about fuel cells is that they have very few moving parts, making them quiet while operating (USDOE, 2024).

NOTE

1 Information in this section from USDOE, 2008. *Hydrogen, Fuel Cells & Infrastructure Technologies Program.* Accessed 12/12/23 @ http://www1.eere.energy.gov/hydrogenandfuelcells/production/basics.html.

REFERENCE

USDOE, 2024. *Fuel Cells.* Washington, DC: U.S. Department of Energy.

12 Small Modular Reactors

12.1 INTRODUCTION

Several four-letter words commonly used in the vernacular are considered vulgar or offensive, so much so that many of them (not to be included here) have been deemed legally indecent under the U.S. Federal Communications Commission (FCC) regulations for TV and radio open-wave broadcasting. Well, other words offend many of us, too. Consider, for example, the words and combinations of words that offend just about everyone, especially those on the receiving end of the addressing. One word gets a lot of attention: radioactive.

This one word gets a lot of attention. However, in this book, radioactive refers to a reliable energy source, which many, including the author, feel is the White Knight we seek. In energy, energy producers, providers, and users, the word nuclear contains all kinds of assorted feelings, interpretations, and current calls for respect and understanding.

Many of us involved in energy science have different views, perceptions, and understandings of nuclear power. Radioactivity is one of the fear words most people think about when nuclear anything is brought up in conversation or otherwise; people have the notion that unchained nuclear power might light them up or worse. Today, when applied for peaceful uses, a standard and current perception of nuclear power is best described by many as a clear-eyed view of what it is and its potential safe use in energy production, including operation without producing carbon footprint and other contaminants that lead to pollution and possibly climate change.

12.2 SMALL MODULAR REACTORS: THE BASICS

Considering our concerns about dwindling energy supplies, the pollution of fossil fuels, and the need for a 24/7 source of reliable energy, again, wind, solar, and battery energy have limited supplies. This limitation is an interruption, a break, or a gap in service.

So, okay, the question is, what is the alternative energy source?

Good question. The common perception of the dangers of nuclear-generated energy is changing. The simple truth is that when we turn our light switches on, we want energy to flow (electron flow—aka electricity) to power our lights, appliances, TVs, power tools, and iPhone and computer chargers.

The growing need for energy is making nuclear power attractive. This grappling with the nation's growing thirst for energy is increased due to the enormous growth in artificial intelligence data centers, which have huge appetites for power and gobble up a large constant flow of water for cooling purposes. This point was amplified in an article by Hauptman (2024) in *USA Today*, where he pointed out that Microsoft

DOI: 10.1201/9781003605430-14

announced plans to reopen the Three Mile Island nuclear power plant to support AI. Increasing demand from tech companies has driven vast energy needs required by artificial intelligence data centers.

Wow! Among the deleterious terms listed earlier, including "nuke," "nuclear," and "radioactive," there are critics who would add "Three Mile Island" to that list.

Why would Three Mile Island be added to the list?

In 1979, the facility's Unit 2 reactor experienced a partial meltdown, which caused it to be shut down. It has been closed ever since. Although the partial collapse of the reactor caused a relatively minor leak of radiation, the incident caused the public to distrust the nuclear power industry—its generator of mistrust is the term radioactive, which conjures up other Hiroshima, Nagasaki, and Chernobyl events.

A recent state poll found that residents favor restarting Three Mile Island by more than a 2-1 margin. This suggests that to continue the so-called "good life," we must develop, process, and distribute more energy and electrical power.

This change in attitude about it being a good idea to tap "safe" nuclear power to provide our increasing need for more energy, along with the benefit of using a clean energy source that leaves no carbon footprint and does not contribute to pollution and climate change, has initiated the change in attitude in using nuclear power. This change in public attitude has increased interest in small modular reactors (SMRs). SMRs offer a lower initial capital investment (costs), greater scalability (i.e., capacity to be changed in scale or size), and location flexibility for sites unable to accommodate more traditional larger reactors. Moreover, they have the potential for enhanced safety and security compared to earlier designs. These can provide both process heat for industry and electricity and off-grid power in developing countries.

12.3 THE BOTTOM LINE

SMRs are good economic tools to help drive trade and industry growth. Additionally, due to their reduced size, there would be less radioactive material in the core, and passive safety features make their operations much safer than earlier designs. They are also 100 times smaller than a solar installation with the same output and 1,000 times smaller than a wind farm, making their environmental footprint small (Rao & Gould, 2023). Moreover, SMRs are an attractive means of producing power in remote locations, and they can serve as a backup energy supplier for numerous operations. Additionally, using SMRs is a critical power station for AI infrastructure.

REFERENCES

Hauptman, M., 2024. Microsoft announced a plan to reopen the Three Mile Island nuclear power plant to support AI. *USA Today*, September 20, 2024.

Rao, V., and Gould, C., 2023. *Carbon-Free Power: Role of Small modular (Nuclear) Reactors*. Research Triangle Location, NC: RTI International.

13 Transformative Energy

13.1 INTRODUCTION

This book about the science of energy cannot overlook the primary energy or power source used in most locations throughout the globe: electricity. Electricity is produced in Nature via lightning. At present, we have not determined how to harness lightning-produced electricity or how to store it. Keep in mind that storing energy is no easy task. Then, there is the naturally produced static electricity. Static electricity is produced when an imbalance between positive and negative charges within a material occurs when electrons (the negatively charged particles in an atom) move from one material to another. Electricity is the flow of electrons in a conductor, usually copper or aluminum, but in electronics, conductors are often silver or gold. Again, it isn't easy to tap static electricity for practical power use, and although it can be stored in certain materials, it isn't easy to use. For practical electrical energy use, it is essential to point out that humans have developed electromagnetic generation, solar generation, and wind generation, which points out an essential characteristic of the electricity we use. Electricity is a transformative source of energy and power, meaning it takes other forms of energy, mainly mechanical generators, wind power, or solar energy, to produce electricity.

13.2 PRINCIPLES OF BASIC ELECTRICITY[1]

People living and working in modern societies generally have little difficulty in recognizing electrical equipment because it is everywhere and (if one pays attention to his or her surroundings) is easy to spot. Despite its great importance in our daily lives, few of us probably stop to think about what life would be like without electricity. Like air and water, we tend to take electricity for granted. But we use electricity for many daily jobs—from lighting, heating, and cooling our homes to powering our televisions and computers. Then there is the workplace. Can we do work without electricity? For example, the typical industrial workplace is outfitted with electrical equipment that:

- generates electricity (a generator or emergency generator);
- stores electricity (batteries);
- changes electricity from one level (voltage or current) to another (transformers);
- transports or transmits and distributes electricity throughout the plant site (wiring distribution systems);
- measures electricity (meters);
- converts electricity into other forms of energy (rotating shafts—mechanical energy, heat energy, light energy, chemical energy, or radio energy);
- protects other electrical equipment (fuses, circuit breakers, or relays);

DOI: 10.1201/9781003605430-15

- operates and controls other electrical equipment (motor controllers);
- converts some condition or occurrence into an electric signal (sensors); and
- converts some measured variable to a representative electrical signal (transducers or transmitters).

13.2.1 Nature of Electricity

The word *electricity* is derived from the Greek word "electron" (meaning AMBER). Amber is a translucent (semitransparent) yellowish fossilized mineral resin. The ancient Greeks used "electric force" to refer to the mysterious forces of attraction and repulsion exhibited by amber when rubbed with cloth. They did not understand the nature of this force. They could not answer the question, "What is electricity?" This question remains unanswered. Today, we often attempt to answer this question by describing the effect, not the force. The standard answer given in physics is that electricity is "the force that moves electrons," which is about the same as defining wind in a sail as "that force which moves a sailboat."

At present, little more is known than the ancient Greeks knew about the fundamental nature of electricity, but we have made tremendous strides in harnessing and using it. As with many other unknown (or unexplainable) phenomena, elaborate theories concerning the nature and behavior of electricity have been advanced and have gained wide acceptance because of their apparent truth—and because they work.

Scientists have determined that electricity behaves constantly and predictably in given situations or when subjected to given conditions. Faraday, Ohm, Lenz, and Kirchoff have described the predictable characteristics of electricity and electric current in specific rules, often called laws. Thus, though electricity has never been clearly defined, its predictable nature and easily used energy form have made it one of the most widely used power sources in modern times.

The bottom line: You can "learn" about electricity by learning the rules or laws that govern its behavior and by understanding the methods of producing, controlling, storing, and using it. Thus, this learning about electricity can be accomplished without ever having determined its fundamental identity.

You are probably scratching your head, puzzled.

I understand the main question running through your brain now: "This is a section in the text about the physics of electricity, and the author can't even explain what electricity is?"

That is correct; we cannot. The point is that no one can definitively define electricity. Electricity is one of those subject areas where the old saying "we don't know what we don't know about it" fits perfectly.

Again, a few theories about electricity have so far stood the test of extensive analysis and much time (relatively speaking, of course). One of the oldest and most generally accepted theories concerning electric current flow (or electricity) is the *electron theory.*

The electron theory states that electricity or current flow results from the flow of free electrons in a conductor. Thus, electricity is the flow of free electrons or simply electron flow. In this section, this is how we define electricity: electricity is the flow of free electrons.

Electrons are extremely tiny particles of matter. To understand electrons and precisely what is meant by "electron flow," we must examine a discussion about the structure of matter.

13.2.2 The Structure of Matter

Matter is anything that has mass and occupies space. To study any matter's fundamental structure or composition, it must be reduced to its essential components. All matter is made of *molecules*, or combinations of *atoms* (Greek: not able to be divided), bound together to produce a given substance, such as salt, glass, or water. For example, if you keep dividing water into smaller and smaller drops, you will eventually arrive at the smallest particle that is still water. That particle is the molecule, which is defined as the *slightest bit of a substance that retains the characteristics of that substance.*

A molecule of water (H_2O) comprises one atom of oxygen and two atoms of hydrogen. If the water molecules were further subdivided, only unrelated oxygen and hydrogen atoms would remain, and the water would no longer exist. Thus, the molecule is the smallest particle to which a substance can be reduced and still be called by the same name. This applies to all substances—solids, liquids, and gases.

Important Point: Molecules are made up of atoms, which are bound together to produce a given substance.

In various combinations, atoms comprise subatomic particles of *electrons, protons,* and *neutrons*. These particles differ in weight (a proton is much heavier than the electron) and charge. We are not concerned with the weights of particles in this section, but the *charge* is significant in electricity. The electron is the fundamental negative charge (–) of electricity. Electrons revolve about the nucleus or center of the atom in paths of concentric *orbits* or shells. The proton is the fundamental positive (+) charge of electricity. Protons are found in the nucleus. The number of protons within an atom's nucleus specifies that atom's atomic number. For example, the helium atom has 2 protons in its nucleus, so the atomic number is 2. The neutron, the fundamental neutral charge of electricity, is also found in the nucleus.

Most of the atom's weight is in the nucleus's protons and neutrons. Whirling around the nucleus are one or more negatively charged electrons. Usually, there is one proton for each electron in the entire atom so that the net positive charge of the nucleus is balanced by the net negative charge of the electrons rotating around the nucleus (see Figure 13.1).

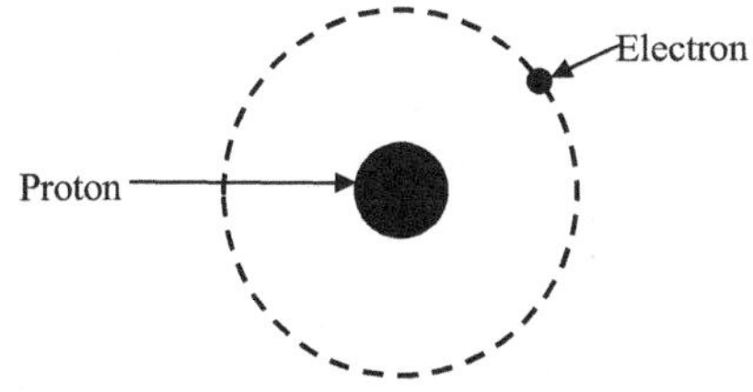

FIGURE 13.1 One proton and one electron = electrically neutral.

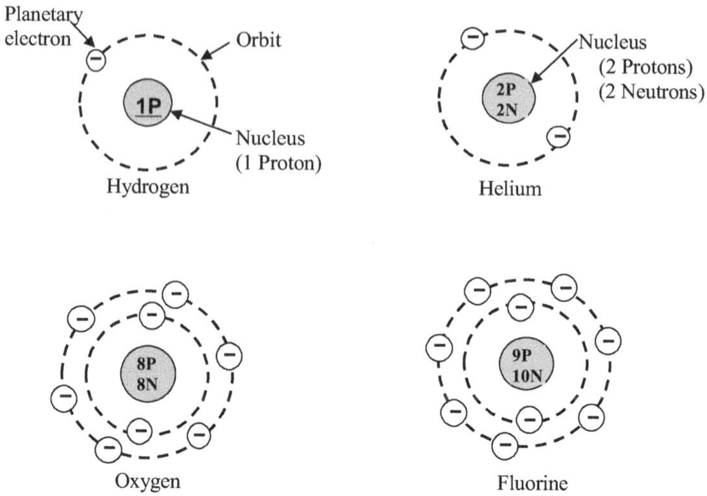

FIGURE 13.2 Atomic structure of sample elements.

Important Point: Most batteries are marked with + and – or even with the abbreviations POS (positive) and NEG (negative). The concept of a positive or negative polarity and its importance in electricity will become clear later. However, for the moment, you need to remember that an electron has a negative charge and that a proton has a positive charge.

Again, the number of protons is usually the same as the number of electrons in an atom. This is an important point because this relationship determines the kind of element (the atom is the smallest particle that makes up an element; an element retains its characteristics when subdivided into atoms) in question. Figure 13.2 presents a simplified illustration of several atoms from different materials, depicting electrons orbiting around the nucleus. For example, hydrogen has a nucleus consisting of 1 proton, around which rotates 1 electron. The helium atom has a nucleus containing 2 protons and 2 neutrons, with 2 electrons encircling the nucleus. These elements are electrically neutral (or balanced) because each has an equal number of electrons and protons. Since each electron's negative (–) charge is equal in magnitude to each proton's positive (+) charge, the two opposite charges cancel.

A balanced (neutral or stable) atom has a certain amount of energy equal to the sum of the energy of its electrons. Electrons, in turn, have different energies called *energy levels.* The energy level of an electron is proportional to its distance from the nucleus. Therefore, the energy levels of electrons in shells are farther from the nucleus than those in shells nearer the nucleus.

When an electric force is applied to a conducting medium, such as a copper wire, electrons in the outer orbits of the copper atoms are forced out of orbit (i.e., liberating or freeing electrons) and impelled along the wire. This electrical force, which forces electrons out of orbit, can be produced in several ways, such as by moving a conductor through a magnetic field, by friction, when a glass rod is rubbed with cloth (silk), or by chemical action, as in a battery.

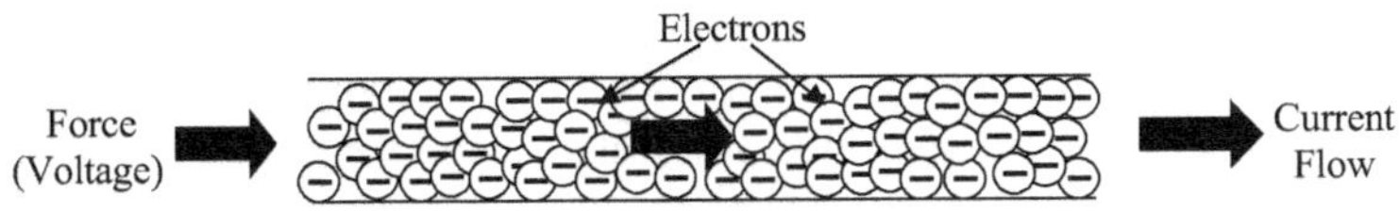

FIGURE 13.3 Electrons flow in a copper wire.

When the electrons are forced from their orbits, they are called *free electrons.* Some of the electrons of certain metallic atoms are so loosely bound to the nucleus that they are relatively free to move from atom to atom. These free electrons constitute the flow of an electric current in electrical conductors.

Important Point: When an electric force is applied to a copper wire, free electrons are displaced from the copper atoms and move along the wire, producing electric current, as shown in Figure 13.3.

If the internal energy of an atom is raised above its normal state, the atom is said to be *excited.* Excitation may be produced by causing the atoms to collide with particles propelled by an electric force, as shown in Figure 13.3. In effect, energy is transferred from the electric source to the atom. The excess energy an atom absorbs may become sufficient to cause loosely bound outer electrons (as shown in Figure 13.3) to leave the atom against the force that acts to hold them within.

Important Point: An atom that has lost or gained one or more electrons is said to be *ionized.* If the atom loses electrons, it becomes positively charged and is referred to as a *positive ion.* Conversely, if the atom gains electrons, it becomes negatively charged and is referred to as a *negative ion.*

13.2.3 Conductors, Semiconductors, and Insulators

Electric current moves quickly through some materials but with incredible difficulty through others. Substances that permit the free movement of many electrons are called conductors. Copper is the most widely used electrical conductor because of its high conductivity (how good a conductor the material is) and cost-effectiveness.

Electrical energy is transferred through a copper or other metal conductor by moving free electrons that migrate from atom to atom inside the conductor. Each electron moves a very short distance to the neighboring atom, which replaces one or more electrons by forcing them out of their orbits. The replaced electrons repeat the process in other nearby atoms until the movement is transmitted throughout the entire length of the conductor. A good conductor is said to have a low opposition, or *resistance,* to the electron (current) flow.

Note: If many electrons flow through a material with only a small force (voltage) applied, we call that material a **conductor**. A convenient way to understand the purpose and function of an electrical conductor is to think of or visualize a pipe or conveyor of water or some other liquid and gaseous substance. As a liquid or gas is conveyed or conducted through a pipe, an electron or electricity is transferred through a conductor. Regarding force applied, it is a pump that pushes the water and voltage that pushes the electrons. It is important to remember that in an electrical circuit, voltage does not move; instead, it is the force that makes electrons move.

Table 13.1 lists many of the metals commonly used as electric conductors. The best conductors appear at the top of the list, with the poorer ones shown last.

Note: The movement of each electron (e.g., in copper wire) takes a minimal amount of time and is almost instantaneous. This is a crucial point to remember when events in an electrical circuit seem to co-occur.

While it is true that electron motion is known to exist to some extent in all matters, some substances, such as rubber, glass, and dry wood, have very few free electrons. In these materials, large amounts of energy must be expended to break the electrons loose from the influence of the nucleus. Substances containing very few free electrons are called *insulators*. Insulators are essential in electrical work because they prevent the current from being diverted from the wires, which helps prevent electrocutions and fires.

Note: If the voltage is large enough, even the best insulators—rubber, plastic, and wood—will break down and allow their electrons to flow.

Table 13.2 lists some materials we often use as insulators in electrical circuits. The list is in a top-to-bottom sequence to show their decreasing ability to withstand high voltages without conducting.

TABLE 13.1
Electrical Conductors

Silver
Copper
Gold
Aluminum
Zinc
Brass
Iron
Tin
Mercury

TABLE 13.2
Materials that Withstand Voltage Without Conducting

Common Insulators
Rubber
Mica
Wax or paraffin
Porcelain
Bakelite
Plastics
Glass
Fiberglass
Dry wood
Air

A material that is neither a good conductor nor a good insulator is called a *semiconductor.* Silicon and germanium are substances that fall into this category. Because of their peculiar crystalline structure, these materials may, under certain conditions, act as conductors and, under other conditions, as insulators. As the temperature rises, however, a limited number of electrons become available for conduction.

13.2.4 Static Electricity

Electricity at rest is often referred to as *static electricity.* More specifically, when two bodies of matter have unequal charges and are near one another, an electric force is exerted between them because of their unequal charges. However, since they are not in contact, their charges cannot equalize. The existence of such an electric force, where current can't flow, is *static electricity.*

However, static electricity at rest will flow if given the opportunity—this is the case because static electricity is an imbalance of negative and positive charges. An example of this phenomenon is often experienced when one walks across a dry carpet and then touches a doorknob—a slight shock is usually felt, and a spark at the fingertips is likely noticed. Another familiar example is "static cling." For example, whenever we rub an air-filled balloon against the hair on our heads and then place the balloon against a wall, the balloon sticks to the wall, defying gravity because of static cling. In the workplace, static electricity is prevented from being built up by properly bonding or grounding equipment to the ground or earth.

13.2.5 Charged Bodies

The fundamental law of charged bodies states: "Like charges repel each other, and unlike charges attract each other." Positive and negative charges, opposite or unlike, tend to move toward each other—attracting each other. In contrast, like bodies tend to repel each other. Electrons repel each other because of their negative charges (like charges repel), and protons repel each other because of their positive charges. Figure 13.4 demonstrates the law of charged bodies.

It is essential to point out another significant part of the fundamental law of charged bodies—the *force of attraction or repulsion between two magnetic poles decreases rapidly as the poles are separated.* More specifically, the force of attraction

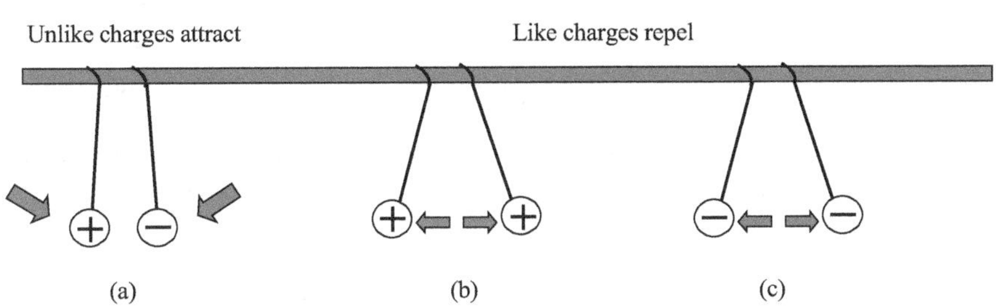

FIGURE 13.4 Reaction between two charged bodies. The opposite charge in (a) attracts. The like charges in (b) and (c) repel each other.

or repulsion varies directly as the product of the separate pole strengths and inversely as the square of the distance separating the magnetic poles, provided the poles are small enough to be considered points.

Let's look at an example:

If you increase the distance between the magnet's two north poles from 2 ft to 4 ft, the force of repulsion between them decreases to one-fourth of its original value. If either pole strength is doubled while the distance remains the same, the force between the poles will be doubled.

13.2.6 Coulomb's Law

Coulomb's Law states that the amount of attracting or repelling force that acts between two electrically charged bodies in free space depends on two things:

a. Their charges
b. The distance between them

Specifically, Coulomb's Law states: "Charged bodies attract or repel each other with a force that is directly proportional to the product of their charges and is inversely proportional to the square of the distance between them."

Note: The number of electrons determines the magnitude of a body's electric charge compared to the number of protons within the body. The symbol for the magnitude of electric charge is Q, expressed in units of *coulombs* (C). A charge of one positive coulomb means a body contains a charge of 6.25×10^{18}. A charge of one negative coulomb, $-Q$, means a body includes a charge of 6.25×10^{18} more electrons than protons.

13.2.7 Electrostatic Fields

The fundamental characteristic of an electric charge is its ability to exert force. The space between and around charged bodies in which their influence is felt is called an *electric field of force*. The electric field is terminated on material objects and extends between positive and negative charges. This region of force can consist of air, glass, paper, or a vacuum. This region of force is referred to as an *electrostatic field*.

When two objects of opposite polarity are brought near each other, the electrostatic field is concentrated in the area between them. The field is generally represented by lines referred to as *electrostatic lines of force*. These imaginary lines are used merely to describe the direction and strength of the field. To avoid confusion, the positive lines of force are always shown as leaving charge; for a negative charge, they are shown as entering. Figure 13.5 illustrates using lines to represent the field of charged bodies.

Note: A charged object will retain its charge temporarily if there is no immediate transfer of electrons to or from it. In this condition, the charge is said to be *at rest*. Remember, electricity at rest is called *static* electricity.

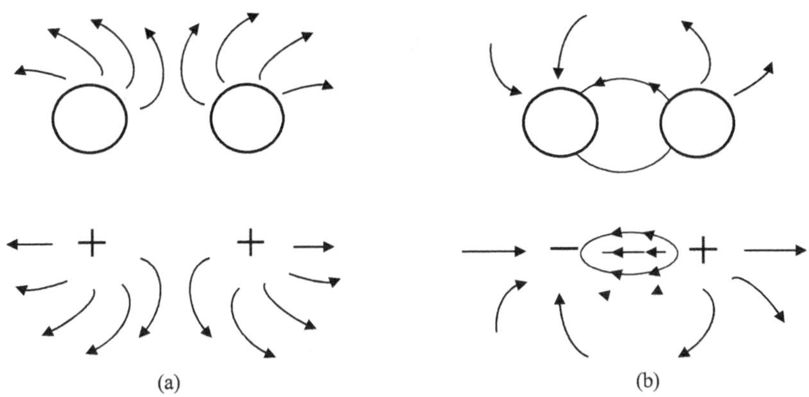

FIGURE 13.5 Electrostatic lines of force. (a) The repulsion of like-charged bodies and their associated fields. (b) The attraction between unlike-charged bodies and their related fields.

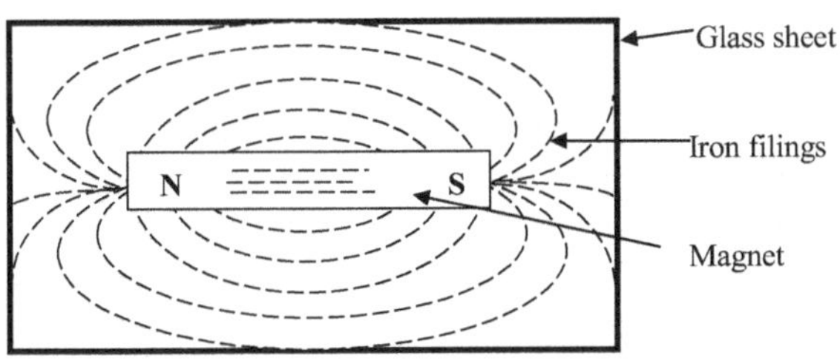

FIGURE 13.6 Magnetic field around a bar magnet. If the glass sheet is tapped gently, the filings will move into a definite pattern that describes the field of force around the magnet.

13.2.8 Magnetism

Most electrical equipment depends directly or indirectly upon magnetism. Magnetism is a phenomenon associated with magnetic fields; that is, it can attract substances such as iron, steel, nickel, or cobalt (metals known as magnetic materials). Correspondingly, a substance is said to be a magnet if it has the property of magnetism. For example, a piece of iron can be magnetized and is thus a magnet.

When magnetized, the piece of iron (note: we will assume a piece of flat bar 6 inches long × 1-inch-wide × 0.5 inches thick; a bar magnet—see Figure 13.6) will have two points opposite each other, which most readily attract other iron pieces. The points of maximum attraction (one on each end) are called the *magnetic poles* of the magnet: the north (N) pole and the south (S) pole. Just as electric charges repel each other and opposite charges attract each other, magnetic poles repel each other, and unlike poles attract each other. Although invisible to the naked eye, its force can be shown to exist by sprinkling small iron filings on a glass covering a bar magnet, as shown in Figure 13.6.

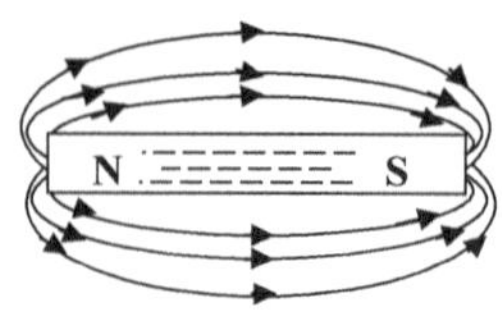

FIGURE 13.7 Magnetic field of force around a bar magnet, indicated by lines of force.

Figure 13.7 shows how the field looks without iron filings. It is shown as lines of force [known as magnetic flux or flux lines; the symbol for magnetic flux is the Greek lowercase letter (phi)] in the field, repelled away from the magnet's north pole and attracted to its south pole.

Note: A *magnetic circuit* is a complete path through which magnetic lines of force may be established under the influence of a magnetizing force. Most magnetic circuits contain magnetic materials to contain the magnetic flux. These circuits are like the *electric circuit* (important point), which is a complete path through which current is caused to flow under the influence of an electromotive force.

There are three types or groups of magnets:

a. **Natural magnets**––These are found in the natural state as a mineral (an iron compound) called magnetite.
b. **Permanent magnets**––These are artificial magnets made of hardened steel or some alloy such as Alinco bars that have been permanently magnetized. Most people are familiar with the permanent horseshoe magnet; this red U-shaped magnet is the universal symbol of magnets, recognized worldwide.
c. **Electromagnets**––These are artificial magnets composed of soft-iron cores around wound coils of insulated wire. When an electric current flows through the coil, the core becomes magnetized. When the current ceases to flow, the core loses most of the magnetism.

13.2.9 Magnetic Materials

Natural magnets are no longer used (they have no practical value) in electrical circuitry because more powerful and conveniently shaped permanent magnets can be produced artificially. Commercial magnets are made from special steels and alloys, which are magnetic materials.

Magnetic materials are attracted or repelled by a magnet and can be magnetized. Iron, steel, and alloy bars are the most common magnetic materials. These materials can be magnetized by inserting the material (in bar form) into a coil of insulated wire and passing a heavy direct current through the coil. The same material may also be magnetized if stroked with a bar magnet. It will then have the same magnetic property as the magnet used to induce the magnetism—namely, there will be two poles of attraction, one at either end. This process produces a permanent magnet by induction—that is, the magnetism is induced in the bar by the influence of the stroking magnet.

Note: Permanent magnets are hard magnetic materials (hard steel or alloys) that retain their magnetism when the magnetizing field is removed. A temporary magnet cannot retain a magnetized state when the magnetizing field is removed.

Even though they are classified as permanent magnets, it is essential to point out that hardened steel and certain alloys are relatively tricky to magnetize and are said to have low permeability because the magnetic lines of force do not easily permeate or distribute themselves readily through the steel.

Note: *Permeability* refers to the ability of a magnetic material to concentrate magnetic flux. Any material that is easily magnetized has high permeability. A measure of permeability for different materials compared to air or vacuum is called *relative* permeability, symbolized by μ or (mu).

Once hard steel and other alloys are magnetized, they retain a large part of their magnetic strength and are called *permanent magnets.* Conversely, relatively easy-to-magnetize materials—such as soft iron and annealed silicon steel—are said to have a high permeability. Such materials retain only a tiny part of their magnetism after removing the magnetizing force and are called *temporary magnets.*

The magnetism that remains in a temporary magnet after the magnetizing force is removed is called *residual magnetism.*

Early magnetic studies classified magnetic materials merely as magnetic and nonmagnetic based on the strong magnetic properties of iron. However, since weak magnetic materials can be necessary in some applications, present studies classify materials into one of three groups: paramagnetic, diamagnetic, and ferromagnetic.

a. **Paramagnetic materials**––These include aluminum, platinum, manganese, and chromium—materials that become only slightly magnetized even under the influence of a strong magnetic field. This slight magnetization is in the same direction as the magnetizing field. Relative permeability is somewhat more than 1 (i.e., considered nonmagnetic materials).
b. **Diamagnetic materials**––These include bismuth, antimony, copper, zinc, mercury, gold, and silver—materials that can also be slightly magnetized when under the influence of a powerful field. Relative permeability is less than 1 (i.e., considered nonmagnetic materials).
c. **Ferromagnetic materials**––These include iron, steel, nickel, cobalt, and commercial alloys, the most important group for applications in electricity and electronics. Ferromagnetic materials are easy to magnetize and have high permeability, ranging from 50 to 3,000.

13.2.10 Magnetic Earth

The Earth is a giant magnet, and its magnetism produces the magnetic field surrounding it. Most people would have no problem understanding or at least accepting this statement. However, if told that the Earth's north magnetic pole is its south magnetic pole and that the south magnetic pole is the Earth's north magnetic pole, they might not accept or understand this statement. But, in terms of magnets, it is true.

Figure 13.8 indicates the Earth's magnetic polarities. The geographic poles are also shown at each end of the earth's rotation axis. The magnetic axis does not

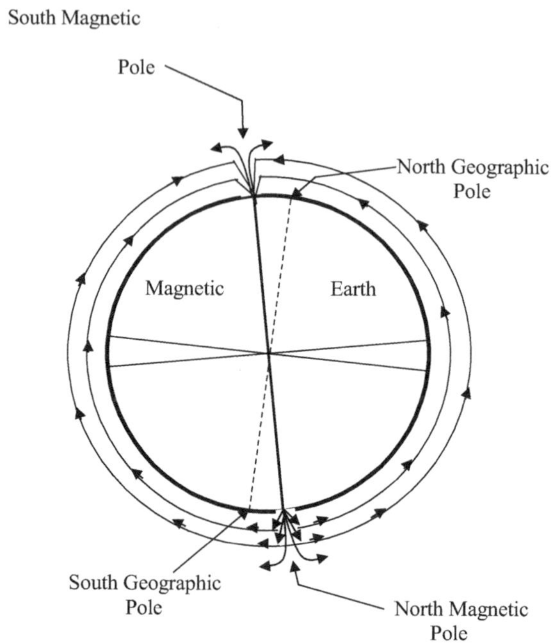

FIGURE 13.8 Earth's magnetic poles.

coincide with the geographic axis; therefore, the magnetic and geographic poles are not at the same place on the Earth's surface.

Recall that magnetic lines of force are assumed to emanate from the magnet's north pole and enter the south pole as closed loops. Because the earth is a magnet, lines of force emanate from its north magnetic pole and enter the south magnetic pole as closed loops. A compass needle aligns itself so that the Earth's lines of force enter at its south pole and leave at its north pole. Because the needle's north pole is defined as the end that points in a northerly direction, it follows that the magnetic pole near the north geographic pole is a south magnetic pole, and vice versa.

13.2.11 Difference in Potential

Because of the force of its electrostatic field, an electric charge can move another charge by attraction or repulsion. The force that causes free electrons to move in a conductor as an electric current may be referred to as follows:

- Electromotive force (emf)
- Voltage
- Difference in potential

When a difference in potential exists between two charged bodies connected by a wire (conductor), electrons (current) flow along the conductor. This flows from the negatively charged body to the positively charged body until the two charges are equalized and the potential difference no longer exists.

Note: The basic unit of potential difference is the *volt* (V). The symbol for the potential difference is V, indicating the ability to force electrons (current flow) to move. Because the volt unit is used, the potential difference is called *voltage.*

13.2.12 Water Analogy

In attempting to train individuals in the concepts of basic electricity, especially regarding the difference of potential (voltage), current, and resistance relationships in a simple electrical circuit, it has been common practice to use what is referred to as the water analogy. We use the water analogy later to explain (in a simple, straightforward fashion) voltage, current, and resistance and their relationships in more detail. Still, for now, we use analogy to explain the basic concept of electricity: the difference between potential and voltage. Because a difference in potential causes current flow (against resistance), it is essential to understand this concept before the concept of current flow and resistance is explained.

Consider the water tanks shown in Figure 13.9—two connected by a pipe and valve. First, the valve is closed, and all the water is in Tank A. Thus, the water pressure across the valve is at maximum. When the valve is opened, the water flows through the pipe from A to B until the water level becomes the same in both tanks. The water then stops flowing in the pipe because there is no longer a difference in water pressure (difference in potential) between the two tanks.

Just as the flow of water through the pipe in Figure 13.9 is directly proportional to the difference in water level in the two tanks, current flow through an electric circuit is directly proportional to the difference in potential across the circuit.

Important Point: A fundamental law of current electricity states that the current is directly proportional to the applied voltage; if the voltage increases, the current increases. If the voltage decreases, the current decreases.

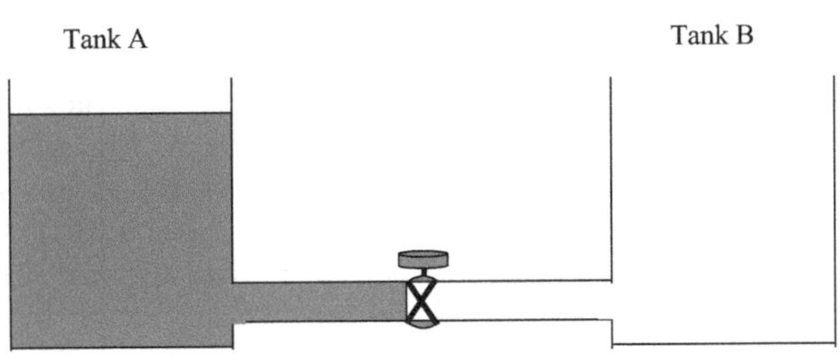

FIGURE 13.9 Water analogy of electric difference of potential

13.2.13 Principal Methods of Producing a Voltage

There are many ways to produce electromotive force or voltage, some more widely used than others. The following list the seven most common methods (USDOE, 1992):

a. **Friction**––The voltage produced by rubbing two materials together (static electricity or electrostatic force). Remember our discussion of static electricity? Let's refresh our memories. For example, have you ever walked across a carpet and felt shocked when you touched a metal doorknob? Your shoe soles built up a charge by rubbing on the carpet, which was transferred to your body. Your body became positively charged, and when you touched the zero-charged doorknob, electrons were transferred to your body until you and the doorknob had equal charges.
b. **Pressure (piezoelectricity)**—The voltage produced by squeezing or applying pressure to crystals of certain substances (e.g., crystals like quartz or Rochelle salts or certain ceramics like barium titanate). When pressure is applied to such substances, electrons can be driven out of orbit in the direction of the force. Electrons leave one side of the material and accumulate on the other, building up positive and negative charges on opposite sides. When the pressure is released, the electrons return to their orbits. Some materials react to bending pressure, while others respond to twisting pressure. This voltage generation is known as the *piezoelectric effect* (*using* Rochelle salts to generate electricity is called pyroelectricity). If external wires are connected while pressure and voltage are present, electrons will flow and current will be produced. If the pressure is constant, the current will flow until the potential difference is equalized. When the force is removed, the material is decompressed and immediately causes an electric force in the opposite direction. The power capacity of these materials is minimal. However, these materials are handy because of their extreme sensitivity to changes in mechanical force. One example is the crystal phonograph cartridge containing a Rochelle salt crystal (potassium sodium tartrate tetrahydrate). A phonograph needle is attached to the crystal. As the needle moves in the grooves of a record, it swings from side to side, applying compression and decompression to the crystal. This mechanical motion applied to the crystal generates a voltage signal to reproduce sound.
c. **Heat (thermoelectricity)**—The voltage produced by heating the joint (junction) where two unlike metals are joined. Some materials readily give up their electrons, and others readily accept electrons. For example, electrons can transfer when two dissimilar metals, like copper and zinc, join. Electrons will leave the copper atoms and enter the zinc atoms. The zinc gets a surplus of electrons and becomes negatively charged. The copper loses electrons and takes on a positive charge. This creates a voltage potential across the junction of the two metals. The heat energy of normal room temperature is enough to make them release and gain electrons, causing a measurable voltage potential. As more heat energy is applied to the junction, more electrons are released and the voltage potential becomes greater.

When heat is removed, the junction cools, the charges dissipate, and the voltage potential decreases. This process is called thermoelectricity. A device like this is generally referred to as a thermocouple.

The thermocouple voltage depends upon the heat energy applied to the junction of the two dissimilar metals. Thermocouples are widely used to measure temperature and as heat-sensing devices in automatic temperature-controlled equipment.

Thermocouple power capacities are minimal compared to other sources but are more significant than those of crystals. A thermocouple can be subjected to higher temperatures than ordinary mercury or alcohol thermometers.

d. **Light (photoelectricity)**—The voltage produced by light (photons) striking photosensitive (light sensitive) substances. When the photons in a light beam hit the surface of a material, they release their energy and transfer it to the atomic electrons of the material. This energy transfer may dislodge electrons from their orbits around the surface of the substance. The photosensitive (light-sensitive) material becomes positively charged upon losing electrons, creating an electric force.

   This phenomenon is called the photoelectric effect that has wide applications in electronics, such as photoelectric cells, photovoltaic cells, optical couplers, and television camera tubes. Three uses of the photoelectric effect are described below:

   - **Photovoltaic**--Light energy in two plates joined together causes one plate to release electrons to the other, building up opposite charges, like a battery.
   - **Photoemission**--The photon energy from a light beam could cause a surface to release electrons in a vacuum tube. A plate would then collect the electrons.
   - **Photoconduction**--Light energy applied to some typically poor materials produces free electrons, making the materials better conductors.

e. **Chemical action**—The voltage produced by a chemical reaction in a battery cell. For example, the voltaic chemical cell wherein a chemical reaction produces and maintains opposite charges on two dissimilar metals that serve as positive and negative terminals. The metals are in contact with an electrolyte solution. Connections between more than one of these cells will produce a battery.

f. **Magnetism**—The voltage produced in a conductor when the conductor moves through a magnetic field, or a magnetic field moves through the conductor in such a manner as to cut the magnetic lines of force of the field. A generator is a machine that converts mechanical energy into electrical energy using the magnetic induction principle. This is one of the most useful and widely employed applications for producing vast electric power.

g. **Thermionic emission**—A thermionic energy converter consists of two electrodes placed near one another in a vacuum. One electrode is typically called the cathode or emitter, and the other is called the anode or plate. Ordinarily, electrons in the cathode are prevented from escaping from the

> surface by a potential energy barrier. When an electron starts to move from the surface, it induces a corresponding positive charge in the material, which tends to pull it back to the surface. To escape the electron, one must somehow acquire enough energy to overcome this energy barrier. Almost none of the electrons can gain enough energy to escape at ordinary temperatures. However, thermal motion dramatically increases the electron energy when the cathode is hot. At sufficiently high temperatures, a considerable number of electrons can escape. The liberation of electrons from a hot surface is called *thermionic emission.*
>
> The electrons that have escaped from the hot cathode form a cloud of negative charges called a space charge near it. If a battery maintains the plate positive for the cathode, the electrons in the cloud are attracted to it. If the potential difference between the electrodes is maintained, there will be a steady current flow from the cathode to the plate.
>
> The simplest example of a thermionic device is a vacuum tube diode, in which the only electrodes are the cathode and plate, or anode. The diode can convert alternating current (AC) flow to pulsating direct current (DC) flow.

In the study of the basic electricity related to renewable energy production, we are most concerned with magnetism (generators powered by hydropower, for example), light (photoelectricity produced by solar cells), and chemistry (chemical energy converted to electricity in batteries) to produce voltage. Friction has little practical applications, though we discussed it earlier in studying static electricity. Pressure and heat have valuable applications, but we do not need to consider them in this section. Magnetism used in generators, solar light-produced electricity, and the chemistry involved with storing electricity in batteries, on the other hand, are, as mentioned, the principal sources of voltage and are discussed at length in this section.

13.2.14 Electric Current

The movement or flow of electrons is called current. To produce current, the electrons must be moved by a potential difference or pressure (voltage).

Note: The terms current, current flow, electron flow, or electron current, etc., may be used to describe the same phenomenon.

For our purposes in this section, electron flow, or current, in an electric circuit moves from a region of less negative potential to an area of more positive potential—from negative to positive.

Note: Current is represented by the letter I. Its basic unit of measurement is the *ampere, or amp* (A). One ampere of current is defined as the movement of one coulomb past a conductor's point during 1 second.

Recall that we used the water analogy to help us understand the potential differences. We can also use the water analogy to help us understand current flow through a simple electric circuit. Consider Figure 13.10, which shows a water tank connected via a pipe to a pump with a discharge pipe. If the water tank contains an amount of water above the level of the pipe opening to the pump, the water exerts pressure (a difference in potential) against the pump. When sufficient water is available for

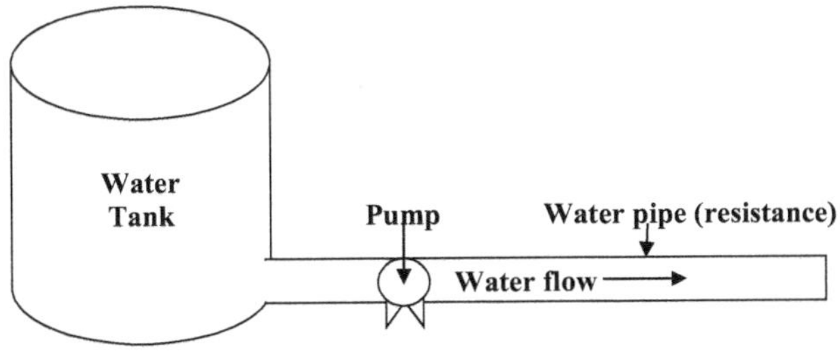

FIGURE 13.10 Water analogy: current flow

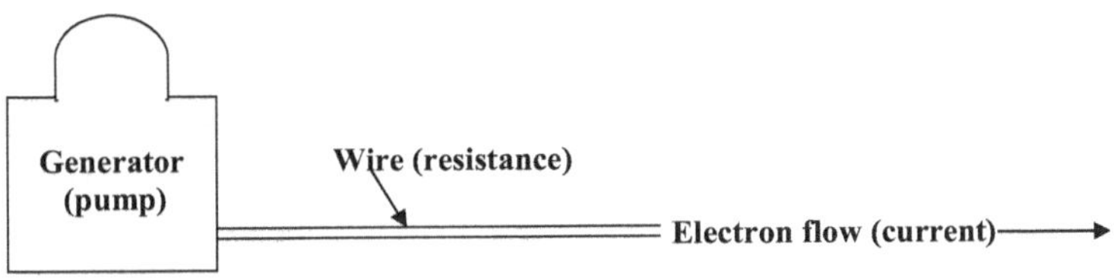

FIGURE 13.11 Simple electric circuit with current flow.

pumping with the pump, water flows through the pipe against the resistance of the pump and pipe. The analogy should be clear with an electric circuit. If a difference of potential exists, current will flow in the circuit.

Another simple way of looking at this analogy is to consider Figure 13.11, where the water tank has been replaced with a generator, the pipe with a conductor (wire), and water flow with the flow of electric current.

Again, Figures 13.10 and 13.11 illustrate that the electrons must be moved by a potential difference to produce current.

Electric current is generally classified into two general types:

- direct current (D-C)
- alternating current (A-C)

Direct current (D-C) only moves through a conductor or circuit in one direction. *Alternating current* (A-C) periodically reverses direction.

13.2.15 Resistance

Earlier, it was pointed out that free electrons, or electric current, could move quickly through a good conductor, such as copper, but an insulator, such as glass, was an obstacle to current flow. In the water analogy shown in Figure 13.10 and the simple electric circuit shown in Figure 13.11, resistance is indicated by either the pipe or the conductor.

Every material offers some resistance to the flow of electric current. Good conductors like copper, silver, and aluminum provide little resistance. Poor conductors or insulators like glass, wood, and paper resist current flow.

Note: The current flow in each circuit depends on voltage and resistance.

Note: The letter R represents resistance. The basic unit in which resistance is measured is the *ohm* (Ω). One ohm is the resistance of a circuit element or circuit that permits a steady current of 1 ampere (1 coulomb per second) to flow when a constant electromotive force (emf) of 1 V is applied to the circuit. Manufactured circuit parts containing definite amounts of resistance are called **resistors**.

The size and type of wires in an electric circuit are chosen to keep the electrical resistance as low as possible. In this way, the current can flow easily through the conductors, as water flows through the pipe between the tanks in Figure 13.10. If the water pressure remains constant, the water flow in the pipe will depend on how far the valve is open. The smaller the opening, the greater the opposition (resistance) to the flow; the smaller the flow rate in gallons per second.

In the electric circuit shown in Figure 13.11, the larger the diameter of the wire, the lower its electrical resistance (opposition) to the flow of current through it. In the water analogy, pipe friction opposes the water flow between the tanks. This friction is like electrical resistance. The resistance of the pipe to the flow of water through it depends upon (1) the length of the pipe, (2) the diameter of the pipe, and (3) the nature of the inside walls (rough or smooth). Similarly, the electrical resistance of the conductors depends upon (1) the length of the wires, (2) the diameter of the wires, and (3) the material of the wires (copper, silver, etc.).

It is important to note that temperature also affects the resistance of electrical conductors to some extent. The resistance increases with temperature in most conductors (copper, aluminum, etc.). Carbon is an exception. In carbon, the resistance decreases as temperature increases.

Important Note: Electricity is a study frequently explained in terms of opposites. The term that is precisely the opposite of resistance is *conductance.* Conductance (G) is the ability of a material to pass electrons. The SI-derived unit of conductance is the Siemens. The commonly used unit of conductance is the *Mho,* which is ohm spelled backward. The relationship between resistance and conductance is reciprocal; a number is obtained by dividing the number into one. If the resistance of a material is known, dividing its value into one will give its conductance. Similarly, if the conductance is known, dividing its value into one will provide its resistance.

13.3 D-C CIRCUITS

An electric circuit includes an **energy source** [source of electromotive force (emf) or voltage; that is, a battery or generator], a conductor (wire), a load, and a means of control (see Figure 13.12). The energy source could be a battery, as in Figure 13.12, or some other voltage production method. The **load** that dissipates the energy could be a lamp, a resistor, or some other device (or devices) that does valuable work, such as an electric toaster, a power drill, a radio, a soldering iron, a laptop computer, or a battery pack for storage of energy as a backup to a renewable energy source such as solar or wind. Again, **conductors** are wires that offer low resistance to current; they connect all the loads in the circuit to the voltage source. No electrical device dissipates energy unless current flows through it. Because conductors, or wires, are

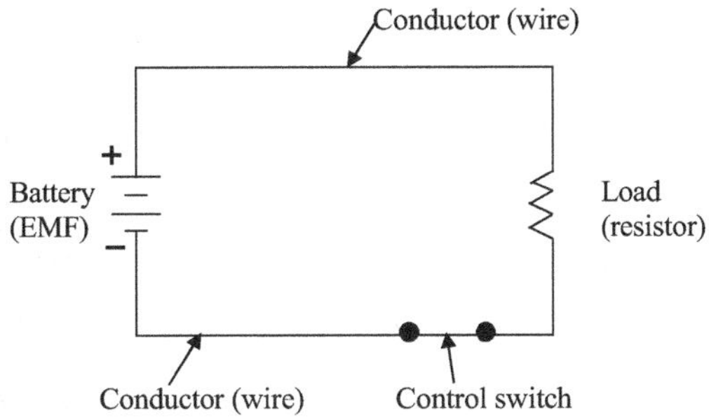

FIGURE 13.12 Simple closed circuit.

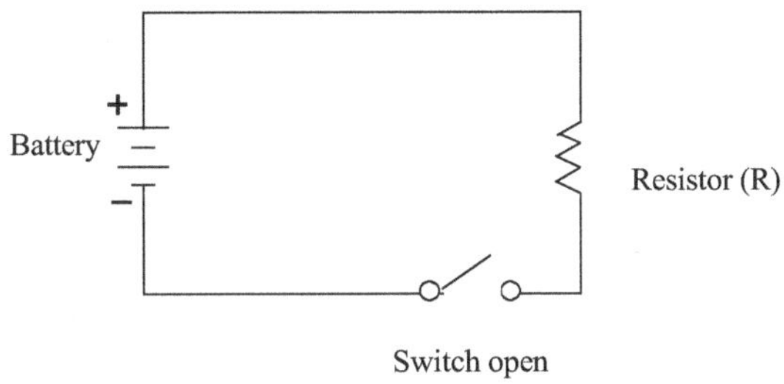

FIGURE 13.13 Open circuit.

not perfect, because of their resistance, they heat up (dissipate energy) and are part of the load. For simplicity, however, we usually think of connecting wiring as having no resistance since assigning a very low resistance value to the wires every time we wanted to solve a problem would be tedious. **Control devices** might be switches, variable resistors, circuit breakers, fuses, or relays.

A complete pathway for current flow, or **closed circuit** (Figure 13.12), is an unbroken path for current from the emf, through a load, and back to the source. A circuit is called **open** (see Figure 13.13) if a break in the circuit (e.g., open switch) does not provide a complete path for current.

Important Point: Current flows from the negative (–) terminal of the battery, as shown in Figures 13.12 and 13.13, through the load to the positive (+) battery terminal and continues by going through the battery from the positive (+) terminal to the negative (–) terminal. If this pathway is unbroken, it is a closed circuit, and current will flow. However, if the path is broken at ANY point, it is an open circuit, and no current flows (see Figure 13.13).

To protect it, a fuse is placed directly into the circuit (see Figure 13.14). A fuse will open the circuit whenever a dangerously large current starts to flow (i.e., a short circuit condition occurs caused by an accidental connection between two points in a circuit that offers very little resistance). A fuse will permit currents smaller than the fuse value to flow but will melt and break or open the circuit if a more significant current flows.

13.3.1 Schematic Representation

The simple circuits shown in Figures 13.12–13.14 are displayed schematically. A *schematic diagram* (usually shortened to "schematic") is a simplified drawing representing the **electrical**, not the physical, situation in a circuit. The symbols used in schematic diagrams are the electrician's "shorthand"; they make the diagrams easier to draw and understand. Consider the symbol in Figure 13.14 used to represent a battery power supply. The symbol is relatively straightforward—but is also very important. For example, by convention, the shorter line in the symbol for a battery represents the **negative** terminal. It is important to remember this because it is sometimes necessary to note the direction of the current flow, which is from negative to positive when you examine the schematic. The battery symbol shown in Figure 13.14 has a single cell, so only one short and one long line are used. The number of lines used to represent a battery varies (and they are not necessarily equivalent to the number of cells), but they are always in pairs, with long and short lines alternating. In the circuit shown in Figure 13.14, the current would flow counterclockwise, in the opposite direction that a clock's hands move. If the long and short lines of the battery symbol (symbol shown in Figure 13.15) were reversed, the current in the circuit shown in Figure 13.14 would flow *clockwise*, that is, in the direction of a clock's hands.

Important Note: In studies of electricity and electronics, many circuits, mainly composed of specially designed resistive components, are analyzed. As previously stated, these components are called **resistors**. The resistive element will consistently be represented by a physical resistor in all subsequent fundamental circuit analyses. However, the resistive component could be any one of several electrical devices.

Remember that in the simple circuits shown in the figures, we have only illustrated and discussed a few symbols used in schematics to represent circuit components. (Other symbols will be introduced as we need them.)

Remembering that a closed loop of wire (conductor) is not necessarily a circuit is also essential. A source of voltage must be included to make it an electric circuit. In any electric circuit where electrons move around a closed loop, current, voltage, and

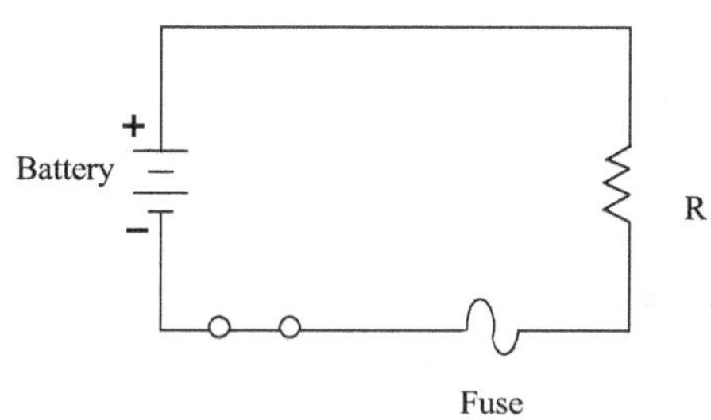

FIGURE 13.14 A simple fused circuit.

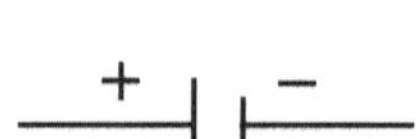

FIGURE 13.15 Schematic symbol for a battery.

resistance are present. The physical pathway for current flow is the circuit. By knowing any two of the three quantities, such as voltage and current, the third (resistance) may be determined. This is done mathematically using Ohm's Law, which is the foundation of electrical theory.

13.4 OHM'S LAW

Ohm's Law defines the relationship between electric circuits' current, voltage, and resistance. It can be expressed mathematically in three ways:

a. The *current* (I) in a circuit is equal to the voltage applied to the circuit divided by the resistance of the circuit. Stated another way, the current in a circuit is DIRECTLY proportional to the applied voltage and INVERSELY proportional to the circuit resistance. Ohm's Law may be expressed as an equation:

$$I = \frac{E}{R} \quad (13.1)$$

where
I = current in amperes (amps)
E = voltage in volts
R = resistance in ohms

b. The *resistance* (R) of a circuit is equal to the voltage applied to the circuit divided by the current in the circuit:

$$R = \frac{E}{I} \quad (13.2)$$

c. The applied *voltage* (E) to a circuit is equal to the product of the current and the resistance of the circuit:

$$E = I \times R = IR \quad (13.3)$$

The third may be easily found if any two of the quantities in equations (13.1)–(13.3) are known. Let's look at an example.

Example 13.1

Problem:

Figure 13.16 shows a circuit with a resistance of 6 Ω and a voltage source of 3 V. How much current flows in the circuit?

Given: E = 3 V
R = 6 Ω
I =?

Solution:

$$I = \frac{E}{R}$$

$$I = \frac{3}{6}$$

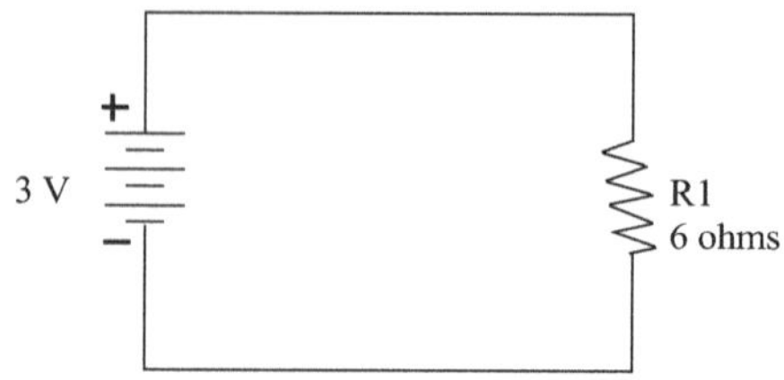

FIGURE 13.16 Determining current in a simple circuit.

I = 0.5 A

To observe the effect of source voltage on circuit current, we use the circuit shown in Figure 13.16 but double the voltage to 6 V.

Example 13.2

Problem:

Given: E = 6 V
R = 6 Ω
I =?

Solution:

$$I = \frac{E}{R}$$

$$I = \frac{6}{6}$$

I = 1 A

Notice that as the source of voltage doubles, the circuit current also doubles.

The Key Point: Circuit current is directly proportional to the applied voltage and will change by the same factor as the voltage.

To verify that current is inversely proportional to resistance, assume the resistor in Figure 13.16 to have a value of 12 Ω.

Example 13.3

Problem:

Given: E = 3 V
R = 12 Ω
I =?

Solution:

$$I = \frac{E}{R}$$

$$I = \frac{3}{12}$$

I = 0.25 A

Comparing this current of 0.25 A for the 12-Ω resistor to the 0.5-A current obtained with the 6-Ω resistor shows that doubling the resistance will reduce the current to one-half the original value—the **point is that circuit** current is inversely proportional to the circuit resistance.

Recall that you can calculate the third if you know any two quantities, E, I, and R. In many circuit applications, current is known, and either the voltage or the resistance will be the unknown quantity. To solve a problem in which current and resistance are known, the basic formula for Ohm's Law must be transposed to solve for E, I, or R.

However, the Ohm's Law equations can be memorized and practiced effectively using an Ohm's Law circle (see Figure 13.17).

To find the equation for E, I, or R when two quantities are known, cover the unknown third quantity with your finger, ruler, piece of paper, etc., as shown in Figure 13.18.

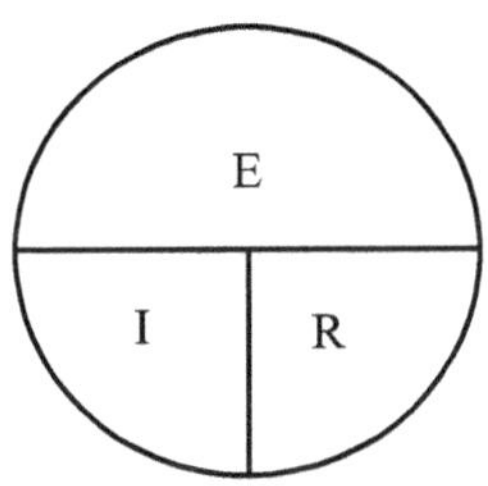

FIGURE 13.17 Ohm's law circle.

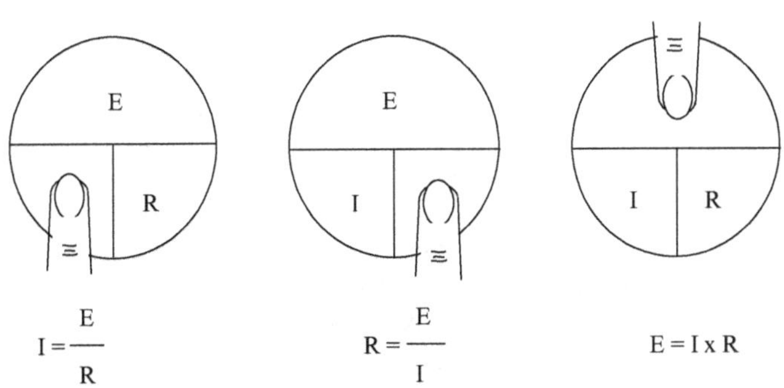

FIGURE 13.18 Putting Ohm's law circle to work.

Example 13.4

Problem: Find I when E = 120 V and R = 40 ohms (Ω).

Solution:

$$I = \frac{E}{R}$$

$$= \frac{120}{40} = 3A$$

Note: In the previous examples, we have demonstrated how the Ohm's Law circle can help us solve simple voltage, current, and amperage problems. However, the beginning student is cautioned not to rely wholly on using this circle when transposing simple formulas but rather to supplement his/her knowledge of the algebraic method. Algebra is an essential tool in solving electrical problems, and knowing how to use it should not be under-emphasized or bypassed after the operator has learned a shortcut method such as the one indicated in this circle.

Example 13.5

Problem:

An electric light bulb draws 0.5 A when operating on a 120-V D-C circuit. What is the resistance of the bulb?

Solution:

The first step in solving a circuit problem is to sketch a schematic diagram of the circuit itself, label each part, and show the known values.

Because I and E are known, we use equation (13.2) to solve for R:

$$R = \frac{E}{I} = \frac{120}{0.5} = 240\Omega$$

13.5 ELECTRIC POWER

Power, whether electrical or mechanical, pertains to the rate at which work is being done, so the power consumption in your plant is related to current flow. A large electric motor or air dryer consumes more power (and draws more current) in each length of time than, for example, an indicating light on a motor controller. **Work** is done whenever a force causes motion. Work is done if a mechanical force is used to lift or move a weight. However, force exerted WITHOUT causing motion, such as the force of a compressed spring acting between two fixed objects, does not constitute work.

Key Point: Power is the rate at which work is done.

13.5.1 Electrical Power Calculations

The electric power *P* used in any part of a circuit equals the current I in that part multiplied by V across that part of the circuit. In equation form,

$$P = EI \tag{13.4}$$

where

P = power, W (Watts)
E = voltage, V
I = current, A

If we know the current I and the resistance R but not the voltage V, we can find the power P by using Ohm's Law for voltage, so that substituting

$$E = IR \tag{13.5}$$

into

$$P = IR \times I = I^2R \tag{13.6}$$

In the same manner, if we know the voltage V and the resistance R but not the current I, we can find P by using Ohm's Law for current so that substituting

$$I = \frac{E}{R}$$

$$P = E\frac{E}{R} = \frac{E^2}{R} \tag{13.7}$$

Key Point: We can calculate the third if we know two quantities.

Example 13.6

Problem:

The current through a 200-Ω resistor in a circuit is 0.25 A. Find the resistor's power rating.

Solution:

Because I and R are known, use equation (13.6) to find P:

$$P = I^2R = (0.25)^2(200) = 0.0625(200) = 12.5 \text{ W}$$

Important Point: The power rating of any resistor used in a circuit should be twice the wattage calculated by the power equation to prevent the resistor from burning out. Thus, the resistor used in Example 13.6 should have a power rating of 25 W.

Example 13.7

Problem:

How many kilowatts of power are delivered to a circuit by a 220-V generator that supplies 30 A to the circuit?

Solution:

$$P = EI = 220(30) = 6{,}600 \text{ W} = 6.6 \text{ Kw}$$

Example 13.8

Problem:

If the voltage across a 30,000-Ω resistor is 450 V, what power is dissipated in the resistor?

Solution:

$$P = \frac{E^2}{R} = \frac{450^2}{30{,}000} = \frac{202{,}500}{30{,}000} = 6.75 \text{ W}$$

This section expressed P as alternate pairs of the other three fundamental quantities, E, I, and R. In practice, you should be able to express any of the three fundamental quantities and P in terms of any two of the others. Figure 13.19 is a summary of twelve basic formulas you should know. The four quantities E, I, R, and P are at the center of the figure.

Adjacent to each quantity are three segments. Note that the fundamental quantity is expressed in terms of two other basic quantities in each segment, and no two segments are alike.

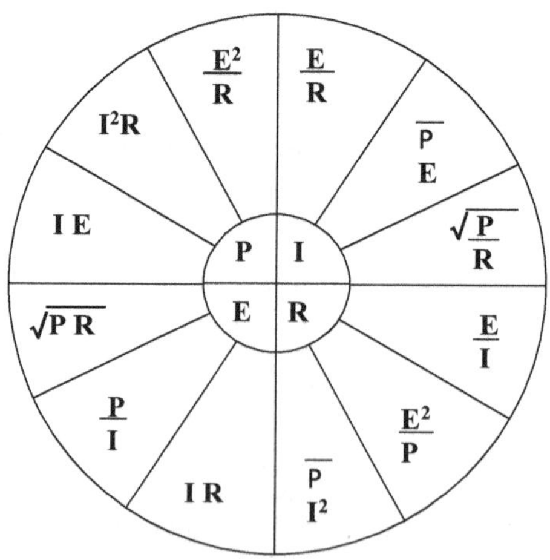

FIGURE 13.19 Ohm's law circle—Summary of basic formulas.

13.6 ELECTRIC ENERGY

Energy (the mechanical definition) is the ability to do work (energy and time are essentially the same and are expressed in identical units). Energy is expended when work is done because it takes energy to maintain a force when that force acts through a distance. The total energy expended to do a certain amount of work equals the working force multiplied by the distance through which the force moved to do the job.

In electricity, the total energy expenditure equals the rate at which work is done multiplied by the time the rate is measured. Essentially, energy W is equal to power P multiplied by time t. Kilowatt-hours (kWh) are units commonly used for large amounts of electric energy or work. The amount of kilowatt-hours is calculated as the product of the power in kilowatts (kW) and the time in hours (h) during which the power is used:

$$\text{kWh} = \text{kW} \times \text{h} \tag{13.8}$$

Example 13.9

Problem:

How much energy is delivered in 4 hours by a generator supplying 12 kW?

Solution:

$$\text{kWh} = \text{kW} \times \text{h}$$

$$= 12(4) = 48$$

Energy delivered = 48 kWh

13.7 SERIES D-C CIRCUIT CHARACTERISTICS

As previously mentioned, an electric circuit consists of a voltage source, the necessary connecting conductors, and the effective load.

If the circuit is arranged so that the electrons have only ONE possible path, the circuit is called a *Series Circuit.* Therefore, a series circuit is defined as a circuit that contains only one path for current flow. Figure 13.20 shows a series circuit having several loads (resistors).

Key Point: A series circuit has only one path for the current to flow.

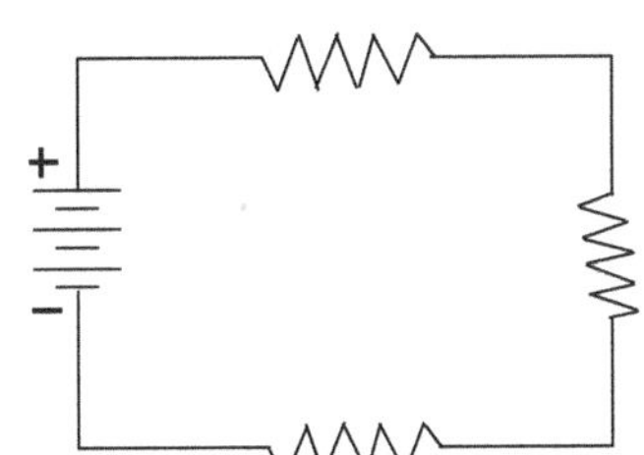

FIGURE 13.20 Series circuit.

13.7.1 D-C Resistance

Referring to Figure 13.20, the current in a series circuit must flow through each resistor inserted into the circuit to complete its electrical path. Thus, each additional resistor offers added resistance. In a series circuit, the total circuit resistance (R_T) equals the sum of the individual resistances. As an equation:

$$R_T = R_1 + R_2 + R_3 \ldots R_n \tag{13.9}$$

where

R_T = total resistance, Ω
R_1, R_2, R_3 = resistance in series, Ω
R_n = any number of additional resistors in equation.

Example 13.10

Problem:

Three resistors of 10, 12, and 25 Ω are connected in series across a battery whose EMF is 110 V (Figure 13.21). What is the total resistance?

Solution:

Given: $R_1 = 10\ \Omega$
$R_2 = 12\ \Omega$
$R_3 = 25\ \Omega$
$R_T = ?$
$R_T = R_1 + R_2 + R_3$
$R_T = 10 + 12 + 25$
$R_T = 47\ \Omega$

Transposition can be used in some circuit applications where the total resistance is known, but a circuit resistor's value must be determined.

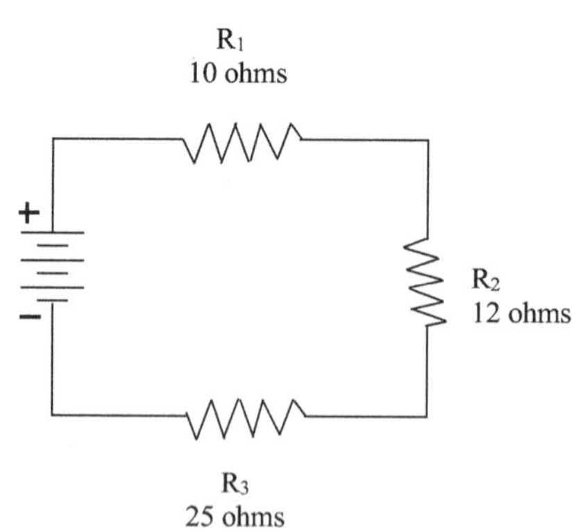

FIGURE 13.21 Solving total resistance in a series circuit.

Example 13.11

Problem:

The total resistance of a circuit containing three resistors is 50 Ω. Two of the circuit resistors are 12 Ω each. Calculate the value of the third resistor.

Solution:

Given: $R_T = 50\ \Omega$
$R_1 = 12\ \Omega$
$R_2 = 12\ \Omega$
$R_3 = ?$
$R_T = R_1 + R_2 + R_3$

Subtracting ($R_1 + R_2$) from both sides of the equation

$R_3 = R_T - R_1 - R_2$
$R_3 = 50 - 12 - 12$
$R_3 = 50 - 24$
$R_3 = 26\ \Omega$

Key Point: When resistances are connected in series, the total resistance in the circuit is equal to the sum of the resistances of all the parts of the circuit.

13.7.2 D-C CURRENT

Because there is only one path for current in a series circuit, the same *current* (I) must flow through each part of the circuit. Thus, to determine the current throughout a series circuit, only the current through one of the parts needs to be known.

Inserting ammeters at various points in a series circuit, as shown in Figure 13.22, can verify that the same current flows through each part. Each meter indicates the same value as the current.

Key Point: In a series circuit, the same current flows in every part of the circuit. *Do not* add the currents in each part of the circuit to obtain I.

13.7.3 D-C VOLTAGE

The *voltage* drop across the resistor in the basic D-C circuit is the total voltage across the circuit and is equal to the applied voltage. The total voltage across a series circuit equals the applied voltage but consists of two or more individual voltage drops. Draw a diagram for this circuit; a source potential (E_T) of 30 V is impressed across a series circuit consisting of two 6-Ω resistors. The circuit's total resistance equals the sum of the two individual resistances or 12 Ω. Using Ohm's Law, the circuit current may be calculated as follows:

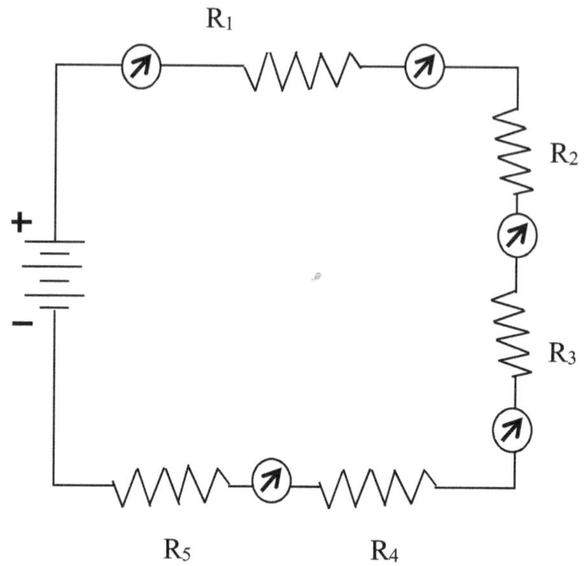

FIGURE 13.22 D-C current in a series circuit.

$$I = \frac{E_T}{R_T}$$

$$I = \frac{30}{12}$$

$$I = 2.5\ A$$

Knowing the value of the resistors to be 6 Ω each and the D-C current through the resistors to be 2.5 A, the voltage drops across the resistors can be calculated. The voltage (E_1) across R_1 is therefore:

$$E_1 = IR_1$$
$$E_1 = 2.5\ A \times 6\ \Omega$$
$$E_1 = 15\ V$$

Since R_2 is the same as R_1 and carries the same current, the voltage drop across R_2 equals 15 V. Adding these two 15-V drops together gives a total drop of 30 V equal to the applied voltage. For a series circuit, then:

$$E_T = E_1 + E_2 + E_3 \ldots E_n \qquad (13.10)$$

where

E_T = total voltage, V
E_1 = voltage across resistance R_1, V

E_2 = voltage across resistance R_2, V
E_3 = voltage across resistance R_3, V

Example 13.12

Problem:

Draw a diagram for a series D-C circuit consisting of three resistors with 10, 20, and 40 Ω, respectively. Find the applied voltage if the current through the 20-Ω resistor is 2.5 A.

Given:

$R_1 = 10\ \Omega$
$R_2 = 20\ \Omega$
$R_3 = 40\ \Omega$
$I = 2.5\ A$

Because the circuit is a D-C series circuit, the same 2.5 A of current flows through each resistor. Using Ohm's Law, the voltage drops across each of the three resistors can be calculated and are:

$E_1 = 25\ V$
$E_2 = 50\ V$
$E_3 = 100\ V$

Once the individual drops are known, they can be added to find the total or applied voltage:

$E_T = E_1 + E_2 + E_3$
$E_T = 25\ V + 50\ V + 100\ V$
$E_T = 175\ V$

Key Point 1: The total voltage (E_T) across a D-C series circuit equals the sum of the voltages across each circuit resistance.

Key Point 2: The voltage drops in a series circuit are directly proportional to the resistance across which they appear. This is the result of having the same current flow through each resistor. Thus, the larger the resistor, the larger the voltage drop across it.

13.7.4 D-C Power

Each resistor in a D-C series circuit consumes *power*. This power is dissipated in the form of heat. Because this power must come from the source, the total power must equal the power consumed by circuit resistance. In a series circuit, the total power equals the **sum** of the powers the individual resistors dissipate. Total power (P_T) is thus equal to:

$$P_T = P_1 + P_2 + P_3 \ldots P_n \quad (13.11)$$

where

P_T = total power, W
P_1 = power used in the first part, W
P_2 = power used in the second part, W
P_3 = power used in the third part, W
P_n = power used in nth part, W

Example 13.13

Problem:

Draw a D-C series circuit diagram consisting of three resistors with values of 5, 15, and 20 Ω. Find the total power dissipation when 120 V is applied to the circuit.

Solution:

Given:

$R_1 = 5\ \Omega$
$R_2 = 15\ \Omega$
$R_3 = 20\ \Omega$
$E = 120$ V

The total resistance is found first.

$R_T = R_1 + R_2 + R_3$
$R_T = 5 + 15 + 20$
$R_T = 40\ \Omega$

Using total resistance and the applied voltage, the circuit current is calculated:

$$I = \frac{E_T}{R_T}$$

$$I = \frac{120}{40}$$

$I = 3$ A

Using the power formula, the individual power dissipation can be calculated:
For resistor R_1:

$P_1 = I^2R_1$
$P_1 = (3)^2 5$
$P_1 = 45$ W

For R_2:

$P_2 = I^2R_2$
$P_2 = (3)^2 15$

$P_2 = 135$ W

For R_3:

$P_3 = I^2R_3$
$P_3 = (3)^2 20$
$P_3 = 180$ W

To obtain total power:

$P_T = P_1 + P_2 + P_3$
$P_T = 45 + 135 + 180$
$P_T = 360$ W

To check the answer, the total power delivered by the source can be calculated:

$P = E \times I$
$P = 3\ A \times 120\ V$
$P = 360$ W

Thus, the total power equals the sum of the individual power dissipation.

Key Point: We found that Ohm's Law can be used for total values in a D-C series circuit and for individual parts. Similarly, the formula for power may be used for total values:

$$P_T = IE_T \tag{13.12}$$

13.7.5 Summary of the Rules for Series D-C Circuits

To this point, we have covered many critical factors governing the operation of basic D-C series circuits. We have laid a strong foundation to build in preparation for the more advanced circuit theory that follows. A summary of the critical factors governing the operation of a D-C series circuit is listed as follows:

a. The same current flows through each part of a series circuit.
b. The total resistance of a series circuit is equal to the sum of the individual resistances.
c. The total voltage across a series circuit equals the sum of the voltage drops.
d. The voltage drop across a resistor in a series circuit is proportional to the size of the resistor.
e. The total power dissipated in a series circuit equals the sum of individual dissipation.

13.7.6 Kirchoff's Voltage Law

Kirchoff's voltage law states that the voltage applied to a closed circuit equals the sum of the voltage drops in that circuit. It should be obvious that this fact was used in the study of series circuits to this point. It was expressed as follows:

Voltage applied = sum of voltage drops

$$E_A = E_1 + E_2 + E_3$$

where

E_A = applied voltage

E_1, E_2, and E_3 = voltage drops

Another way of stating Kirchoff's Law is that the algebraic sum of the instantaneous EMFs and voltage drops around any closed circuit is zero.

Kirchoff's Law can solve circuit problems, which would be difficult and often impossible with only knowledge of Ohm's Law. When Kirchoff's Law is properly applied, an equation can be set up for a closed loop, and the unknown circuit values may be calculated.

13.7.7 Ground

The ground is used to denote a common electrical point of zero potential. The reference point of a circuit is always considered to be at zero potential. Because the earth (ground) is said to have zero potential. The common ground for many electrical/electronic equipment is the metal chassis. The value of ground is noted when considering its contribution to the economy, simplification of schematics, and ease of measurement. When completing each electrical circuit, common points of a circuit at zero potential are connected directly to the metal chassis, thereby eliminating a large amount of connecting wire.

Note: Most voltage measurements used to check proper circuit operation in electronic equipment are taken concerning the ground. One meter lead is attached to the ground, and the other is moved to various test points.

13.7.8 Open and Short Circuits

A circuit is *open* if a break does not provide a complete path for DC or AC, for example, a circuit switch is open or off or a blown fuse. A fuse is placed directly into the circuit to protect a circuit. A fuse will open the circuit whenever a dangerously large current flows. A fuse will permit currents smaller than the fuse value to flow but will melt and break or open the circuit if a more significant current flows. A dangerously large current will flow when a "short circuit" occurs. A short circuit is usually caused by an accidental connection between two points in a circuit that offers very little resistance and passes an abnormal amount of current. A short circuit often occurs because of improper wiring or broken insulation.

13.8 PARALLEL D-C CIRCUITS

The principles we applied to solving simple D-C series circuit calculations to determine voltage, current, and resistance reactions can also be used in D-C parallel and series-parallel circuits.

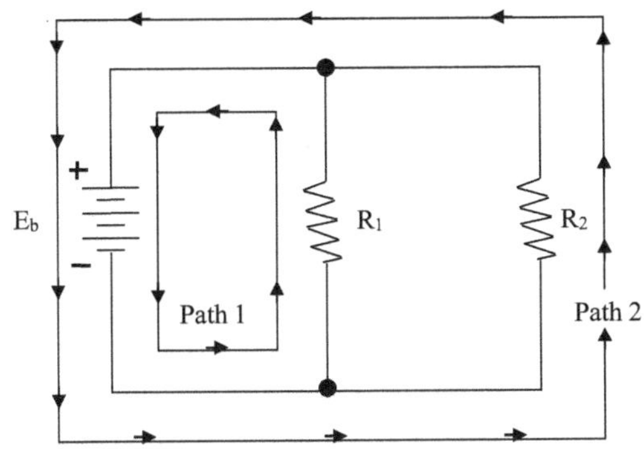

FIGURE 13.23 Basic parallel circuit.

13.8.1 Parallel Circuit Characteristics

A *parallel circuit* has two or more components connected across the same voltage source (see Figure 13.23). Recall that there is only one path for current flow in a series circuit. As additional loads (resistors, etc.) are added to the circuit, the total resistance increases, and the total current decreases. This is *not the case* in a **parallel** circuit. Each load (or branch) is connected directly across the voltage source in a parallel circuit. In Figure 13.23, commencing at the voltage source (E_b) and tracing counterclockwise around the circuit, two complete and separate paths can be identified where current can flow. One path is traced from the source through resistance R_1 and back to the source; the other through resistance R_2 and back to the source.

13.8.2 Voltage in Parallel Circuits

Recall that in a series circuit, the source voltage divides proportionately across each resistor in the circuit. In a parallel circuit, the same voltage is present across all the resistors of a parallel group. This voltage is equal to the applied voltage (E_b) and can be expressed in equation form as:

$$E_b = E_{R1} = E_{R2} = E_{Rn} \tag{13.13}$$

We can verify equation (13.13) by taking voltage measurements across the resistors of a parallel circuit. Each voltmeter indicates the same amount of voltage; that is, the voltage across each resistor is the same as the applied voltage.

Key Point: In a parallel circuit, the voltage remains the same.

Example 13.14

Problem:

Assume that the current through a resistor in a parallel circuit is 4.0 milliamperes (mA) and that the resistor's value is 40,000 Ω. Determine the potential (voltage) across the resistor.

Solution:

Given:

$R_2 = 40\ k\Omega$
$I_{R2} = 4.0\ mA$

Find:
$E_{R2} =?$
$E_b =?$

Select proper equation
$E = IR$
Substitute known valúes:

$E_{R2} = I_{R2} \times R_2$
$E_{R2} = 4.0\ mA \times 40{,}000\ \Omega$
[use the power of tens]
$E_{R2} = (4.0 \times 10^{-3}) \times (40 \times 10^{3})$
$E_{R2} = 4.0 \times 40$

Resultant:

$E_{R2} = 160\ V$

Therefore:

$E_b = 160\ V$

13.8.3 Current in Parallel Circuits

Important Point: Ohm's Law states that the current in a circuit is inversely proportional to the circuit resistance. This fact, essential as a fundamental building block of electrical theory, also helps explain current flow in parallel circuits.

In a series circuit, a single current flows. Its value is determined in part by the total resistance of the circuit. However, the source current in a parallel circuit is divided among the available paths according to the value of the resistors in the circuit. Ohm's Law remains unchanged. For a given voltage, current varies inversely with resistance. The total current I_T is equal to the sum of all branch currents:

$$I_T = I_1 = I_2 = I_3 \tag{13.14}$$

This formula applies to any number of parallel branches, whether the resistances are equal or unequal.

By Ohm's Law, each branch current equals the applied voltage divided by the resistance between the two points where the voltage is applied. With the same applied voltage, any branch with less resistance allows more current through it than a branch with higher resistance.

$$I_2 = \frac{V}{R_2} = \frac{120}{15} = 8\text{ A}$$

$$I_3 = \frac{V}{R_3} = \frac{120}{12} = 10\text{ A}$$

Now, find the total current:

$$I_T = I_1 + I_2 + I_3$$
$$= 8 + 8 + 10 = 26\text{ A}$$

13.8.4 Parallel Circuits and Kirchoff's Current Law

The division of current in a parallel network follows a definite pattern. This pattern is described by *Kirchoff's Current Law*, which is stated as follows:

The algebraic sum of the currents entering and leaving any junction of conductors equals zero. This can be stated mathematically as

$$I_a + I_b + \ldots + I_n = 0 \qquad (13.15)$$

where I_a, I_b +, etc., are the currents entering and leaving the junction. Currents entering the junction are assumed to be positive, and currents leaving the junction are considered negative. When solving a problem using equation (13.15), the currents must be placed into the equation with the proper polarity.

13.8.5 Parallel Resistance

Unlike series circuits where total resistance (R_T) is the sum of the individual resistances, in a parallel circuit, the total resistance is *not* the sum of the individual resistances.

We can use Ohm's Law to find total resistance in a parallel circuit. We use the equation

$$R = \frac{E}{I}$$

or

$$R_T = \frac{E_s}{I_t}$$

where

R_T = total resistance of all the parallel branches across the voltage source E_S

I_T = sum of all the branch currents

Key Point: When two equal-value resistors are connected in parallel, they present a total resistance equivalent to a single resistor of one-half the value of either of the original resistors.

Example 13.15

Problem:

Five 50-Ω resistors are connected in parallel. What is the equivalent circuit resistance?

Solution:

$$R_{eq} = \frac{R}{N} = \frac{50}{5} = 10\ \Omega$$

What about parallel circuits containing resistance of unequal value? How is equivalent resistance determined?

Key Point: Remember, in parallel circuits, the equivalent resistance will always be smaller than the resistance of any branch.

13.8.5.1 The Reciprocal Method

When circuits in which resistors of unequal value are connected in parallel are encountered, the equivalent resistance may be computed using the *reciprocal method.*

Note: A *reciprocal* is an inverted fraction; the reciprocal of the fraction 3/4, for example, is 4/3. We consider a whole number to be a fraction with one as the denominator, so the reciprocal of a whole number is that number divided into 1. For example, the reciprocal of R_t is $1/R_t$.

The formula gives equivalent resistance in parallel:

$$\frac{1}{R_T} = \frac{1}{R_1} + \frac{1}{R_2} + \frac{1}{R_3} + \ldots \frac{1}{R_n} \tag{13.16}$$

where

R_T = total resistance in parallel
R_1, R_2, R_3, and R_n = branch resistances

Note: When resistances are connected in parallel, the total resistance is always **less** than the slightest resistance of any single branch.

13.8.5.2 Product Over the Sum Method

When any two unequal resistors are parallel, it is often easier to calculate the total resistance by multiplying the two resistances and dividing the product by the sum of the resistances:

$$R_T = \frac{R_1 \times R_2}{R_1 + R_2} \tag{13.17}$$

where

R_T = total resistance in parallel
R_1 and R_2 = two resistors in parallel

Example 13.16

Problem:

What is the equivalent resistance of a 20-Ω and a 30-Ω resistor connected in parallel?

Solution:

Given: $R_1 = 20$
$R_2 = 30$

$$R_T = \frac{R_1 \times R_2}{R_1 + R_2}$$

$$R_T = \frac{20 \times 30}{20 + 30}$$

$$R_T = 12\ \Omega$$

13.8.6 Power in Parallel Circuits

As in the series circuit, the total *power* consumed in a parallel circuit equals the sum of the individual resistors.

Note: Because power dissipation in resistors consists of heat loss, power dissipation is additive regardless of how the resistors are connected to the circuit.

$$P_T = P_1 + P_2 + P_3 + \ldots + P_n \tag{13.18}$$

where

P_T = total power
P_1, P_2, P_3, and P_n = branch powers

The equation can also calculate the total power

$$P_T = EI_T \tag{13.19}$$

where

P_T = total power
E = voltage source across all parallel branches
I_T = total current

The power dissipated in each branch is equal to EI and equal to V^2/R.

Note: In parallel and series arrangements, the sum of the individual values of power dissipated in the circuit equals the total power generated by the source. The circuit arrangements cannot change the fact that all power in the circuit comes from the source.

13.9 SERIES-PARALLEL CIRCUITS

To this point, we have discussed series and parallel D-C circuits. However, the maintenance operator will seldom encounter a circuit that consists solely of either type of circuit. Most circuits consist of both series and parallel elements. A circuit of this type is referred to as a *series-parallel circuit* or as a combination circuit. The solution of a series-parallel (combination) circuit is simply a matter of application of the laws and rules discussed before this point.

13.10 CONDUCTORS

Recall that electric current moves quickly through some materials but with incredible difficulty through others. Three good electrical conductors are copper, silver, and aluminum (generally, we can say that most metals are good conductors). At present, copper is the material of choice used in electrical conductors. Under special conditions, certain gases are also used as conductors (e.g., neon gas, mercury vapor, and sodium vapor are used in various lamps).

The function of the wire conductor is to connect a source of applied voltage to a load resistance with a minimum IR voltage drop in the conductor so that most of the applied voltage can produce current in the load resistance. Ideally, a conductor must have a very low resistance (e.g., a typical value for a conductor—copper—is less than 1 Ω/10 ft).

13.11 ELECTROMAGNETISM

Earlier, we discussed the fundamental theories concerning simple magnets and magnetism. Those discussions dealt mainly with forms of magnetism that were not directly related to electricity—permanent magnets, for instance. Further, only brief mention was made of those forms of magnetism directly related to electricity—producing electricity with magnetism, for example.

Anatomy and physiology are so closely related that medical students cannot study one at a time without involving the other. A similar relationship holds for the electrical field; magnetism and basic electricity are so closely related that one cannot be studied at length without involving the other. This close fundamental relationship is continually borne out in the study of generators, transformers, battery packs, and motors. To be proficient in electricity, we must become familiar with the general relationships that exist between magnetism and electricity as follows:

a. Electric current flow will always produce some form of magnetism.
b. Magnetism is the most used means for producing or using electricity.
c. Magnetic influences cause the peculiar behavior of electricity under certain conditions.

13.11.1 Magnetic Field Around a Single Conductor

In 1819, Hans Christian Oersted, a Danish scientist, discovered that a field of magnetic force exists around a single wire conductor carrying an electric current. In Figure 13.24, a wire is passed through a piece of cardboard and connected through a switch to a dry cell. With the switch open (no current flowing), if we sprinkle iron filings on the cardboard and tap it gently, the filings will fall back haphazardly. If we close the switch, the current will flow in the wire. If we tap the cardboard again, the magnetic effect of the current in the wire will cause the filings to fall back into a definite pattern of concentric circles with the wire as the center of the circles. Every section of the wire has this force field around it in a plane perpendicular to the wire, as shown in Figure 13.24.

The ability of the magnetic field to attract bits of iron (as demonstrated in Figure 13.24) depends on the number of lines of force present. The strength of the magnetic field around a wire carrying a current depends on the current since it is the current that produces the field. The greater the current, the greater the strength of the field. A large current will create many lines of force extending far from the wire, while a small current will produce only a few lines close to the wire, as shown in Figure 13.25.

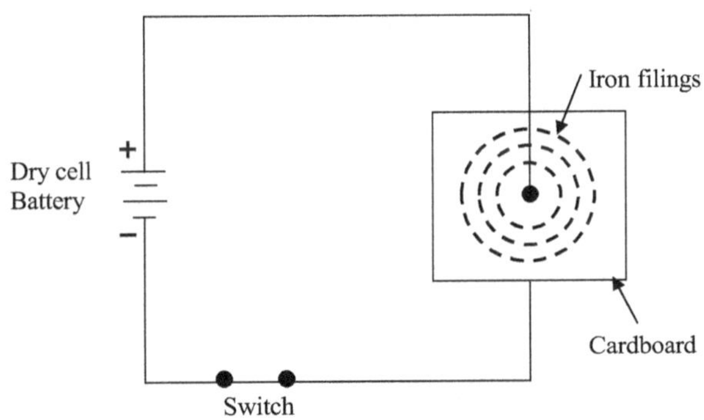

FIGURE 13.24 A circular pattern of magnetic force exists around a wire carrying an electric current.

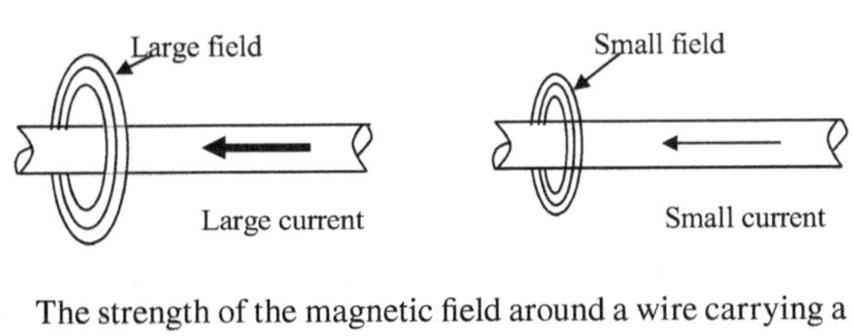

FIGURE 13.25 The strength of the magnetic field around a wire carrying a current depends on the amount of current.

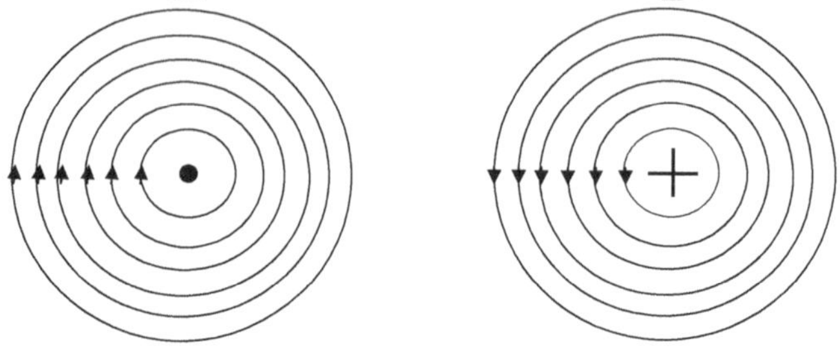

FIGURE 13.26 Magnetic field around a current-carrying conductor.

13.11.2 Polarity of a Single Conductor

The relation between the direction of the magnetic lines of force around a conductor and the direction of current flow along the conductor may be determined using the **left-hand rule for a conductor**. If the conductor is grasped in the left hand with the thumb extended in the direction of electron flow (– to +), the fingers will point in the direction of the magnetic lines of force. This is the same direction that the north pole of a compass would point if the compass were placed in the magnetic field.

Important Note: Arrows are generally used in electric diagrams to denote the direction of current flow along the length of the wire. A unique view of the arrow is used where cross sections of wire are shown. A cross-sectional view of a conductor that is carrying current toward the observer is illustrated in Figure 13.26a. The direction of the current is indicated by a dot, which represents the head of the arrow. A conductor carrying current away from the observer is illustrated in Figure 13.26b. The direction of the current is indicated by a cross, which represents the arrow's tail.

13.11.3 Field Around Two Parallel Conductors

When two parallel conductors carry current in the same direction, the magnetic fields tend to encircle both conductors, drawing them together with a force of attraction, as shown in Figure 13.27a. Two parallel conductors carrying currents in opposite directions are shown in Figure 13.27b. The field around one conductor is opposite to the field around the other. The resulting lines of force are crowded together in the space between the wires and tend to push the wires apart. Therefore, two parallel adjacent conductors carrying currents in the same direction attract each other, and two parallel conductors carrying currents in opposite directions repel each other.

13.11.4 Magnetic Field of a Coil

The magnetic field around a current-carrying wire exists at all points along its length. Bending the current-carrying wire into a single loop has two results. First,

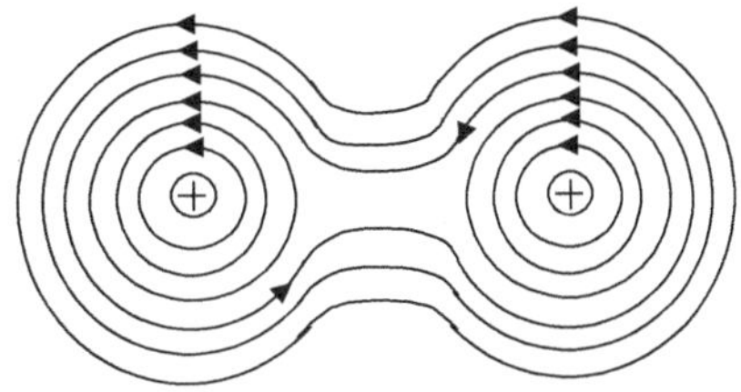

(a) Current flowing in the same direction

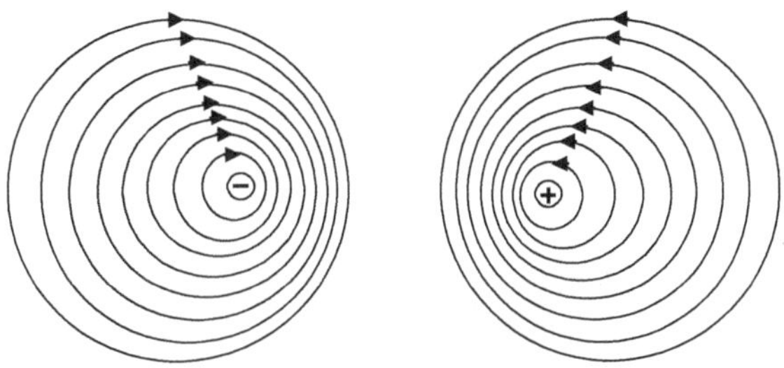

(b) Currents flowing in opposite directions

FIGURE 13.27 Magnetic field around two parallel conductors.

the magnetic field consists of more dense concentric circles in a plane perpendicular to the wire, although the total number of lines is the same as for the straight conductor. Second, all the lines inside the loop are in the same direction. When this straight wire is wound around a core, it becomes a coil, and the magnetic field assumes a different shape. When current is passed through the coiled conductor, the magnetic field of each turn of wire links with the fields of adjacent turns. The combined influence of all the turns produces a two-pole field like a simple bar magnet. One end of the coil will be a north pole, and the other will be a south pole.

13.11.5 Polarity of an Electromagnetic Coil

The magnetic field around a straight conductor depends on the direction of current flow through that conductor. Thus, a reversal of current flow through a conductor causes a reversal in the direction of the produced magnetic field. It follows that a reversal of the current flow through a coil also causes a reversal of its two-pole field. This is true because that field is the product of the linkage between the individual turns of wire on the coil. Therefore, if the field of each turn is reversed, the total field (coils' field) is also reversed.

When the direction of electron flow through a coil is known, its polarity may be determined using the **left-hand rule for coils**. This rule is illustrated in Figure 13.28 and is stated as follows: Grasping the coil in the left hand, with the fingers "wrapped around" in the direction of electron flow, the thumb will point toward the north pole.

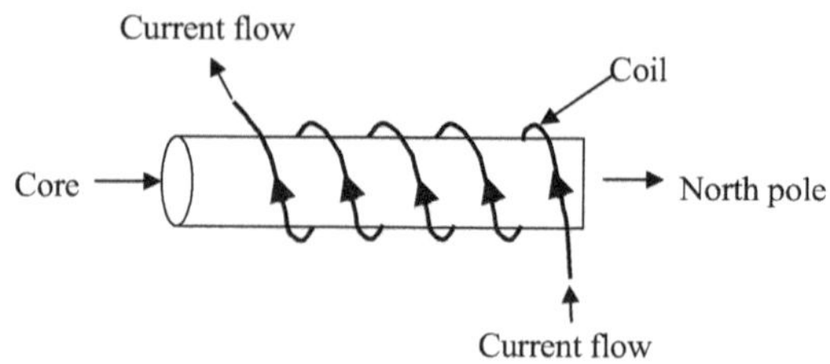

FIGURE 13.28 Current-carrying coil.

13.11.6 Strength of an Electromagnetic Field

The strength or intensity of the magnetic field of a coil depends on several factors:

- The *number of turns* of the conductor.
- The *amount of current flows* through the coil.
- The ratio of the coil's length to its width.
- The type of material in the core.

13.11.7 Magnetic Units

The current flow in the electric circuit is like the law establishing flux in the magnetic circuit. The *magnetic flux* (phi) is like the current in Ohm's Law formula and comprises the total number of lines of force in the magnetic circuit. The Maxwell is the unit of flux—that is, 1 line of force is equal to 1 Maxwell.

Note: The maxwell is often called simply a line of force, induction, or line.

The *strength* of a magnetic field in a coil of wire depends on how much current flows in the coil turns; the more current, the stronger the magnetic field. Also, the more turns, the more concentrated the lines of force are. The *force* that produces the flux in the magnetic circuit (comparable to electromotive force in Ohm's Law) is known as *magnetomotive force* or mmf. The practical unit of magnetomotive force is the **ampere-turn** (At). In equation form,

$$F = \text{ampere-turns} = NI \quad (13.20)$$

where

F = magnetomotive force, At
N = number of turns
I = current, A.

Example 13.17

Problem:

Calculate the ampere-turns for a coil with 2,000 turns and a 5-mA current.

Solution:

Use equation (13.20) and substitute $N = 2{,}000$ and $I = 5\times10^{-3}$ A.

$$NI = 2{,}000\left(5\times10^{-3}\right) = 10 \text{ At}$$

The *intensity* of magnetizing force per unit of length is designated as H and sometimes expressed as Gilberts per centimeter. This is expressed as follows:

$$H = \frac{NI}{L} \tag{13.21}$$

where

H = magnetic field intensity, ampere-turns per meter (At/m)
NI = ampere-turns, At
L = length between poles of the coil, m

Note: Equation (13.21) is for a solenoid. H is the intensity of an air core. With an iron core, H is the intensity through the entire core, and L is the length or distance between poles of the iron core.

13.11.8 Properties of Magnetic Materials

13.11.8.1 Permeability

When the core of an electromagnet is made of annealed sheet steel, it produces a stronger magnet than a cast iron core. This is the case because the magnetizing force of the coil more readily acts upon annealed sheet steel than the hard cast iron. Simply put, soft sheet steel is said to have more excellent *permeability* because of its greater ease with which magnetic lines are established.

Recalling that permeability is the relative ease with which a substance conducts magnetic lines of force. The permeability of air is arbitrarily set at 1. The permeability of other substances is the ratio of their ability to conduct magnetic lines to that of air. The permeability of nonmagnetic materials, such as aluminum, copper, wood, and brass, is essentially unity, or the same as that of air.

Important Note: The permeability of magnetic materials varies with the degree of magnetization, being more negligible for high flux density values. *Reluctance*—analogous to resistance, which is the opposition to the production of flux in a material—is inversely proportional to permeability. Iron has high permeability and, therefore, low reluctance. Air has low permeability and, hence, high reluctance.

13.11.9 Electromagnets

An *electromagnet* is composed of a coil of wire wound around a usually soft iron core because of its high permeability and low hysteresis. When direct current flows through the coil, the core becomes magnetized with the same polarity that the coil would have without the core. If the current is reversed, the coil's and core's polarity is reversed.

Electromagnets are essential in electricity because magnetism can be "turned on" or "turned off" at will. A good example is the starter solenoid (an electromagnet) in automobiles and power boats. In a car or boat, an electromagnet is part of a relay that connects the battery to the induction coil, generating the high voltage needed to start the engine. The starter solenoid isolates this high voltage from the ignition switch. When no current flows in the coil, it is an "air core," but a movable soft-iron core does two things when the coil is energized—first, the magnetic flux increases because the soft-iron core is more permeable than the air core. Second, the flux is more concentrated. All this concentration of magnetic lines of force in the soft-iron core results in a perfect magnet when current flows in the coil. But soft iron loses its magnetism quickly when the current is shut off. The effect of soft iron is the same whether it is movable, as in some solenoids, or permanently installed in the coil. An electromagnet consists of a coil and a core; it becomes a magnet when current flows through the coil.

The ability to control the action of magnetic force makes an electromagnet very useful in many circuit applications, many of which are discussed in this manual.

13.12 A-C THEORY

Because voltage is induced in a conductor when lines of force are cut, the induced electromotive force (emf) amount depends on the number of lines cut in a unit of time. To induce an emf of 1 V, a conductor must cut 100,000,000 lines of force per second. To obtain this significant number of "cuttings," the conductor is formed into a loop and rotated on an axis at incredible speed (see Figure 13.29). The two sides of the loop become individual conductors in series, each side cutting lines of force and inducing twice the voltage that a single conductor would induce. In commercial generators, the number of "cuttings" and the resulting emf are increased by

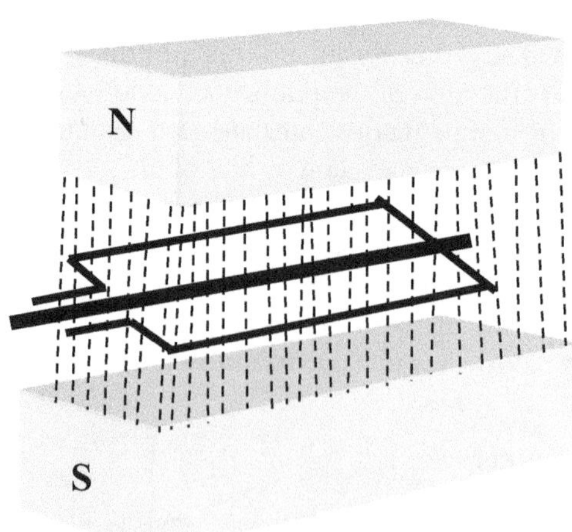

FIGURE 13.29 Loop rotating in a magnetic field produces an A-C voltage.

(1) increasing the number of lines of force by using more magnets or more potent electromagnets, (2) using more conductors or loops, and (3) rotating the loops faster. [Note: both A-C and D-C generators are covered later].

How an A-C generator operates to produce an A-C voltage and current is a fundamental concept today, taught in elementary and middle school science classes. Of course, we accept technological advances as commonplace today—we surf the internet, text our friends, watch cable television, use our cell phones, and take outer space flight as a given—and consider producing the electricity that makes all these technologies possible as our right. These technologies are on the bottom shelf for us today. We have them available for us, so we use them. It's no big deal. It's not worth thinking about. This point of view was not held initially, especially by those who broke ground in developing technology and electricity.

In groundbreaking years of electric technology development, the geniuses of the science of electricity (including George Simon Ohm) performed their technological breakthroughs in faltering steps. We tend to forget that those first faltering steps of scientific achievement in electricity were performed with crude and mostly homemade apparatus. (It sounds like the more contemporary young garage-and-basement inventors who came up with the first fundamental, user-friendly microcomputer and software packages, and hopefully the renewable energy innovations of tomorrow.)

Indeed, electricity innovators had to fabricate nearly all the laboratory equipment used in their experiments. At the time, the only convenient source of electrical energy available to these early scientists was the voltaic cell, invented some years earlier. Because cells and batteries were the only power sources available, some early electrical devices were designed to operate from **direct current**.

Thus, initially, direct current was used extensively. However, when electricity became widespread, certain disadvantages of using direct current became apparent. In a direct-current system, the supply voltage must be generated at the level required by the load. To operate a 240-V lamp, for example, the generator must deliver 240 V. A 120-V lamp could not be operated from this generator by any convenient means. A resistor could be placed in series with the 120-V lamp to drop the extra 120 V, but the resistor would waste an amount of power equal to that consumed by the lamp.

Another disadvantage of direct-current systems is the large amount of power lost due to the transmission wires' resistance to carry current from the generating station to the consumer. Operating the transmission line at very high voltage and low current could significantly reduce this loss. However, this is not a practical solution in a D-C system since the load must also operate at high voltage. As a result of the difficulties encountered with direct current, practically all modern power distribution systems use **alternating current (A-C).**

Unlike D-C voltage, A-C voltage can be stepped up or down by a transformer device (discussed later). Transformers permit the transmission lines to be operated at high voltage and low current for maximum efficiency. Then, at the consumer end, the voltage is stepped down to whatever value the load requires by using a transformer. Due to its inherent advantages and versatility, alternating current has replaced direct current in all but a few commercial power distribution systems.

13.12.1 Basic A-C Generator

As shown in Figure 13.29, an A-C voltage and current can be produced when a conductor loop rotates through a magnetic field and cuts lines of force to generate an induced A-C voltage across its terminals. This describes the basic principle of operation of an alternating current generator or **alternator**. An alternator converts mechanical energy into electrical energy. It does this by utilizing the principle of **electromagnetic induction**. The essential components of an alternator are an armature, about which many turns of the conductor are wound, which rotates in a magnetic field, and some means of delivering the resulting alternating current to an external circuit. [Note: We cover generator construction in more detail; in this section, we concentrate on the theory of operation].

13.12.2 Cycle

An A-C voltage changes in magnitude and periodically reverses in polarity (see Figure 13.30). The zero axis is a horizontal line across the center. The vertical variations on the voltage wave show the changes in magnitude. The voltages above the horizontal axis have positive (+) polarity, while voltages below the horizontal axis have negative (–) polarity.

Note: To bring the important points presented up to this point into finer focus and to expand the presentation, Figure 13.30 is provided with an accompanying explanation.

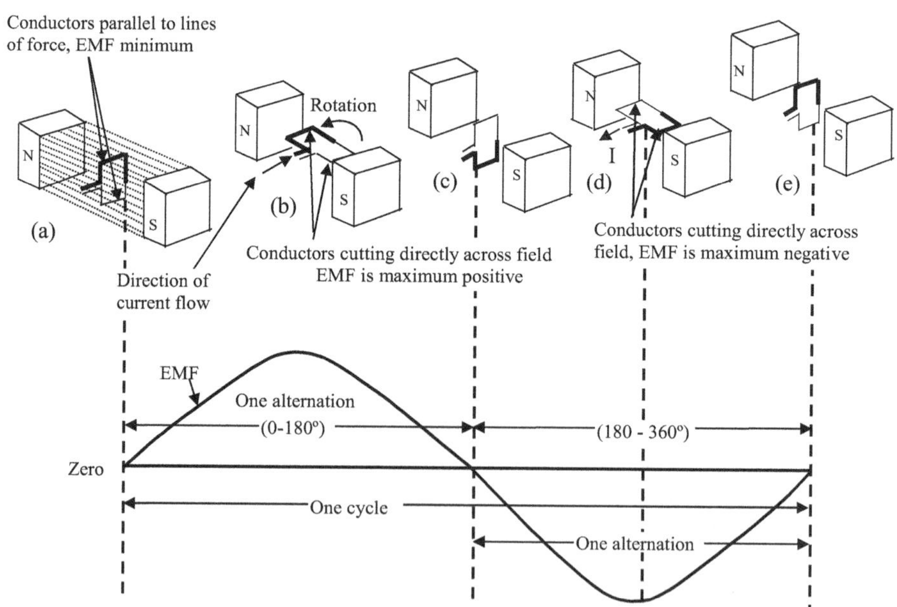

FIGURE 13.30 An A-C voltage sine wave and alternating-current generator.

Figure 13.30 shows a suspended loop of wire (conductor or armature) being rotated (moved) in a counterclockwise direction through the magnetic field between the poles of a permanent magnet. The loop has been divided into thick and thin halves for easy explanation. Notice that in part (A), the thick half is moving along (parallel to) the lines of force. Consequently, it is cutting none of these lines. The same is true of the thin half, moving in the opposite direction. Because the conductors are not cutting any lines of force, no EMF is induced. As the loop rotates toward the position shown in part (Bb), it cuts more and more lines of force per second because it is cutting more directly across the field (lines of force) as it approaches the position shown in (b). The induced voltage is most significant at position (b) because the conductor cuts directly across the field.

As the loop continues to be rotated toward the position shown in part (c), it cuts fewer and fewer lines of force per second. The induced voltage decreases from its peak value. Eventually, the loop moves again in a plane parallel to the magnetic field, and no voltage (zero voltage) is induced. The loop has now been rotated through half a circle (one alternation, or 180°). The sine curve shown in the lower part of Figure 13.30 shows the induced voltage at every instant of rotation of the loop. Notice that this curve contains 360 or two alternations.

Important Point: Two complete alternations in a period are called a *cycle.*

In Figure 13.30, if the loop is rotated at a steady rate, and if the strength of the magnetic field is uniform, the number of cycles per second (cps), or **Hertz**, and the voltage will remain at fixed values. Continuous rotation will produce a series of sine-wave voltage cycles, or, in other words, an A-C voltage. In this way, mechanical energy is converted into electrical energy.

13.12.3 Frequency, Period, and Wavelength

The *frequency* of an alternating voltage or current is the number of complete cycles occurring each second. It is indicated by the symbol *f* and is expressed in hertz (Hz). One cycle per second equals 1 Hz. Thus, 60 cycles per second (cps) equals 60 Hz. A frequency of 2 Hz (Figure 13.31a) is twice the frequency of 1 Hz (Figure 13.31a).

The period is the amount of time it takes to complete one cycle. It is indicated by the symbol T for time and expressed in seconds (seconds). Frequency and period are reciprocals.

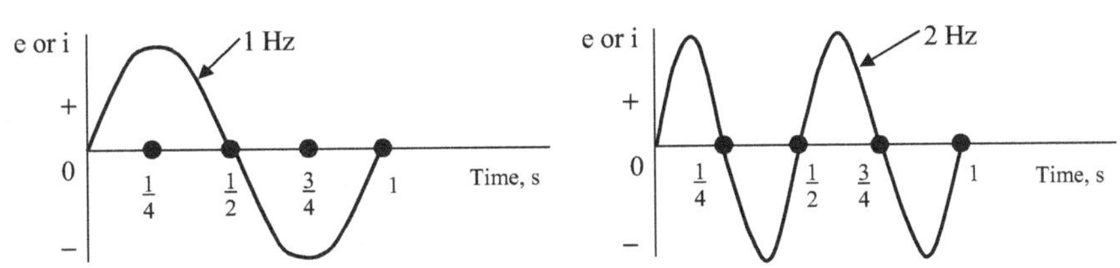

FIGURE 13.31 Comparison of frequencies.

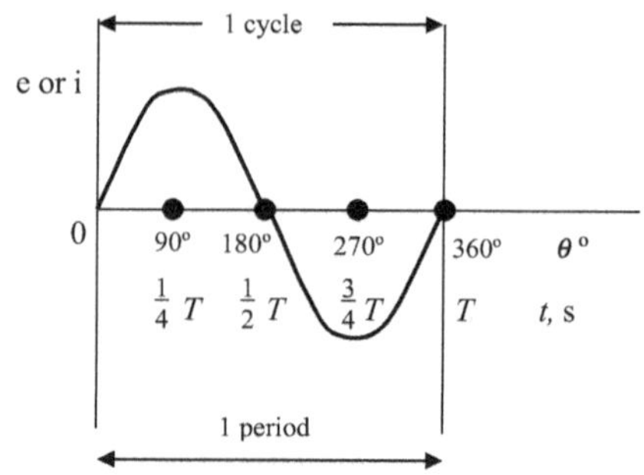

FIGURE 13.32 Relationship between electrical degrees and time.

$$f=\frac{1}{T} \tag{13.22}$$

$$T=\frac{1}{f} \tag{13.23}$$

Important Point: The higher the frequency, the shorter the period.

The angle of 360° represents the time for one cycle, or the period T.

The sine wave's horizontal axis is either electrical degrees or seconds (see Figure 13.32).

The *wavelength* is the length of one complete wave or cycle. It depends upon the frequency of the periodic variation and its transmission velocity. It is indicated by the symbol λ (Greek lowercase lambda). This can be expressed as a formula:

$$\lambda=\frac{\text{Velocity}}{\text{Frequency}} \tag{13.24}$$

13.13 CHARACTERISTIC VALUES OF A-C VOLTAGE AND CURRENT

Because an A-C sine wave voltage or current has many instantaneous values throughout the cycle, it is convenient to specify magnitudes for comparing one wave with another. The peak, average, or root-mean-square (rms) values can be determined (see Figure 13.33). These values apply to current or voltage.

One of a sine wave's most frequently measured characteristics is its *amplitude.* Unlike D-C measurement, the amount of alternating current or voltage in a circuit can be measured in various ways. One measurement method measures the maximum amplitude of either the positive or the negative alternation. The value of current or voltage obtained is called the *peak voltage* or the *peak current.* An oscilloscope must be used to measure the peak value of current or voltage. The peak value is described in Figure 13.33.

A second method of indicating the amplitude of a sine wave consists of determining the total voltage or current between the positive and negative peaks. This

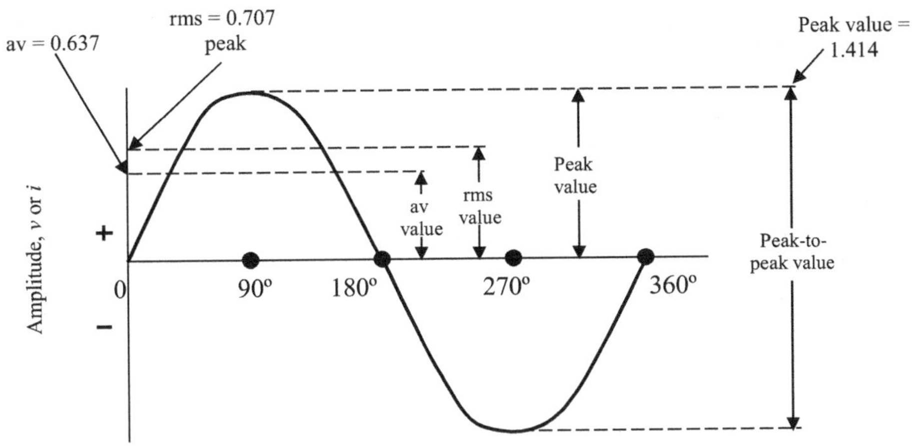

FIGURE 13.33 Amplitude values for A-C sine wave.

current or voltage value is *peak-to-peak* (see Figure 13.33). Because both alternations of a pure sine wave are identical, the peak-to-peak value is twice the peak value. Peak-to-peak voltage is usually measured with an oscilloscope, although some voltmeters calibrate a unique scale in peak-to-peak volts.

The *instantaneous value* of a sine wave of voltage for any angle of rotation is expressed by the formula:

$$e = E_m \times \sin\theta \qquad (13.25)$$

where

e = instantaneous voltage
E_m = maximum or peak voltage
$\sin\theta$ = sine of the angle at which e is desired.

Similarly, the equation for the instantaneous value of a sine wave of current would be:

$$i = I_m \times \sin\theta \qquad (13.26)$$

where

i = instantaneous current
I_m = maximum or peak current
$\sin\theta$ = sine of the angle at which i desired.

Note: The instantaneous value of voltage constantly changes as the armature of an alternator moves through a complete rotation. Because current varies directly with voltage, according to Ohm's Law, the instantaneous changes in current also result in a sine wave whose positive and negative peaks and intermediate values can be plotted precisely as we plotted the voltage sine wave. However, instantaneous values do not help solve most A-C problems, so practical value is used.

The *effective value* of an A-C voltage or current of a sine waveform is defined in terms of the equivalent heating effect of a direct current. The heating effect is independent of the direction of current flow.

Important Point: Because all instantaneous values of induced voltage are somewhere between zero and E_M (maximum or peak voltage), the practical value of a sine wave voltage or current must be greater than zero and less than E_M (the maximum or peak voltage).

The alternating current of the sine waveform, having a maximum value of 14.14 A, produces the same amount of heat in a circuit having a resistance of 1 Ω as a direct current of 10 A. Because this is true, we can calculate, having a maximum value of 14.14 A, out a constant value to convert any peak value to a corresponding effective value. X represents this constant in the simple equation below. Solve for X to three decimal places.

$$14.14X = 10$$

$$X = 0.707$$

The effective value is also called the *root-mean-square (rms)* value because it is the square root of the average of the squared values between zero and maximum. The effective value of an A-C current is stated in terms of an equivalent D-C current. The phenomenon used as the standard comparison is the heating effect of the current.

Important Point: Anytime an A-C voltage or current is stated without qualifications, it is assumed to be an effective value.

Converting from effective to peak or vice versa using a standard equation is often necessary. Figure 13.33 shows that the peak value of a sine wave is 1.414 times the effective value; therefore, the equation we use is:

$$E_m = E \times 1.414 \tag{13.27}$$

where

E_m = maximum or peak voltage
E = effective or RMS voltage

and

$$I_m = I \times 1.414 \tag{13.28}$$

where

I_m = maximum or peak current
I = effective or RMS current.

Occasionally, converting a peak current or voltage value to an effective value is necessary. This is accomplished by using the following equations:

$$E = E_m \times 0.707 \tag{13.29}$$

where

E = effective voltage
E_m = maximum or peak voltage

$$I = I_m \times 0.707 \tag{13.30}$$

where

I = effective current
I_m = maximum or peak current

Because the positive alternation is identical to the negative alternation, the *average value* of a complete cycle of a sine wave is zero. However, it is necessary to compute the average value of one alternation in certain circuits. Figure 13.33 shows that the average value of a sine wave is 0.637 × peak value, and therefore:

$$\text{Average Value} = 0.637 \times \text{Peak Value} \tag{13.31}$$

or

$E_{avg} = E_m \times 0.637$ where

E_{avg} = average voltage of one alternation
E_m = maximum or peak voltage

similarly

$$I_{avg} = I_m \times 0.637 \tag{13.32}$$

where

I_{avg} = average current in one alternation
I_m = maximum or peak current.

Table 13.3 lists the various values of sine wave amplitude used to multiply in the conversion of A-C sine wave voltage and current.

TABLE 13.3
A-C Sine Wave Conversion Table

Multiply the Value	By	To Get the Value
Peak	2	Peak-to-peak
Peak-to-peak	0.5	Peak
Peak	0.637	Average
Average	1.637	Peak
Peak	0.707	RMS (effective)
RMS (effective)	1.414	Peak
Average	1.110	RMS (effective)
RMS (effective)	0.901	Average

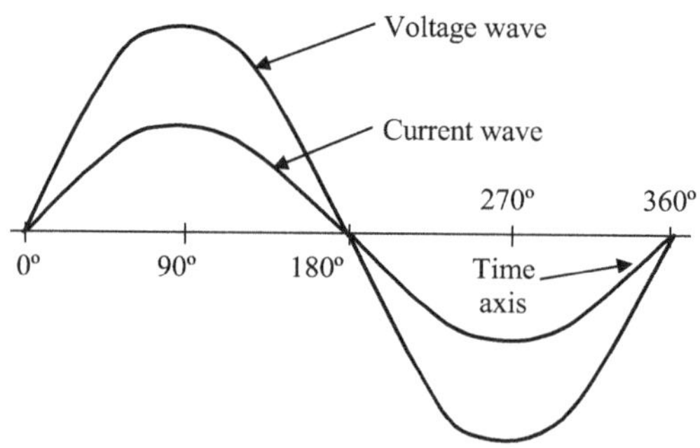

FIGURE 13.34 Voltage and current waves in phase.

13.14 RESISTANCE IN A-C CIRCUITS

If a sine wave of voltage is applied to a resistance, the resulting current will also be a sine wave. This follows Ohm's Law, which states that the current is directly proportional to the applied voltage. Figure 13.34 shows a sine wave of voltage and the resulting sine wave of current superimposed on the same time axis. Notice that as the voltage increases in a positive direction, the current increases along with it. When the voltage reverses direction, the current reverses direction. At all times, the voltage and current pass through the same relative parts of their respective cycles simultaneously. When two waves, such as those shown in Figure 13.34, are precisely in step, they are said to be **in phase**. To be in phase, the two waves must go through their maximum and minimum points simultaneously and in the same direction.

In some circuits, several sine waves can be in phase. Thus, it is possible to have two or more voltage drops in phase with each other and in phase with the circuit current.

Note: It is important to remember that Ohm's Law for DC circuits applies only to AC circuits with resistance.

Voltage waves are not always in phase. For example, Figure 13.35 shows a voltage wave E_1 considered to start at 0° (time 1). As voltage wave E_1 reaches its positive peak, a second voltage wave E_2 rises (time 2). Because these waves do not go through their maximum and minimum points at the same instant of time, a **phase difference** exists between the two waves. The two waves are said to be out of phase. For the two waves in Figure 13.35, this phase difference is 90°.

In the preceding section, we discussed the essential concepts of **phase** and **phase difference**. Another important phase concept is the *phase angle*. The angular difference at a given instant is the phase angle between two waveforms of the same frequency. For example, the phase angle between waves B and A (see Figure 13.36) is 90°. Take an instant of time at 90°. The horizontal axis is shown in angular units of time. Wave B starts at maximum value and reduces to zero at 90°, while Wave A begins at zero and increases to a maximum at 90°. Wave B reaches its maximum

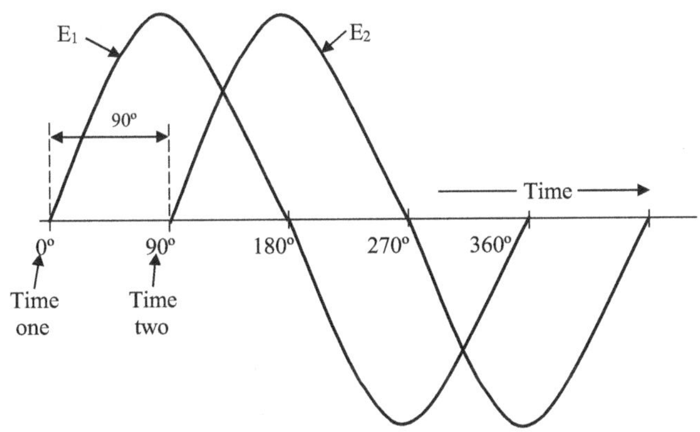

FIGURE 13.35 Voltage waves 90° out of phase.

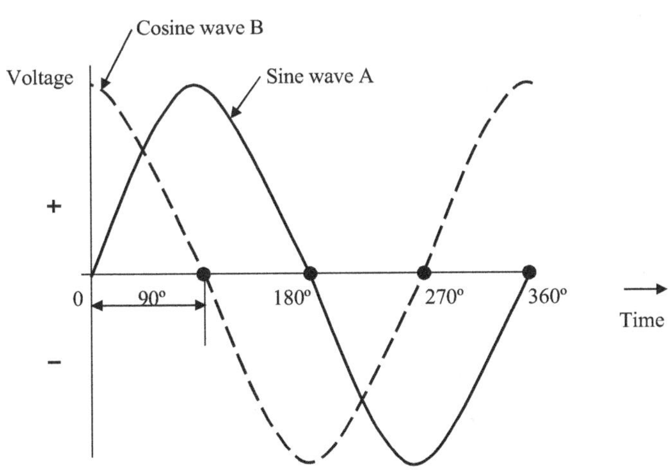

FIGURE 13.36 Wave B leads wave A by a phase angle of 90°.

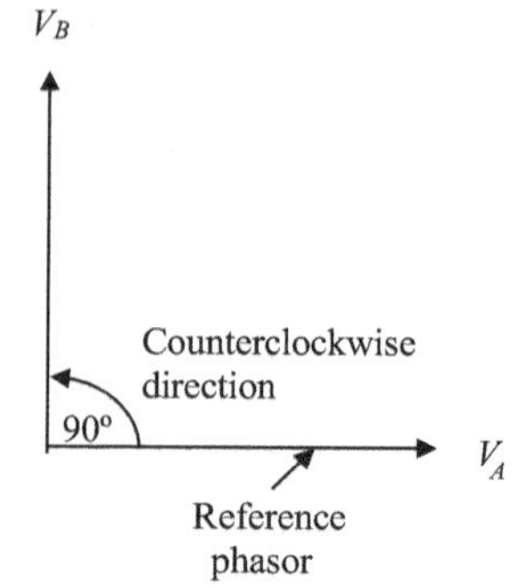

FIGURE 13.37 Phasor diagram.

value of 90° ahead of wave A, so wave B **leads** wave A by 90° (and wave A **lags** wave B by 90°). This 90-phase angle between Waves B and A is maintained throughout the complete and successive cycles. At any instant, wave B has the value that wave A will have 90° later. Wave B is a cosine wave displaced 90° from wave A, a sine wave.

Important Point: The amount by which one wave leads or lags another is measured in degrees.

To compare phase angles or phases of alternating voltages or currents, it is more convenient to use vector diagrams corresponding to the voltage and current waveforms. A *vector* is a straight line denoting a given quantity's magnitude and direction. Magnitude is denoted by the length of the line drawn to scale, and the direction is indicated by the arrow at one end of the line, together with the angle the vector makes with a horizontal reference vector.

Note: Since different directions represent time as a phase relationship in electricity, an electrical vector is called a **phasor**. The voltage and current co-occur or are in phase in an A-C circuit containing only resistance. To indicate this condition using phasors, all that is necessary is to draw the phasors for the voltage and current in the same direction. The length of the phasor indicates the value of each.

Figure 13.37 shows a vector, or phasor, diagram. Vector VB is vertical to show the phase angle of 90° concerning vector VA, which is the reference. Since lead angles are shown in the counterclockwise direction from the reference vector, V_B leads V_A by 90°.

13.15 INDUCTANCE

To this point, we have learned the following key points about magnetic fields:

- A field of force exists around a wire carrying a current.
- This field has concentric circles around the wire, in planes perpendicular to the wire, and with the wire at the center of the circles.
- The strength of the field depends on the current. Large currents produce large fields; small currents produce small fields.
- When lines of force cut across a conductor, a voltage is induced in the conductor.

Moreover, to this point, we have studied circuits that have been resistive (i.e., resistors presented the only opposition to current flow). Two other phenomena—inductance and capacitance—exist in D-C circuits to some extent, but they are major players in A-C circuits. Both inductance and capacitance present a kind of opposition to current flow called **reactance**, which we will cover later. Before we examine reactance, however, we must first study inductance and capacitance.

Inductance is the characteristic of an electrical circuit that makes itself evident by opposing the starting, stopping, or changing of current flow. A simple analogy can be used to explain inductance. We are all familiar with how difficult it is to push a heavy load (a cart full of heavy materials). It takes more work to start the load than it does to keep it moving. This is because the load possesses the property of **inertia**. Inertia is the characteristic of mass that opposes a **change** in velocity. Therefore, inertia can hinder us in some ways and help us in others. Inductance exhibits the same effect on current in an electric circuit as inertia does on the velocity of a mechanical object. The effects of inductance are sometimes desirable and sometimes undesirable.

Important Point: Inductance is the characteristic of an electrical conductor that opposes a **change** in current flow.

Because inductance is the property of an electric circuit that opposes any **change** in the current through that circuit if the current increases, a self-induced voltage opposes this change and delays the growth; on the other hand, if the current decreases, a self-induced voltage tends to aid (or prolong) the current flow, slowing the reduction. Thus, current can neither increase nor decrease as fast in an inductive circuit as in a purely resistive circuit.

This effect becomes significant in A-C circuits because it affects the **phase** relationships between voltage and current. Earlier, we learned that voltages (or currents) can be out of phase if induced in separate alternator armatures. In that case, the voltage and current generated by each armature were in phase. When inductance is a factor in a circuit, the voltage and current generated by the **same** armature are out of phase. We shall examine these phase relationships later. We aim to understand the nature and effects of inductance in an electric circuit.

The unit for measuring inductance, L, is the *Henry* (named for the American physicist, Joseph Henry), abbreviated H and generally written in lower case, henry. Figure 13.38 shows the schematic symbol for an inductor. An inductor has an inductance of 1 Henry if an emf of 1 V is induced in the inductor when the current through the inductor changes at the rate of 1 A/s. The relation between the induced voltage, inductance, and rate of change of current for time is stated mathematically as

$$E = L\frac{\Delta I}{\Delta t} \tag{13.33}$$

FIGURE 13.38 Schematic symbol for an inductor.

where

E = induced emf in volts
L = inductance in henry
ΔI = change in amperes occurring in Δt seconds.

Note: The symbol Δ (Delta) means "a change in…".

The henry is a large unit of inductance used with relatively large inductors. The unit employed with small inductors is the millihenry (mH). For still smaller inductors, the unit of inductance is the microhenry (μH).

As previously explained, current flow in a conductor always produces a magnetic field surrounding or linking with it. The magnetic field changes when the current changes, and an electromagnetic field (EMF) is induced in the conductor. This EMF is called a self-induced EMF because it is induced in the conductor carrying the current.

Note: Even a perfectly straight length of conductor has some inductance.

The direction of the induced emf has a definite relation to the direction in which the field that induces the emf varies. When a circuit's current increases, the circuit's linking flux increases. This flux cuts across the conductor and induces an emf in the conductor in such a direction to oppose the increase in current and flux. This emf is sometimes referred to as **counterelectromotive force** (cemf). The two terms are used synonymously throughout this manual. Likewise, when the current decreases, an EMF is induced in the opposite direction and opposes the decrease in current.

Important Point: The effects just described are summarized by **Lenz's Law**, which states that the induced emf in any circuit is always in a direction opposed to the impact that produced it.

Shaping a conductor so that the electromagnetic field around each portion of the conductor cuts across some other portion of the same conductor increases the inductance. This is shown in its simplest form in Figure 13.39a. A loop of conductor is looped so that two portions of the conductor lie adjacent and parallel to one another. These portions are labeled as Conductor 1 and Conductor 2. When the switch is closed, the electron flow through the conductor establishes a typical concentric field around **all** portions of the conductor. The field is shown in a single plane (for simplicity) that is perpendicular to both conductors. Although the field originates simultaneously in both conductors, it is considered as originating in Conductor 1, and its effect on Conductor 2 will be noted. With increasing current, the field expands outward, cutting across a portion of Conductor 2. The dashed arrow shows the resultant induced emf in Conductor 2. Note that it opposes the battery current and voltage, according to Lenz's Law.

In Figure 13.39b, the same section of Conductor 2 is shown, but with the switch opened and the flux collapsing.

Important Point: From Figure 13.39, the critical point to note is that the voltage of self-induction opposes both **changes** in current. It delays the initial buildup of current by opposing the battery voltage and the current breakdown by exerting an induced voltage in the same direction that the battery voltage acted.

Four major factors affect the self-inductance of a conductor or circuit.

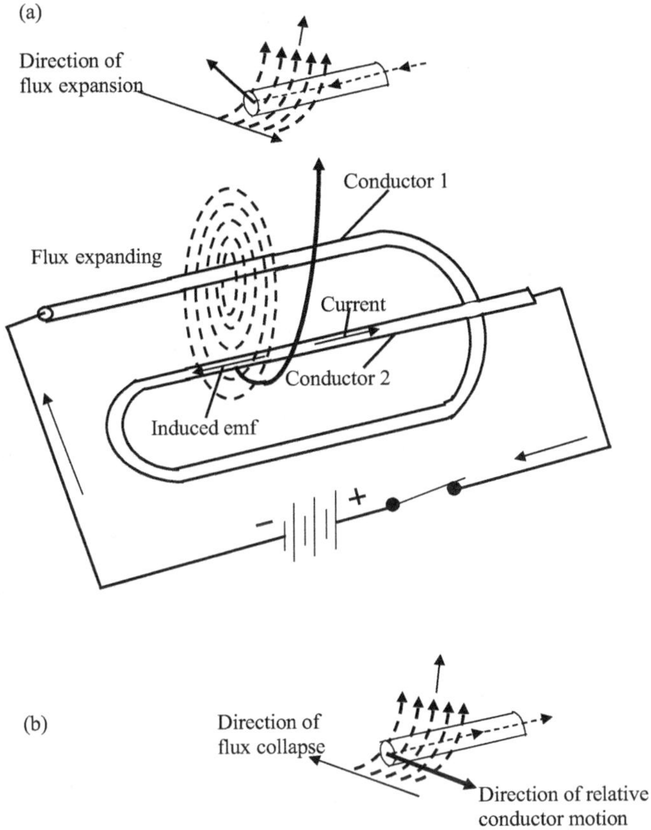

FIGURE 13.39 Self-inductance.

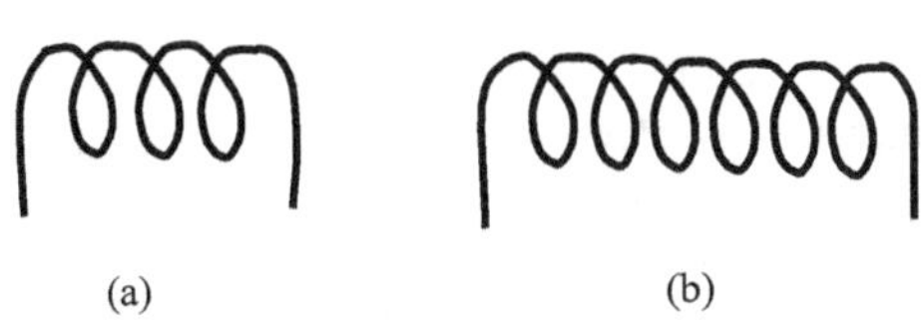

FIGURE 13.40 (a) Few turns, low inductance. (b) More turns, higher inductance.

1. **Number of turns**—Inductance depends on the number of wire turns. Wind more turns to increase inductance and take turns off to decrease it. Figure 13.40 compares the inductance of two coils made with different numbers of turns.
2. **Spacing between turns**—Inductance depends on the spacing between turns or the inductor's length. Figure 13.41 shows two inductors with the same number of turns. The first inductor's turns have a wide spacing. The second inductor's turns are close together. The second coil, though shorter,

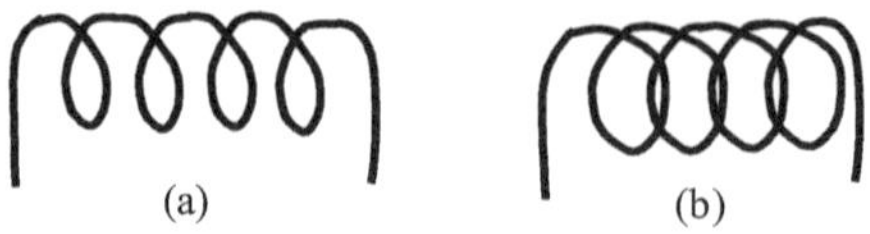

FIGURE 13.41 (a) Wide spacing between turns, low inductance. (b) Close spacing between turns, higher inductance.

FIGURE 13.42 (a) Small diameter, low inductance. (b) Larger diameter, higher inductance.

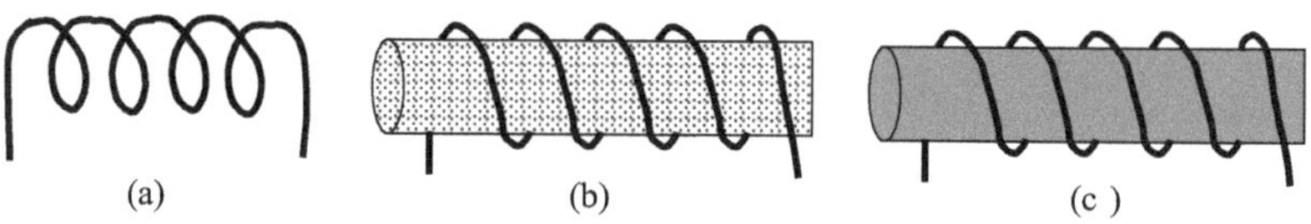

FIGURE 13.43 (a) Air core, low inductance; (b) powdered iron core, higher inductance; and (c) soft iron core, highest inductance.

has a more considerable inductance value because of its close spacing between turns.

3. **Coil diameter**—Coil diameter, or cross-sectional area, is highlighted in Figure 13.42. The larger-diameter inductor has more inductance. Both coils shown have the same number of turns, and the spacing between turns is the same. The first inductor has a small diameter, and the second one has a larger diameter. The second inductor has more inductance than the first one.
4. **Type of core material—Permeability**, as pointed out earlier, measures how easily a magnetic field goes through a material. Permeability also tells us how much stronger the magnetic field will be with the material inside the coil. Figure 13.43 shows three identical coils. One has an air core, a powdered iron core in the center, and a soft iron core. This figure illustrates the effects of core material on inductance. The inductance of a coil is affected by the current magnitude when the core is a magnetic material. When the core is air, the inductance is independent of the current.

Key Point: A coil's inductance increases rapidly as the number of turns increases. It also increases as the coil is shorter, the cross-sectional area is larger, or the core's permeability increases.

Suppose a battery is connected across a pure inductance. In that case, the current builds up to its final value at a rate determined by the battery voltage and the internal resistance. The current buildup is gradual because of the cemf generated by the self-inductance of the coil. When the current starts to flow, the magnetic lines of force move out, cut the turns of wire on the inductor, and build up a cemf that opposes the emf of the battery. This opposition causes a delay in the time it takes the current to build up to steady value. When the battery is disconnected, the lines of force collapse, again cutting the turns of the inductor and building up an emf that tends to prolong the current flow.

Although the analogy is not exact, electrical inductance resembles mechanical inertia. A boat moves on the water's surface when a constant force is applied. At this instant, its rate of change in speed (acceleration) is the greatest, and all the applied force is used to overcome the boat's inertia. After a while, the speed of the boat increases (its acceleration decreases), and the applied force is used up to overcome the friction of the water against the hull. As the speed levels off and the acceleration becomes zero, the applied force equals the opposing friction force at this speed, and the inertia effect disappears. In the case of inductance, electrical inertia must be overcome.

13.15.1 Mutual Inductance

When the current in a conductor or coil changes, the varying flux can cut across any nearby conductor or coil, thus inducing voltages in both. Therefore, a varying current in L1 induces a voltage across L_1 and L_2 (see Figures 13.44 and 13.45 for the

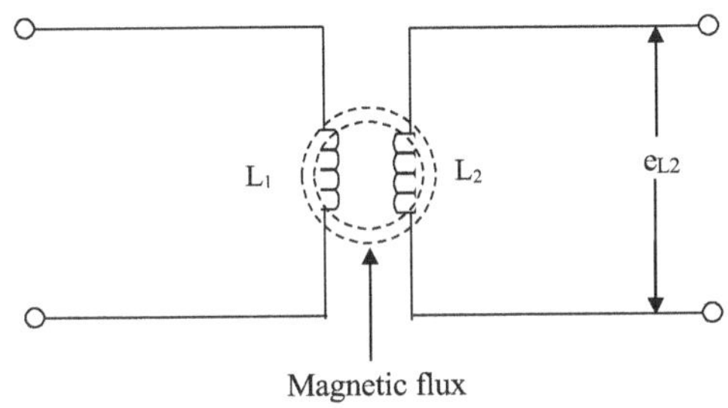

FIGURE 13.44 Mutual inductance between L_1 and L_2.

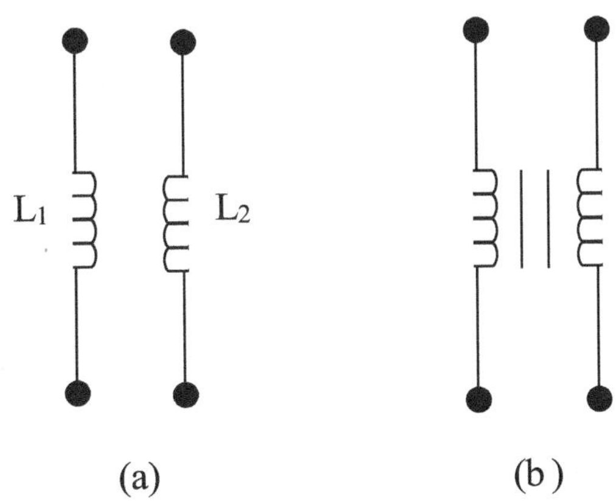

FIGURE 13.45 (a) Schematic symbol for two coils (air core) with mutual inductance and (b) two coils (iron core) with mutual inductance.

schematic symbol for two coils with mutual inductance). When the induced voltage e_{L2} produces current in L_2, its varying magnetic field induces a voltage in L_1. Hence, the two coils, L_1 and L_2, have mutual inductance because the current change in one coil can induce a voltage in the other. The unit of mutual inductance is the henry, and the symbol is L_M. Two coils have L_M of 1 H when a current change of 1 A/s in one coil induces 1 E in the other coil.

The factors affecting the mutual inductance of two adjacent coils depend upon

- physical dimensions of the two coils,
- number of turns in each coil,
- distance between the two coils,
- relative positions of the axes of the two coils, and
- the permeability of the cores.

Important Point: The amount of mutual inductance depends on the relative position of the two coils. If the coils are separated considerably, the flux standard for both is small, and the mutual inductance is low. Conversely, if the coils are close together so that nearly all the flow of one coil links the turns of the other, mutual inductance is high. The mutual inductance can be significantly increased by mounting the coils on a common iron core.

13.15.2 Calculation of Total Inductance

Note: In the study of advanced electrical theory, knowing the effect of mutual inductance in solving for total inductance in both series and parallel circuits is necessary. However, for our purposes, we do not attempt to make these calculations in this manual. Instead, we discuss the basic total inductance calculations the maintenance operator should know.

If inductors in series are located far enough apart or well shielded to make the effects of mutual inductance negligible, the total inductance is calculated in the same manner as for resistances in series; we merely add them:

$$L_t = L_1 + L_2 + L_3 \ldots (\text{etc.}) \tag{13.34}$$

Example 13.18

Problem:

If a series circuit contains three inductors whose values are 40 μH, 50 μH, and 20 μH, what is the total inductance?

Solution:

$$L_t = 40\ \mu\text{H} + 50\ \mu\text{H} + 20\ \mu\text{H}$$

$$= 110\ \mu\text{H}$$

In a parallel circuit containing inductors (without mutual inductance), the total inductance is calculated in the same manner as for resistances in parallel:

$$\frac{1}{L_t} = \frac{1}{L_1} + \frac{1}{L_2} + \frac{1}{L_2} + \frac{1}{L_3} + \ldots \text{ (etc.)} \qquad (13.35)$$

Example 13.19

Problem:

A circuit contains three shielded inductors in parallel. The values of the three inductances are 4, 5, and 10 mH. What is the total inductance?

Solution:

$$\frac{1}{L_t} = \frac{1}{4} + \frac{1}{5} + \frac{1}{10}$$

$$= 0.25 + 0.2 + 0.1$$

$$= 0.55$$

$$L_t = \frac{1}{0.55}$$

$$= 1.8 \text{ mH}$$

13.16 CAPACITANCE

No matter how complex the electrical circuit is, it is composed of no more than three basic electrical properties: resistance, inductance, and capacitance. Accordingly, a thorough understanding of these three basic properties is a necessary step toward the knowledge of electrical equipment. The last of the basic three, capacitance, is covered here.

In the previous section, we learned that inductance opposes any change in current. *Capacitance* is the property of an electric circuit that opposes any voltage change in a circuit. If the applied voltage increases, capacitance opposes the change and delays the voltage increase across the circuit. If the applied voltage is decreased, capacitance maintains the higher original voltage across the circuit, thus delaying the decrease.

Capacitance is also defined as the property of a circuit that enables energy to be stored in an electric field. Natural capacitance exists in many electric circuits. However, in this manual, we are concerned only with the capacitance designed into the circuit using devices called *capacitors*.

Key Point: The most noticeable effect of capacitance in a circuit is that voltage can neither increase nor decrease rapidly in a capacitive circuit as it can in a circuit that does not include capacitance.

13.16.1 The Capacitor

A *capacitor*, or condenser, is a manufactured electrical device that consists of two conducting plates of metal separated by an insulating material called a *dielectric* (see Figure 13.46). [Note: the prefix "di--" means "through" or "across"].

The schematic symbol for a capacitor is shown in Figure 13.47.

There is a short current pulse when a capacitor is connected to a voltage source. A capacitor stores this electric charge in the dielectric (it can be charged and discharged, as we shall see later). To form a capacitor of any appreciable value, however, the area of the metal pieces must be pretty large, and the thickness of the dielectric must be quite small.

Key Point: A capacitor is essentially a device that stores electrical energy.

The capacitor is used in several ways in electrical circuits. It may block D-C portions of a circuit since it is effectively a barrier to direct current (but not to A-C current). It may be part of a tuned circuit—one such application is tuning a radio to a particular station. It may be used to filter A-C out of a D-C circuit. Most of these are advanced applications beyond this manual's scope; however, a basic understanding of capacitance is necessary for the fundamentals of A-C theory.

The two plates of the capacitor shown in Figure 13.48 are electrically neutral since each plate contains as many protons (positive charge) as electrons (negative charge). Thus, the capacitor has *no charge.*

Now, a battery is connected across the plates (see Figure 13.49a). When the switch is closed (see Figure 13.49b), the negative charge on plate A is attracted to the battery's positive terminal. This movement of charges will continue until the difference in charge between plates A and B equals the battery's electromotive force (voltage).

The capacitor is now *charged.* Because almost none of the charge can cross the space between plates, the capacitor will remain in this condition even if the battery

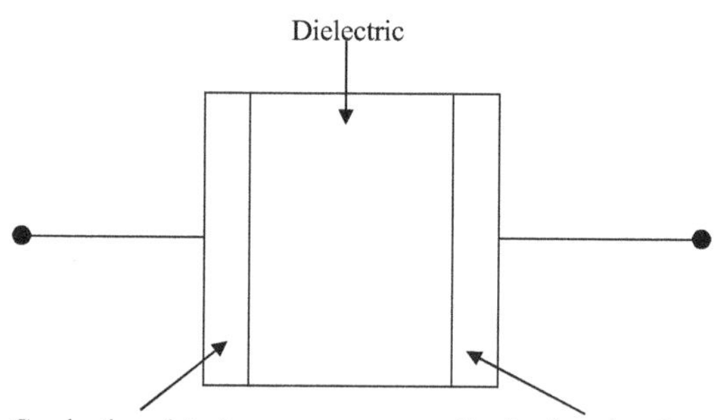

FIGURE 13.46 Capacitor.

FIGURE 13.47 (a) Schematic for a fixed capacitor and (b) variable capacitor.

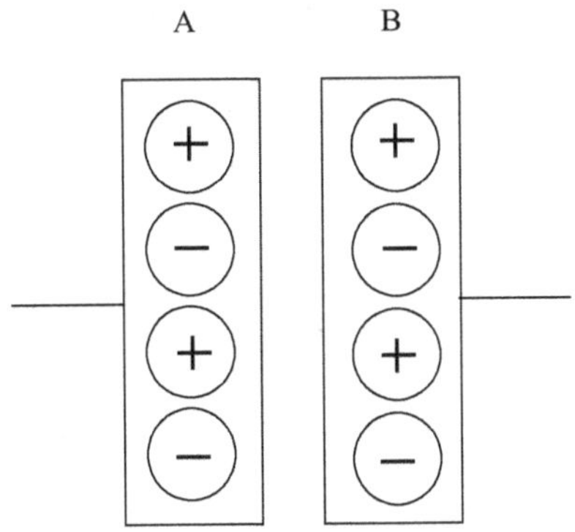

FIGURE 13.48 Two plates of a capacitor with a neutral charge.

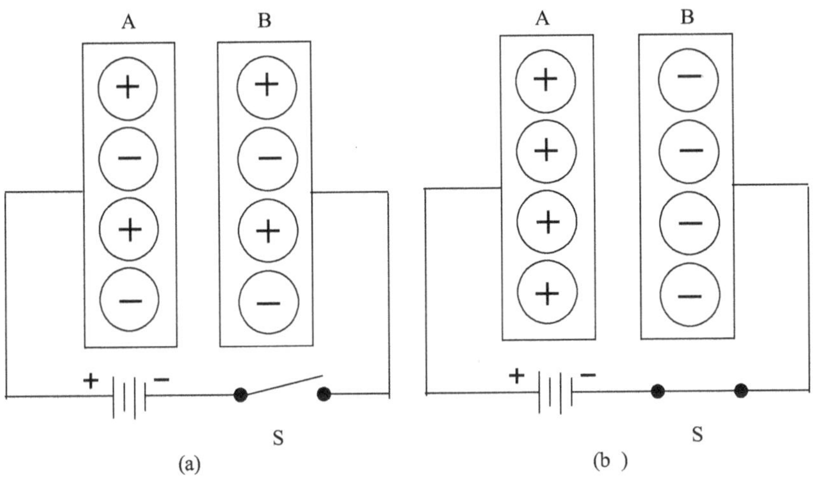

FIGURE 13.49 (a) Neutral capacitor and (b) charged capacitor.

is removed (see Figure 13.49a). However, if a conductor is placed across the plates (see Figure 13.50b), the electrons find a path back to plate A, and the charges on each plate are again neutralized. The capacitor is now *discharged.*

Important Point: In a capacitor, electrons cannot flow through the dielectric because it is an insulator. Because it takes a definite quantity of electrons to charge ("fill up") a capacitor, it is said to have **capacity**. This characteristic is referred to as *capacitance.*

13.16.2 Dielectric Materials

Similar to the phenomenon of permeability in magnetic circuits, various materials differ in their ability to support electric flux (lines of force) or serve as dielectric material for capacitors. Materials are rated in terms of their ability to support electric

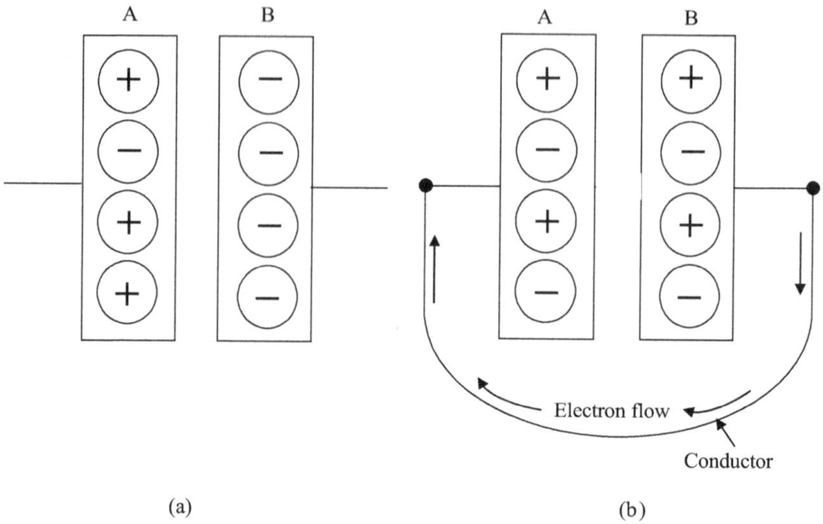

FIGURE 13.50 (a) Charged capacitor and (b) discharging a capacitor.

TABLE 13.4
Dielectric Constants

Material	Constant
Vacuum	1.0000
Air	1.0006
Paraffin paper	3.5
Glass	5–10
Quartz	3.8
Mica	3–6
Rubber	2.5–35
Wood	2.5–8
Porcelain	5.1–5.9
Glycerin (15°C)	56
Petroleum	2
Pure water	81

flux in terms of a number called a *dielectric constant.* Other factors being equal, the higher the value of the dielectric constant, the better the dielectric material. Dry air is the standard (the reference) by which other materials are rated. Dielectric constants for some common materials are given in Table 13.4.

Note: From Table 13.4, it is obvious that "pure" water is the best dielectric. Keep in mind that the keyword is "pure." Water capacitors are used today in high-energy applications in which differences in potential are measured in thousands of volts.

13.16.3 Unit of Capacitance

Capacitance equals the amount of charge that can be stored in a capacitor divided by the voltage applied across the plates:

$$C = \frac{Q}{E} \tag{13.36}$$

where

C = capacitance, F (farads)
Q = amount of charge, C (coulombs)
E = voltage, V

Example 13.20

Problem:

What is the capacitance of two metal plates separated by one centimeter of air if 0.002 C of charge is stored when a potential of 300 V is applied to the capacitor?

Solution:

Given: Q = 0.001 C
F = 200 V

$$C = \frac{Q}{E}$$

Converting to the power of ten

$$C = \frac{10 \times 10^{-4}}{2 \times 10^{2}}$$

$$C = 5 \times 10^{-6}$$

$$C = 0.000005 \text{ F}$$

Note: Although the capacitance obtained in Example 13.21 appears small, many electronic circuits require capacitors of much smaller value. Consequently, the farad is a cumbersome unit that is far too large for most applications. The microfarad is a more convenient unit, one-millionth of a farad (1×10^{-6} F). The symbols used to designate microfarads are μF.

Equation (13.36) can be rewritten as follows:

$$Q = CE \tag{13.37}$$

$$E = \frac{Q}{C} \tag{13.38}$$

Important Point: From Equation (13.37), do not deduce the mistaken idea that capacitance depends on charge and voltage. Capacitance is determined entirely by physical factors. The symbol used to designate a capacitor is (C). The unit of capacitance is the farad (F). The farad is that capacitance that will store one coulomb of charge in the dielectric when the voltage applied across the capacitor terminals is one volt.

13.16.4 Factors Affecting the Value of Capacitance

The capacitance of a capacitor depends on three main factors: plate surface area, distance between plates, and the dielectric constant of the insulating material.

- **Plate surface area**—Capacitance varies directly with place surface area. We can double the capacitance value by doubling the capacitor's plate surface area. Figure 13.51 shows a capacitor with a small surface area and another with a large surface area.

Adding more capacitor plates can increase the plate surface area. Figure 13.52 shows alternate plates connecting to opposite capacitor terminals.

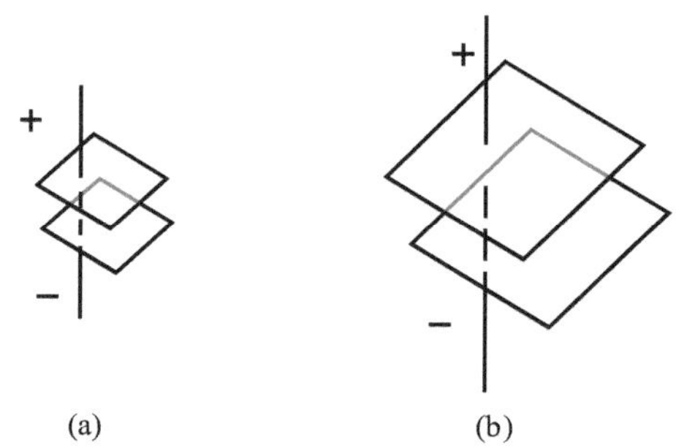

FIGURE 13.51 (a) Small plates, small capacitance. (b) Larger plates, higher capacitance.

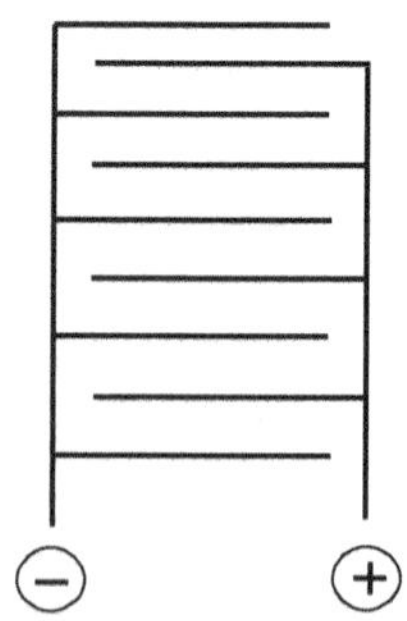

FIGURE 13.52 Several sets of plates connected to produce a capacitor with more surface area.

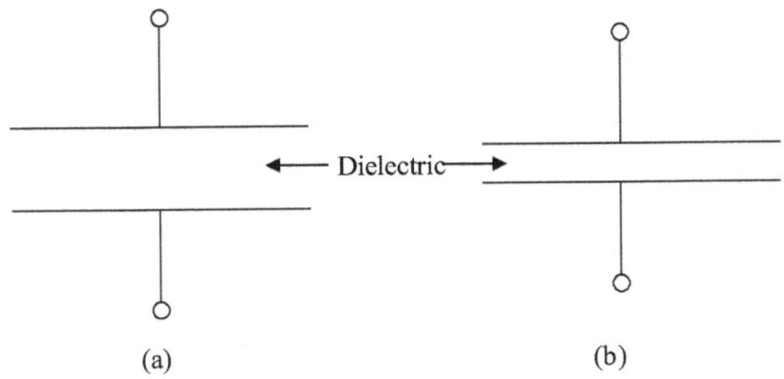

FIGURE 13.53 (a) Wide plate spacing, small capacitance. (b) Narrow plate spacing, larger capacitance.

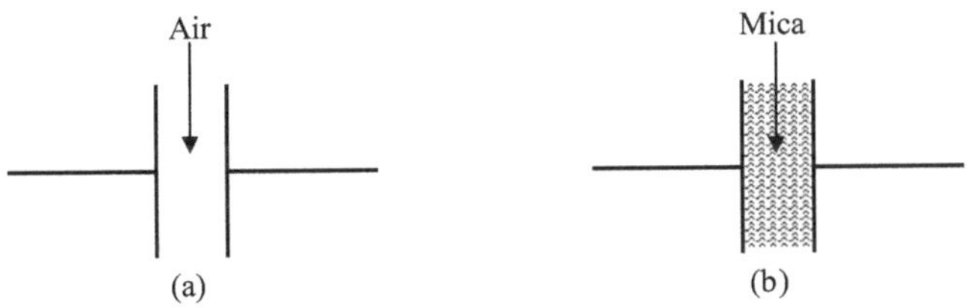

FIGURE 13.54 (a) Low capacitance and (b) Higher capacitance.

- **Distance between plates**—Capacitance varies inversely with the distance between plate surfaces. The capacitance increases when the plates are closer together. Figure 13.53 shows capacitors with the same plate surface area, but with different spacing.
- **Dielectric constant of the insulating material**—An insulating material with a higher dielectric constant produces a higher capacitance rating. Figure 13.54 shows two capacitors. Both have the same plate surface area and spacing. Air is the dielectric in the first capacitor, and mica is the dielectric in the second. Mica's dielectric constant is 5.4 times greater than air's dielectric constant. The mica capacitor has 5.4 times more capacitance than the air-dielectric capacitor.

13.16.5 Voltage Rating of Capacitors

There is a limit to the voltage that may be applied across any capacitor. If too large a voltage is applied, it will overcome the resistance of the dielectric and a current will be forced through it from one plate to the other, sometimes burning a hole in the dielectric. In this event, a short circuit exists and the capacitor must be discarded. The maximum voltage that may be applied to a capacitor is the *working voltage* and must never be exceeded.

The working voltage of a capacitor depends on (1) the type of material used as the dielectric and (2) the thickness of the dielectric. As a margin of safety, the capacitor should be selected so that its working voltage is at least 50% greater than the highest voltage to be applied. For example, if a capacitor is expected to have a maximum of 200 V applied, its working voltage should be at least 300 V.

13.16.6 Capacitors in Series and Parallel

Like resistors or inductors, capacitors may be connected in series, in parallel, or as a series-parallel combination. Unlike resistors or inductors, however, total capacitance in series, parallel, or series-parallel combinations is found differently. Simply put, the rules are not the same for calculating total capacitance. This difference is explained as follows:

Parallel capacitance is calculated like series resistance, and series capacitance is calculated like parallel resistance.

For example:

When capacitors are connected in series (see Figure 13.55), the total capacitance C_T is

$$\text{Series}: \frac{1}{C_T} = \frac{1}{C_1} + \frac{1}{C_2} + \frac{1}{C_3} + \ldots \frac{1}{C_n} \tag{13.39}$$

Example 13.21

Problem:

Find the total capacitance of a 3-μF, a 5-μF, and a 15-μF capacitor in series.

Solution:

Write equation (13.39) for three capacitors in the series:

$$\frac{1}{C_T} = \frac{1}{C_1} + \frac{1}{C_2} + \frac{1}{C_3}$$

$$= \frac{1}{3} + \frac{1}{5} + \frac{1}{15} \quad \frac{9}{15} = \frac{3}{5} = \frac{5}{3} = 1.7\ \mu\text{F}$$

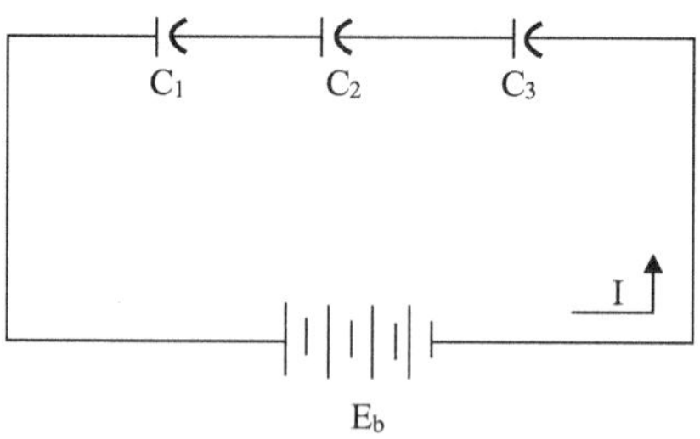

FIGURE 13.55 Series capacitive circuit.

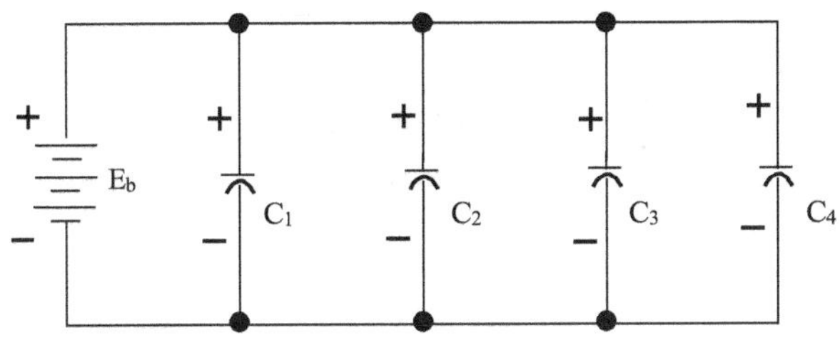

FIGURE 13.56 Parallel capacitive circuit.

When capacitors are connected in *parallel* (see Figure 13.56), the total capacitance C_T is the sum of the individual capacitances:

$$\text{Parallel: } C_T = C_1 + C_2 + C_3 + \ldots + C_n \qquad (10.37)$$

Example 13.22

Problem:

Determine the total capacitance in a parallel capacitive circuit:

Given: $C_1 = 2\mu F$
$C_2 = 3\mu F$
$C_3 = 0.25\mu F$

Solution:

Write equation (13.39) for three capacitors in parallel:

$$\begin{aligned} C_T &= C_1 + C_2 + C_3 \\ &= 2 + 3 + 0.25 \\ &= 5.25\ \mu F \end{aligned}$$

13.16.7 Types of Capacitors

Capacitors used for commercial applications are divided into two major groups—fixed and variable—and named according to their dielectric. The most common are air, mica, paper, and ceramic capacitors, plus the electrolytic type. Table 13.5 compares these types.

The fixed capacitor has a set capacitance value determined by its construction. The construction of the variable capacitors allows a range of capacitances. Within this range, the desired value of capacitance is obtained by some mechanical means, such as by turning a shaft (as in turning a radio tuner knob, for example) or adjusting a screw to adjust the distance between the plates.

The electrolytic capacitor consists of two metal plates separated by an electrolyte. The electrolyte, either paste or liquid, is in contact with the negative terminal,

TABLE 13.5
Comparison of Capacitor Types

Dielectric	Construction	Capacitance Range
Air	Meshed plates	10–400 pF
Mica	Stacked plates	10–5,000 pF
Paper	Rolled foil	0.001-1μF
Ceramic	Tubular	0.5–1,600 pF
	Disk	0.002-0.1 μF
Electrolytic	Aluminum	5–1,000 μF
	Tantalum	0.01–300 μF

forming the negative electrode. The dielectric is a thin film of oxide deposited on the positive electrode, an aluminum sheet. Electrolytic capacitors are polarity sensitive (i.e., they must be connected in a circuit according to their polarity markings) and are used where a large amount of capacitance is required.

13.17 INDUCTIVE AND CAPACITIVE REACTANCE

We have learned that the inductance of a circuit opposes any change of current flow in that circuit and that capacitance opposes any change of voltage. These reactions are not crucial in D-C circuits because they are momentary and occur only when a circuit is first closed or opened. In A-C circuits, these effects become significant because the direction of current flow is reversed many times each second, and the opposition presented by inductance and capacitance is, for practical purposes, constant.

In purely resistive circuits, either D-C or A-C, the term for opposition to current flow is resistance. When the effects of capacitance or inductance are present, as often in A-C circuits, the opposition to current flow is called *reactance.* The total opposition to current flow in circuits with both resistance and reactance is called *impedance.*

13.17.1 Inductive Reactance

To understand the reactance of a typical coil, we need to review exactly what occurs when an A-C voltage is impressed across it:

1. The A-C voltage produces an alternating current.
2. When a current flows in a wire, lines of force are produced around the wire.
3. Large currents produce many lines of force; small currents produce only a few.
4. As the current changes, the number of lines of force will change. The field of force will expand and contract as the current increases and decreases, as shown in Figure 13.57.

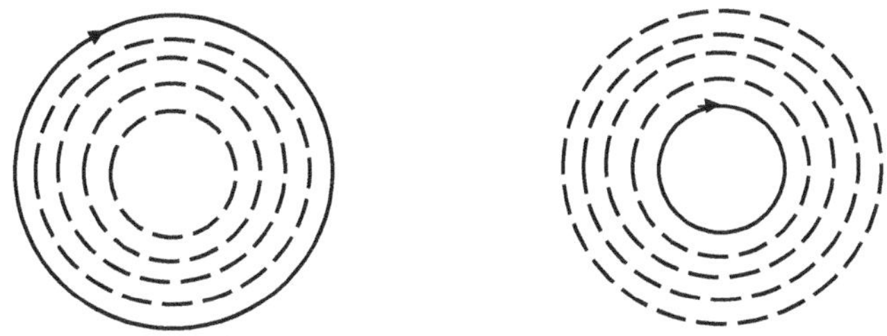

FIGURE 13.57 An A-C current producing a moving (expanding and collapsing) field. In a coil, this moving field cuts the coil's wires.

5. As the field expands and contracts, the lines of force must cut across the wires that form the coil turns.
6. These cuttings induce an emf in the coil.
7. This emf acts in the direction to oppose the original voltage and is called a *counter*, or back, *emf.*
8. The effect of this cemf is to reduce the original voltage impressed on the coil. The net effect will be to reduce the current below, which would flow if there were no cuttings or cemf.
9. In this sense, the cemf is a resistance in reducing the current.
10. We don't do this because it would be more convenient to consider the current-reducing effect of a cemf as several ohms of effectual resistance. Instead, since a cemf is not actually a resistance but merely **acts** as a resistance, we use the term *reactance* to describe this effect.

Important Point: The **reactance** of a coil is the number of ohms of resistance that the coil **seems** to offer due to a cemf induced in it. Its symbol is *X* to differentiate it from the D-C resistance R.

The inductive reactance of a coil depends primarily on (1) the coil's inductance and (2) the frequency of the current flowing through the coil. The value of the reactance of a coil is proportional to its inductance and the frequency of the A-C circuit in which it is used.

The formula for inductive reactance is

$$X_L = 2\pi fL \tag{13.40}$$

Since $2\pi = 2(3.14) = 6.28$, equation (13.40) becomes

$$X_L = 6.28fL$$

where

X_L = inductive reactance, Ω
f = frequency, Hz

L = inductance, H

The third can be found if any two quantities are known in equation (13.41).

$$L = \frac{X_L}{6.28\ f} \quad (13.41)$$

$$f = \frac{X_L}{6.28\ L} \quad (13.42)$$

Example 13.23

Problem:

The circuit's frequency is 60 Hz, and the inductance is 20 mH. What is X_L?

Solution:

$X_L = 2\pi fL$
$= 6.28 \times 60 \times 0.02$
$= 7.5\ \Omega$

Example 13.24

Problem:

A 30-mH coil is in a circuit operating at a frequency of 1,400 kHz. Find its inductive reactance.

Solution:

Given: L = 30 mH
f = 1,400 kHz
Find = X_L =?
Step 1: Change units of measurement.
30 mH = 30×10^{-3} H
1,400 kHz = $1{,}400 \times 10^3$
Step 2: Find the inductive reactance.
$X_L = 6.28fL$
$X_L = 6.28 \times 1{,}400 \times 10^3 \times 30 \times 10^{-3}$
$X_L = 263{,}760\ \Omega$

Example 13.25

Problem:

Given: L = 400 μH
f = 1,500 Hz
Find X_L =?

Solution:

$X_L = 2\pi fL$
$= 6.28 \times 1{,}500 \times 0.0004$
$= 3.78\ \Omega$

Key Point: Inductive reactance must also vary if frequency or inductance varies. A coil's inductance does not vary appreciably after manufacturing unless it is designed as a variable inductor. Thus, frequency is generally the only variable affecting a coil's inductive reactance. The coil's inductive reactance will vary directly with the applied frequency.

13.17.2 Capacitive Reactance

Previously, we learned that as a capacitor is charged, electrons are drawn from one plate and deposited on the other. As more and more electrons accumulate on the second plate, they begin to act as an opposing voltage, which attempts to stop the flow of electrons just as a resistor would do. This opposing effect is called the *reactance* of the capacitor and is measured in ohms. The basic symbol for reactance is X, and the subscript defines the type of reactance. In the symbol for inductive reactance, X_L, the subscript L refers to inductance. Following the same pattern, the symbol for capacitive reactance is X_C.

Key Point: *Capacitive reactance,* X_C, is the opposition to the flow of A-C current due to the capacitance in the circuit.

Factors affecting capacitive reactance, X_C, are

- the size of the capacitor
- frequency

The larger the capacitor, the greater the number of electrons that may be accumulated on its plates. However, because the plate area is large, the electrons do not accumulate in one spot but spread out over the entire area of the plate and do not impede the flow of new electrons onto the plate. Therefore, a large capacitor offers a small reactance. If the capacitance were small, as in a capacitor with a small plate area, the electrons could not spread out and would attempt to stop the flow of electrons coming to the plate. Therefore, a small capacitor offers a large reactance. The reactance is therefore **inversely** proportional to the capacitance.

If an A-C voltage is impressed across the capacitor, electrons are accumulated first on one plate and then on the other. If the frequency of the changes in polarity is low, the time available to accumulate electrons will be large. This means that a large number of electrons will be able to accumulate, which will result in a large opposing effect, or a large reactance. If the frequency is high, the time available to accumulate electrons will be small. This means that there will be only a few electrons on the plates, which will result in a small opposing effect, or a small reactance. The reactance is, therefore, *inversely* proportional to the frequency.

The formula for capacitive reactance is

$$X_C = \frac{1}{2\pi fC} \tag{13.43}$$

with C measured in farads.

13.18 GENERATORS

13.18.1 D-C Generators

A *D-C generator* is a rotating machine that converts mechanical energy into electrical energy. This conversion is accomplished by rotating an armature, which carries conductors, in a magnetic field, thus inducing an emf in the conductors. As stated previously, for an EMF to be induced in the conductors, a relative motion must always exist between the conductors and the magnetic field so that conductors cut through the field. In most D-C generators, the armature is the rotating member, and the field is the stationary member. A mechanical force is applied to the shaft of the rotating member to cause relative motion. Thus, when mechanical energy is put into the machine as a mechanical force or twist on the shaft, causing the shaft to turn at a certain speed, electrical energy in the form of voltage and current is delivered to the external load circuit.

Important Point: Mechanical power must be constantly applied to the shaft as long as the generator supplies electrical energy to the external load circuit.

To gain a basic understanding of the operation of a D-C generator, consider the following explanation:

A simple D-C generator consists of an armature coil with a single turn of wire (see Figure 13.58a ad b). [**Note:** The armature coils used in large D-C machines are usually wound in their final shape before being put on the armature. The sides of the preformed coil are placed in the slots of the laminated armature core]. This armature coil cuts across the magnetic field to produce voltage. If a complete path is present, the current will move through the circuit in the direction the arrows show (see Figure 13.58a). In this coil position, commutator segment 1 is in contact with brush 1, while commutator segment 2 is in contact with brush 2. As the armature rotates half turn clockwise, the contacts between the commutator segments and the brushes are reversed (see Figure 13.58). Currently, segment 1 is in contact with brush 2, and segment 2 is in contact with brush 1. Because of this commutator action, that side of the armature coil, in contact with either of the brushes, always cuts across the

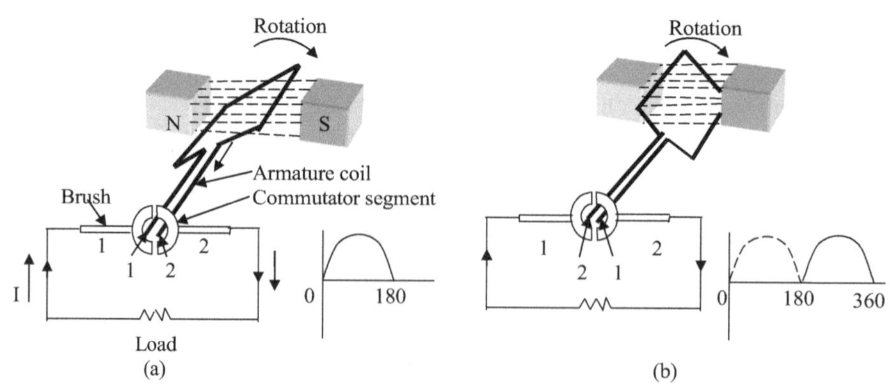

FIGURE 13.58 Basic operation of a D-C generator.

magnetic field in the same direction. Thus, brushes 1 and 2 have constant polarity, and a **pulsating D-C current** is delivered to the external load circuit.

Note: In D-C generators, the voltage induced in individual conductors is A-C. The commutator, which rotates in contact with carbon brushes, converts it to D-C (rectified) so that the current generated is in one direction or direct current.

There are several different types of D-C generators. They take their names from the type of field excitation used (i.e., they are classified according to how the field windings are connected to the armature circuit). For example, when the generator's field is excited (or supplied) from a separate D-C source (such as a battery) other than its armature, it is called a **separately excited D-C generator** (see Figure 13.59).

A **shunt generator** (self-excited) has its field windings connected in series with a rheostat across the shunt with the load, as shown in Figure 13.60. The shunt generator is widely used in industry.

A **series generator** (self-excited) has its field windings connected in series with the armature and load, as shown in Figure 13.61. Series generators are seldom used.

Compound generators (self-excited) contain both series and shunt field windings, as shown in Figure 13.62. Compound generators are widely used in industry.

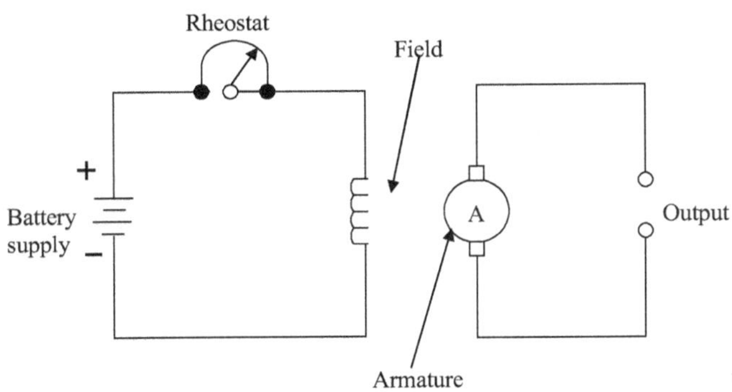

FIGURE 13.59 Separately excited D-C generator.

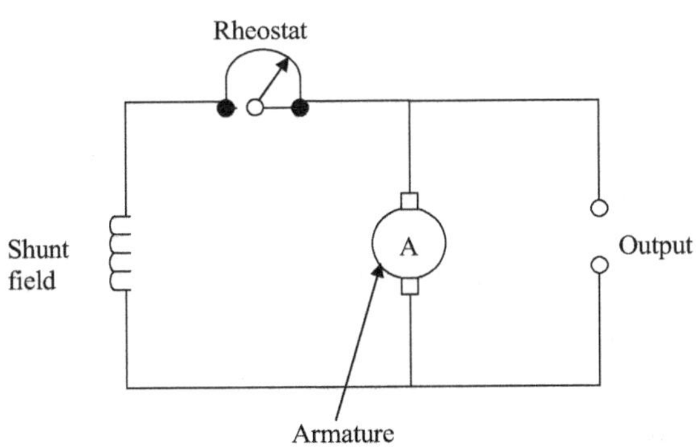

FIGURE 13.60 D-C shunt generator.

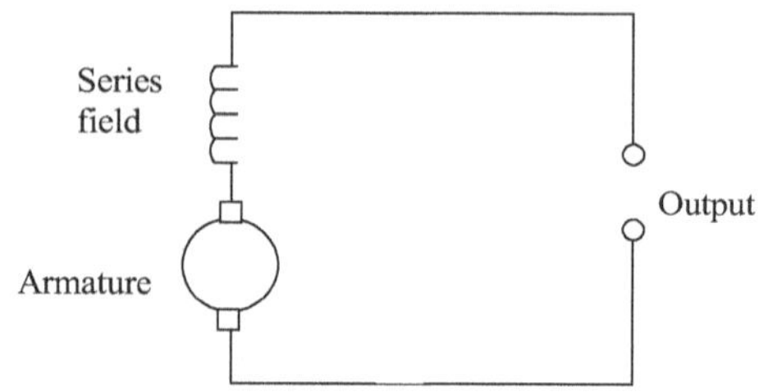

FIGURE 13.61 D-C series generator.

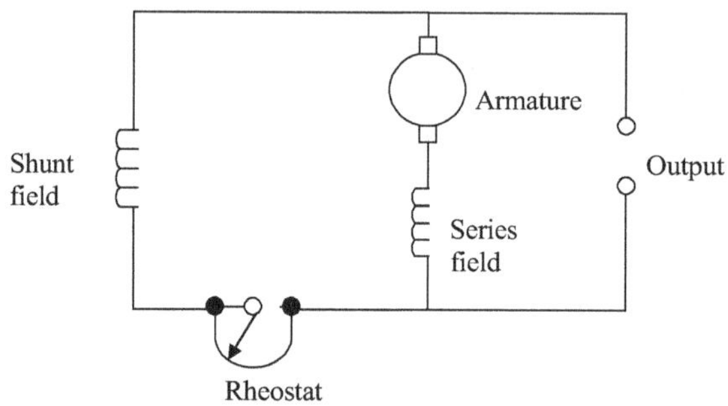

FIGURE 13.62 D-C compound generator.

Note: As central generating stations increased in size along with their number and power distribution distances, D-C generating systems were replaced by A-C generating systems to reduce power transmission costs. This was due to the high-power losses in long D-C transmission lines.

13.18.2 A-C Generators

Most electric power utilized today is generated by *alternating-current generators* (also called **alternators**). They are made in many different sizes, depending on their intended use. Regardless of size, however, all generators operate on the same basic principle: magnetic field cutting through conductors or conductors passing through a magnetic field. They are (1) a group of conductors in which the output voltage is generated and (2) a second group of conductors through which direct current is passed to obtain an electromagnetic field of fixed polarity. The conductors in which the electromagnetic field originates are always called field windings.

In addition to the armature and field, there must be motion between them. To provide this, A-C generators are built in two major assemblies: the **stator** and the **rotor**. The rotor rotates inside the stator.

The revolving-field A-C generator (see Figure 13.63) is the most widely used type. In this type of generator, direct current from a separate source is passed through windings on the rotor using slip rings and brushes. [**Note:** Slip rings and brushes are adequate for the A-C field supply because the power level in the field is much smaller than in the armature circuit]. This maintains a rotating electromagnetic field

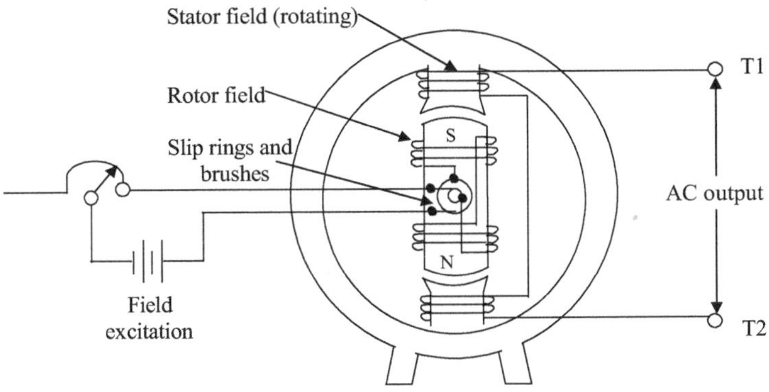

FIGURE 13.63 Essential parts of a rotating field A-C generator.

of fixed polarity. The rotating magnetic field, following the rotor, extends outward and cuts through the armature windings embedded in the surrounding stator. As the rotor turns, A-C voltages are induced in the windings since magnetic fields of first one polarity and the other cut through them. Because the output power is taken from stationary windings, the output may be connected through fixed output terminals T1 and T2 in Figure 13.63. This is advantageous because there are no sliding contacts, and the whole output circuit is continuously insulated.

Important Point: In A-C generators, frequency and electromagnetic wave cycles per second depend on how fast the rotor turns and the number of electromagnetic field poles. The voltage generated depends on the speed of the rotor, the number of coils in the armature, and magnetic field strength.

13.19 MOTORS

Electric motors perform almost endless tasks when operating industrial and domestic equipment. An *electric motor* is a machine that changes electrical energy to mechanical energy to do the work. [**Note:** Recall that a generator does just the opposite; that is, a generator changes mechanical energy to electrical energy].

Previously, we pointed out that a magnetic field is produced around the wire when a current passes through a wire. If this magnetic field passes through a stationary magnetic field, the fields either repel or attract, depending on their relative polarity. If both are positive or negative, they repel. If they are opposite polarity, they attract.

Applying this basic information to motor design, an electromagnetic coil, the armature rotates on a shaft. The armature and shaft assembly are called the rotor. The rotor is assembled between the poles of a permanent magnet, and each end of the rotor coil (armature) is connected to a commutator mounted on the shaft. A commutator comprises copper segments insulated from the shaft and each other by an insulating material. As the electromagnet poles in the rotating armature pass the stationary permanent magnet poles, they are repelled, continuing the motion. As the opposite poles are near each other, they attract, continuing the motion.

13.19.1 D-C Motors

The construction of a D-C motor is essentially the same as that of a D-C generator. However, it is essential to remember that the D-C generator converts mechanical energy into electrical energy back into mechanical energy. A DC generator can operate as a motor when an appropriate DC voltage source is applied across the usual electrical terminals.

Various types of D-C motors exist depending on how the field coils are connected. Each has advantageous characteristics under given load conditions.

Shunt motors (see Figure 13.64) have field coils connected in parallel with the armature circuit. With constant potential applied, this type of motor develops variable torque at an essentially constant speed, even under changing load conditions. Such loads are found in machine-shop equipment such as lathes, shapes, drills, and milling machines.

Series motors (see Figure 13.65) have field coils connected in series with the armature circuit. With constant potential applied, this type of motor develops variable torque, but its speed varies widely under changing load conditions. The speed is low under heavy loads but becomes excessively high under light loads. Series motors are commonly used to drive electric hoists, winches, cranes, and certain types of vehicles (e.g., electric trucks). Also, series motors are used extensively to start internal combustion engines.

Compound motors (see Figure 13.66) have one set of field coils in parallel with the armature circuit and another set in series with the armature circuit. This type of

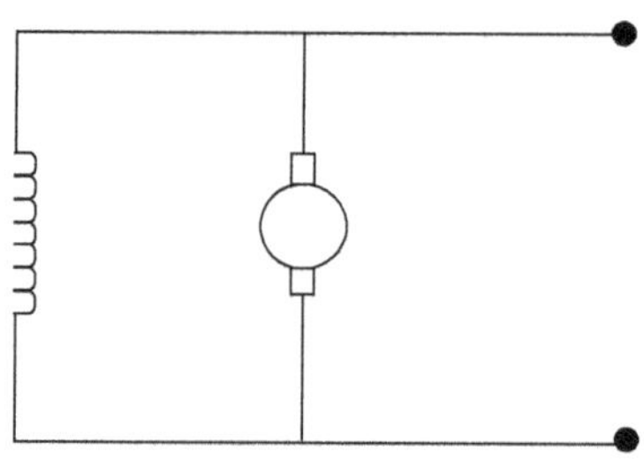

FIGURE 13.64 D-C shunt motor.

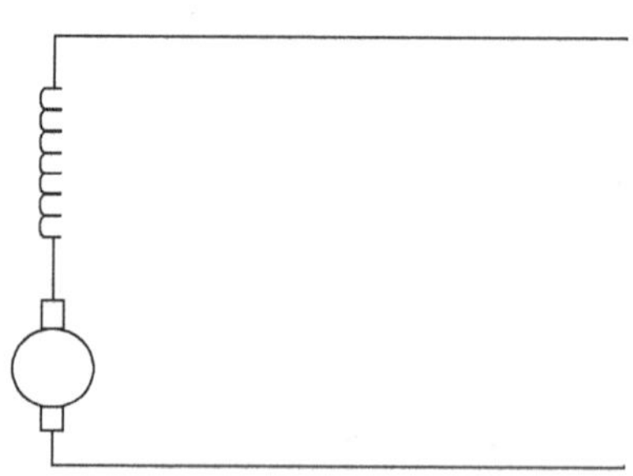

FIGURE 13.65 D-C series motor.

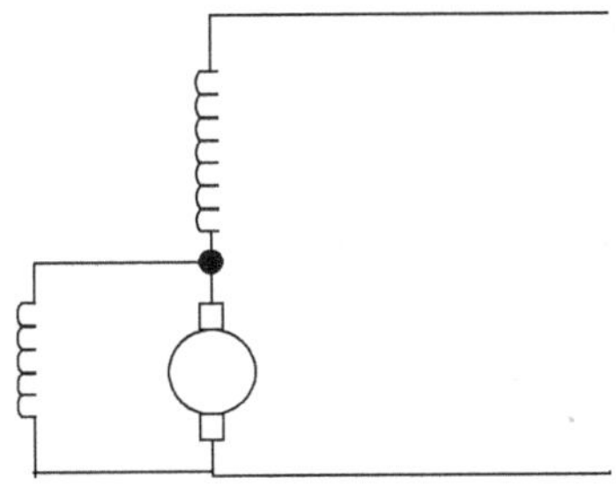

FIGURE 13.66 D-C compound motor.

motor is a compromise between shunt and series motors. It develops an increased starting torque over that of the shunt motor and has less variation in speed than the series motor.

The speed of a D-C motor is variable. It is increased or decreased by a rheostat connected in series with the field or parallel with the rotor. Interchanging either the rotor or field winding connections reverses direction.

13.19.1.1 Brushless D-C Motors

Brushless D-C motors can be deployed in any area currently fulfilled by brushed D-C motors. In renewable energy applications, the brushless D-C motor is industrial machinery. Brushless D-C motors are commonly found in consumer devices such as computer hard drives, CD players, and PC cooling fans.

The brushless D-C motor is a synchronous motor powered by D-C current that uses permanent magnets (usually rare earth-type magnets) on the spinning rotor or various winding styles. Instead of a mechanical commutation system based on brushes, it has a commutation system that a solid-state controller electronically controls. The commutator and brush system form a set of electrical switches, each opening and closing in sequence, so electrical power always flows through the armature coil closest to the stationary stator.

13.19.2 A-C Motors

A-C voltage can be easily transformed from low voltages to high voltages or vice versa, and it can be moved over a much greater distance without too much loss in efficiency. Most of the power-generating systems today, therefore, produce alternating currents. Thus, it follows that most electrical motors today are designed to operate on alternating current. However, there are other advantages to using A-C motors besides the wide availability of A-C power. In general, A-C motors are less expensive than D-C motors. Most types of A-C motors do not employ brushes and commutators. This eliminates many maintenance and wear problems and eliminates dangerous sparking.

A-C motors are manufactured in many different sizes, shapes, and ratings for use in an even more significant number of jobs. They are designed for use with either polyphase or single-phase power systems.

This text cannot possibly cover all aspects of A-C motors. Consequently, it will deal mainly with the operating principles of the two most common types: induction and synchronous motors.

13.19.2.1 Induction Motors (Polyphase)

The induction motor is the most used type of A-C motor because of its simple, rugged construction and good operating characteristics. It consists of the **stator** (stationary part) and the **rotor** (rotating part).

The most essential type of polyphase induction motor is the three-phase motor.

Important Note: A three-phase (3-θ) system is a combination of three single-phase (1-θ) systems. In a 3-balanced system, the power comes from an AC generator that produces three separate but equal voltages, each of which is out of phase with the other voltages by 120. Although 1-circuits are widely used in electrical systems, most generation and distribution of A-C current is 3-θ.

The driving torque of D-C and A-C motors is derived from the reaction of current-carrying conductors in a magnetic field. The magnetic field is stationary in the D-C motor, and the armature rotates with its current-carrying conductors. The current is supplied to the armature through a commutator and brushes.

In *induction motors*, the rotor currents are supplied by electromagnet induction. The stator windings, connected to the A-C supply, contain two or more out-of-time-phase currents, which produce the corresponding mmf. These mmf establish a rotating magnetic field across the air gap. This magnetic field rotates continuously at a constant speed regardless of the load on the motor. The stator winding corresponds to the armature winding of a D-C motor or the primary winding of a transformer. The rotor is not connected electrically to the power supply.

The induction motor derives its name from mutual induction (or transformer action) between the stator and the rotor under operating conditions. The magnetic revolving field produced by the stator cuts across the rotor conductors, inducing a voltage in the conductors. This voltage causes rotor current to flow. Hence, motor torque is developed by the interaction of the rotor current and the magnetic revolving field.

13.19.2.2 Synchronous Motors

Like induction motors, *synchronous motors* have stator windings that produce a rotating magnetic field. But unlike the induction motor, the synchronous motor requires a separate D-C source from the field. It also requires unique starting components. These include a salient-pole field with starting grid winding. The rotor of the conventional type synchronous motor is essentially the same as that of the salient-pole A-C generator. The stator windings of induction and synchronous motors are essentially the same.

In operation, the synchronous motor rotor locks into step with the rotating magnetic field and rotates at the same speed. If the rotor is pulled out of step with the rotating stator field, no torque is developed, and the motor stops. Because a synchronous motor develops torque only when running at synchronous speed, it is not self-starting and needs some device to bring the rotor to synchronous speed. For example, a synchronous motor may rotate with a D-C motor on a standard shaft.

After the motor is brought to synchronous speed, the A-C current is applied to the stator windings. The D-C starting motor now acts as a D-C generator, which supplies D-C field excitation for the rotor. The load can then be coupled to the motor.

13.19.2.3 Single-Phase Motors

Single-phase (1-θ) motors are so called because their field windings are connected directly to a single-phase source. These motors are used extensively in fractional horsepower sizes and in commercial and domestic applications. The advantages of using single-phase motors in small sizes are that they are less expensive to manufacture than other types and eliminate the need for 3-phase A-C lines. Single-phase motors are used in fans, refrigerators, portable drills, grinders, etc.

A single-phase induction motor with only one stator winding and a cage rotor is like a 3-phase induction motor with a cage rotor, except that the single-phase motor has no magnetic revolving field at the start and hence no starting torque. However, suppose the rotor is brought up to speed by external means. In that case, the induced currents in the rotor will cooperate with the stator currents to produce a revolving field, which causes the rotor to continue running in the direction it was started.

Several methods are used to provide the single-phase induction motor with starting torque. These methods identify the motor as split-phase, capacitor, shaded-pole, and repulsion-start induction motor.

Another class of single-phase motors is the A-C series (universal) type. Only the more commonly used types of single-phase motors are described. These include the (1) split-phase motor, (2) capacitor motor, (3) shaded-pole motor, (4) repulsion-start motor, and (5) A-C series motor.

13.19.2.4 Split-Phase Motor

The *split-phase motor* (see Figure 13.67) has a stator of slotted laminations containing a starting and running winding.

Note: If two stator windings of unequal impedance are spaced 90 electrical degrees apart but connected in parallel to a single-phase source, the field produced will appear to rotate. This is the principle of **phase splitting**.

The starting winding has fewer turns and smaller wire than the running winding, hence has higher resistance and less reactance. The primary winding occupies the lower half of the slots, and the starting winding occupies the upper half. When

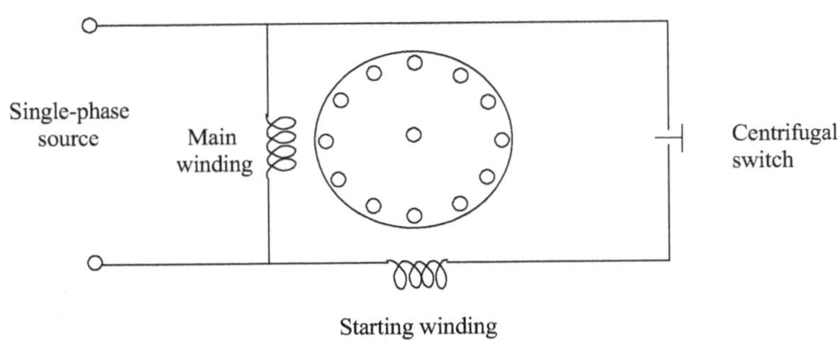

FIGURE 13.67 Split-phase motor.

the same voltage is applied to both windings, the current in the main winding lags behind the current in the starting winding. The angle θ between the main and starting windings is enough phase difference to provide a weak rotating magnetic field to produce a starting torque. When the motor reaches a predetermined speed, usually 75% of synchronous speed, a centrifugal switch mounted on the motor shaft opens, disconnecting the starting winding.

Because they have a low starting torque, fractional-horsepower split-phase motors are used in various equipment, such as washers, oil burners, ventilating fans, and woodworking machines. The direction of rotation of the split-phase motor can be reversed by interchanging the starting winding leads.

13.19.2.5 Capacitor Motor

The *capacitor motor* is a modified split-phase motor, with a capacitor in series with the starting winding (see Figure 13.68). It operates with an auxiliary winding and a series capacitor permanently connected to the line (see Figure 13.68). The capacitance in series may be one value for starting and another for running. As the motor approaches synchronous speed, the centrifugal switch disconnects one section of the capacitor.

If the starting winding is cut out after the motor has increased in speed, the motor is called a **capacitor-start motor**. If the starting winding and capacitors are designed to be left in the circuit continuously, the motor is called a **capacitor-run motor**. Capacitor motors drive grinders, drill presses, refrigerator compressors, and other loads that require relatively high starting torque. Interchanging the starting winding leads may reverse the capacitor motor's rotation direction.

13.19.2.6 Shaded-Pole Motor

A *shaded-pole* motor employs a salient-pole stator and a cage rotor. The projecting poles on the stator resemble those of D-C machines, except that the entire magnetic circuit is laminated, and a portion of each pole is split to accommodate a short-circuited coil called a **shading coil** (see Figure 13.69). The coil is usually a single band or strap of copper. The effect of the coil is to produce a slight sweeping motion of the field flux from one side of the pole piece to the other as the field pulsates. This subtle shift in the magnetic field produces a small starting torque. Thus, shaded-pole motors are self-starting. This motor is generally manufactured in small sizes, up to 1/20 horsepower, for driving small fans, small appliances, and clocks.

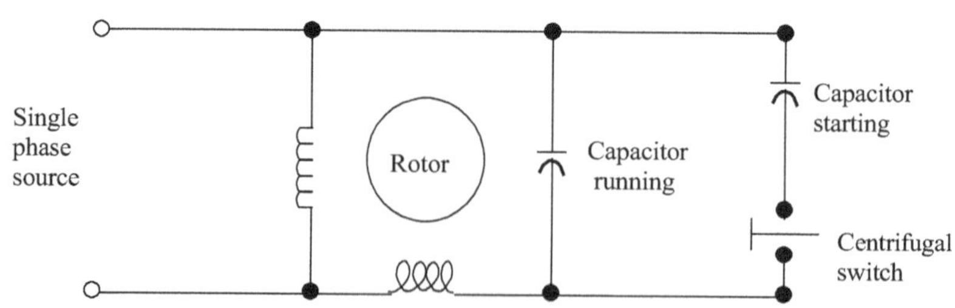

FIGURE 13.68 Capacitor motor.

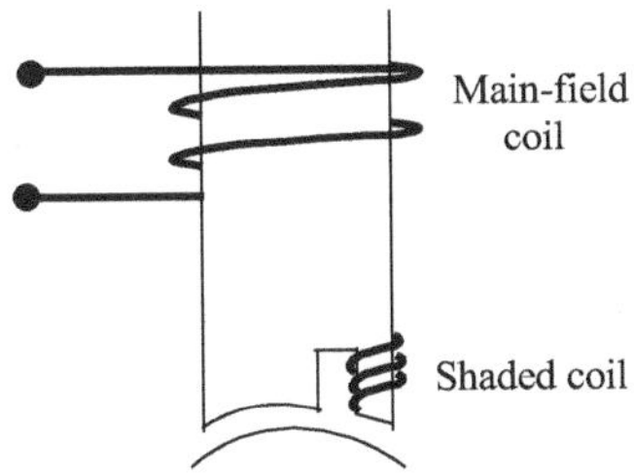

FIGURE 13.69 Shaded pole.

During that part of the cycle, when the central pole flux is increasing, the flux cuts the shading coil, and the resulting induced EMF and current in the shading coil tend to prevent the flux from rising readily through it. Thus, the greater portion of the flux rises in that portion of the pole that is not near the shading coil. When the flux reaches its maximum value, the rate of change of flux is zero, and the voltage and current in the shading coil are also zero. Currently, the flux is distributed more uniformly over the entire pole face. Then, as the main flux decreases toward zero, the induced voltage and current in the shading coil reverse their polarity, and the resulting mmf tends to prevent the flux from collapsing through the iron in the shading coil region. The result is that the main flux rises in the unshaded portion of the pole and later in the shaded portion. This action is equivalent to sweeping the field across the pole face toward the shaded pole. This moving field cuts the rotor conductors, and the force exerted on them causes the rotor to turn toward the sweeping field. The shaded-pole method of starting is used in tiny motors, up to about 1/25 hp, for driving small fans, appliances, and clocks.

13.19.2.7 Repulsion-Start Motor

Like a DC motor, the *repulsion-start motor* has a form-wound rotor with a commutator and brushes. The stator is laminated and contains a distributed single-phase winding. In its simplest form, the stator resembles that of the single-phase motor. In addition, the motor has a centrifugal device that removes the brushes from the commutator and places a short-circuiting ring around the commutator. This action occurs at about 75% of synchronous speed. After that, the motor operates with the characteristics of the single-phase induction motor. This type of motor is made in sizes ranging from 1/2 to 15 hp and is used in applications requiring a high starting torque.

13.19.2.8 A-C Series Motor

The *A-C series motor* will operate on either A-C or D-C circuits. When an ordinary D-C series motor is connected to an A-C supply, the current drawn by the motor is low due to the high series-field impedance. The result is low running torque. A-C series motors are built with as few turns as possible to reduce the field reactance to a minimum. Armature reaction is overcome using **compensating windings** (see Figure 13.70) in the pole pieces.

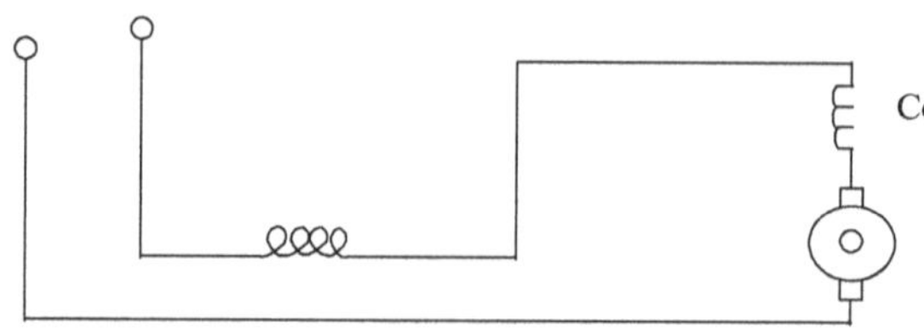

FIGURE 13.70 A-C series motor.

As with D-C series motors, the speed increases to a high value with a decrease in load in A-C series motors. The torque is high for high armature currents, so the motor has a good starting torque. A-C series motors operate more efficiently at low frequencies.

Fractional horsepower A-C series motors are called **universal motors**. They do not have compensating windings and are used extensively to operate fans and portable tools such as drills, grinders, and saws.

13.20 A-C VERSUS D-C

In the 1880s and early 1890s, there was a personal war called the "War of Currents" between two American geniuses: Thomas Edison, the American inventor who used D-C extensively to operate fans and portable tools, and the businessman who established the first investor-owned electric utility in 1882, basing its infrastructure on D-C power, and George Westinghouse, the American entrepreneur and engineer who introduced the rival A-C distribution system. Westinghouse came out on top of the argument for one crucial reason: A-C power can be transformed; it can be stepped up and stepped down in the voltage delivered to customers. Edison's side of the argument was doomed because D-C power is straight-line power; it cannot be stepped up and down without huge voltage losses along the transmission lines. A-C power can be sent and transmitted in steady distribution to a point to counter voltage drops due to transmission line resistance. A-C power can then be stepped up to continue its transmission and then stepped down or further up to suit the user's needs. This steady supply of adjustable electricity is possible because of transformers.

A *Transformer* is an electric control device (with no moving parts) that raises or lowers voltage or current in an electric distribution system. A transformer consists of two electrically insulated coils and wound on a common core (see Figure 13.71). Magnetic coupling is used to transfer electric energy from one coil to another. The coils which **receive** energy from A-C sources are called the **primary**. The coil that delivers energy to A-C loads is called the **secondary**. The core of transformers used at low frequencies is generally made of magnetic material, usually laminated sheet steel. Cores of transformers used at higher frequencies are made of powdered iron and ceramics or nonmagnetic materials. Some coils are wound on nonmagnetic hollow forms such as cardboard or plastic, so the core material is air.

In operation, an alternating current will flow when an A-C voltage is applied to the primary coil of a transformer. This current produces a field of force that changes as the current changes. The magnetic core carries the changing magnetic field to the

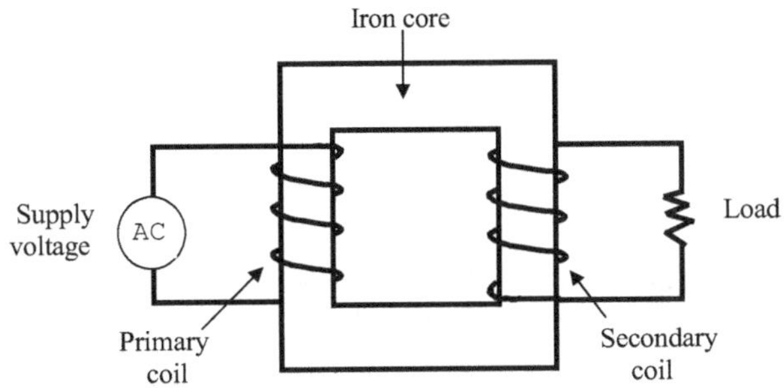

FIGURE 13.71 Basic transformer.

secondary coil, where it cuts across the turns of that coil. In this way, an A-C voltage in one coil is transferred to another, even though there is no electrical connection between them. The number of lines of force available in the primary is determined by the primary voltage and the number of turns on the primary—each turn producing a given number of lines. If there are many turns on the secondary, each line of force will cut many turns of wire and induce a high voltage. If the secondary contains only a few turns, there will be few cuttings and low induced voltage. The secondary voltage, then, depends on the number of secondary turns compared to the number of primary turns. If the secondary voltage has twice as many turns as the primary voltage, the secondary voltage will be twice as large as the primary voltage. If the secondary has half as many turns as the primary, the secondary voltage will be one-half as large as the primary voltage.

Important Point: The voltage on a transformer's coils is directly proportional to the number of turns on the coils.

A voltage ratio of 1:4 (read as "1 to 4") means that for each volt on the primary, there are 4 V on the secondary. This is called a **step-up** transformer. A step-up transformer receives a low voltage on the primary and a high voltage from the secondary. A voltage ratio of 4:1 (read as "4 to 1") means that for 4 V on the primary, there is only 1 V on the secondary. This is called a **step-down** transformer. A step-down transformer receives a high voltage on the primary and delivers a low voltage from the secondary.

13.21 THE BOTTOM LINE

Energy transformation is the process of converting energy from one form to another. Because energy can neither be created nor destroyed, energy transformation takes place. Energy is available from several sources. No matter what form of energy we use, it will be transformed. Energy transformation into electricity is the frontrunner of energy transformers at present. AI energy applications are increasing in leaps and bounds in this transformation process. The paradox, the irony, is that not only is AI used in electricity, but it is AI that consumes vast amounts of electricity in their growing data centers. Nuclear power plants must be reactivated, and new ones must be built to address this ever-increasing need for energy transformation to generate electricity. The author believes that the endless flow of electrons in a conductor to a load is the future.

NOTE

1 Adaptation from Spellman, F.R., and Drinan, J.E., 2001. *Electricity.* Boca Raton, FL: CRC Press.

REFERENCES

Spellman, F.R., and Drinan, J.E., 2001. *Electricity.* Boca Raton, FL: CRC Press.
USDOE, 1992. *DOE Fundamentals Handbook: Electrical Science Volume 1 of 4.* Washington, DC: U.S. Department of Energy.

14 When Water and Electricity Mix

14.1 SUMMARY

The future lies in electron flow: electricity moving through a conductor feeding a load. The age of fossil fuels, which provide many of our present energy needs, is ending. With the eventual decline and depletion of fossil fuels, renewable sources, such as wind, sun, and hydro, are progressing to fill the massive gap in our energy needs that is occurring and widening. The truth be told, the energy landscape is evolving rapidly. Electricity production is increasing rapidly, and finding the means to store it is the latest technological challenge. This energy storage issue is a current problem and is presently the only barrier preventing even more exponential growth in electrical energy production.

Batteries are currently the primary storage devices for electricity. However, electricity can also be stored in ultra- or supercapacitors. The author believes that the ultimate storage is water, as in conventional hydropower or the flow provided by pumped storage hydro. Water itself does not produce electricity. Water flow offers the force to turn a turbine generator in its electromagnetic field, which produces electron flow and electricity in a conductor.

In Chapter 2, it was pointed out that there are several different ways to produce electricity:

a. **Friction**—The voltage produced by rubbing two materials together (static electricity or electrostatic force).
b. **Pressure (piezoelectricity)**—The voltage produced by squeezing or applying pressure to crystals of certain substances (e.g., crystals like quartz or Rochelle salts or certain ceramics like barium titanate).
c. **Heat (thermoelectricity)**—The voltage produced by heating the joint (junction) where two unlike metals are joined.
d. **Light (photoelectricity)**—The voltage produced by light (photons) striking photosensitive (light-sensitive) substances.
e. **Chemical action**—The voltage produced by a chemical reaction in a battery cell.
f. **Thermionic emission**—A thermionic energy converter consists of two electrodes placed near one another in a vacuum.
g. **Magnetism**—The voltage produced in a conductor when the conductor moves through a magnetic field, or when a magnetic field moves through the conductor in such a manner as to cut the magnetic lines of force of the field.

DOI: 10.1201/9781003605430-16

At present, we are most concerned with magnetism (generators powered by hydropower, for example), light (photoelectricity produced by solar cells), and chemistry (chemical energy converted to electricity in batteries) to produce voltage. Friction has few practical applications. Pressure and heat have valuable applications but are limited. Magnetism used in generators, solar light-produced electricity, and the chemistry involved with storing electricity in batteries, on the other hand, are the principal and practical sources of voltage. Recall that voltage is pressure. Voltage does not move or flow, and when someone shocked by electricity says, "The voltage reached out and got me." This is an incorrect statement. Voltage does not move. Instead, voltage or pressure forces electron flow in a conductor to power up a load, whatever the load, and provides the electric shock.

You might wonder where the wind is. That is where the wind-generated electricity is. Well, wind is only the motive source, the prime mover that turns the turbine blades. The turbine generates electricity via magnetism, the primary energy source.

Generating electricity is one thing; using the current flow is another, and the bottom line is the availability of the current flow.

Availability?

When you turn on your electric lights or appliances, you expect them to come on and function as designed.

This points to the most vital factor related to electric power: its accessibility and availability. Storing electricity for future use is an issue that is being investigated from many different angles and studied using experimental technologies. When we think about electricity storage, it is common to think about storage batteries. The problem with batteries is that they hold an electrical charge until it is depleted unless there is some constant or available charging source.

14.2 THE BOTTOM LINE

Artificial intelligence and smart grid technology are being integrated to optimize energy delivery based on demand patterns. This is important in transitioning from fossil fuels to renewable energy in the supply mix.

We all learned early on that water and electricity do not mix. This mix involves using water to power turbo-generators for electricity production.

The bottom line is that AI can make sense of data from large and scattered information sources much better and faster than humans using conventional approaches. Moreover, AI is key to ensuring the reliability and sustainability of today's energy systems, as well as creating and scaling future energy systems.

Glossary[1]

Absorber: The material readily absorbs photons to generate charge carriers (free electrons or holes) in a photovoltaic device.

Acid Hydrolysis: A chemical process in which acid converts cellulose or starch to sugar.

Albedo: The ratio of light is reflected by a surface to the light falling on it.

Alcohol: A general class of hydrocarbons that contain a hydroxyl group (OH). There are many types of alcohol (Butanol, ethanol, methanol).

Alcohol Fuels: Alcohol can be blended with gasoline for use as transportation fuel. It may be produced from a wide variety of organic feedstock. The common alcohol fuels are methanol and ethanol. Methanol may be produced from coal, natural gas, wood, and organic waste. Ethanol is commonly made from agricultural plants, primarily corn, which contains sugar.

Absorber: The material readily absorbs photons to generate charge carriers (free electrons or holes) in a photovoltaic device.

Acid Hydrolysis: A chemical process in which acid converts cellulose or starch to sugar.

Albedo: The ratio of light is reflected by a surface to the light falling on it.

Alcohol: A general class of hydrocarbons that contain a hydroxyl group (OH). There are many types of alcohol (Butanol, ethanol, methanol).

Alcohol Fuels: Alcohol can be blended with gasoline for use as transportation fuel. It may be produced from a wide variety of organic feedstock. The common alcohol fuels are methanol and ethanol. Methanol may be produced from coal, natural gas, wood, and organic waste. Ethanol is commonly made from agricultural plants, primarily corn, which contains sugar.

Alkaline Fuel Cell (AFC): In a hydrogen/oxygen fuel cell type, the electrolyte is concentrated potassium hydroxide (KOH), and hydroxide ions (OH) are transported from the cathode to the anode.

Alternating Current (AC): An electric current that reverses its direction at regularly recurring intervals, usually 50 or 60 times per second.

Alternator: A device that turns the rotation of a shaft into alternating current (A-C).

Ambient: The natural condition of the environment at any given time.

Ampere (Amp): A unit of electrical current can be thought of as the rate of water flowing through a pipe (L/min).

Ampere-Hour: Amperes times hour measures energy production over time and battery capacity.

Amorphous Silicon: An alloy of silica and hydrogen, with a disordered, noncrystalline internal atomic arrangement that can be deposited in thin layers (a few micrometers in thickness) by several deposition methods to produce thin-film photovoltaic cells on glass, metal, or plastic substrates.

Anaerobic Digestion: A biochemical process by which organic matter is decomposed of bacteria in the absence of oxygen, producing methane and other byproducts.

Anemometer: A wind speed measurement device sends data to the controller and surveys the wind site.

Angle of Attack: In wind turbine operation, the airflow angle is relative to the blade.

Angle of Incidence: The angle a sun ray makes with a line perpendicular to the surface. For example, a surface that directly faces the sun has a solar incidence of zero, but if the surface is parallel to the sun (for example, sunrise striking a horizontal rooftop), the incidence is 90°.

Anion: A negatively charged ion is an ion that is attracted to the anode.

Annual Removals: The net volume of growing stock trees removed from the inventory during a specified year by harvesting, cultural operations such as timber land improvement, or land clearing.

Annualized Growth Rate: It is calculated as follows: $(x_n/x_1)1/n$
where x is the value under consideration and n is the number of periods.

Anode: The electrode at which oxidation (a loss of electrons) takes place. The anode is the negative terminal for fuel cells and other galvanic cells; the anode is the positive terminal for electrolytic cells.

Aquifer: Water-bearing stratum of permeable sand, rock, or gravel.

Asexual Reproduction: The naturally occurring ability of some plant species to reproduce asexually through seeds, meaning the embryos develop without a male gamete. This ensures the seeds will produce plants identical to the mother plant.

Availability Factor: A percentage represents the number of hours a generating unit is available to produce power (regardless of the amount of power) in each period, compared to the number of hours.

Azimuth Angle: The angle between the authentic south and the point on the horizon is directly below the sun.

B

Bagasse: Sugar mills often burn the fibrous material remaining after the juice is extracted from sugarcane as an energy source.

Baseload Plants: Electricity-generating units that are operated to meet the constant or minimum load on the system. The energy cost from such units is usually the lowest available to the system.

Binary Cycle: Binary geothermal systems heat a secondary fluid, which drives the power turbine, using the extracted hot water or steam.

Biobased Product: The term "biobased product, " as defined by the Farm Security and Rural Investment Act (FSRIA), means a product determined by the U.S. Secretary of Agriculture to be a commercial or industrial product (other than food or feed) that is composed, in whole or in significant part, of biological products or renewable domestic agriculture materials (including plant, animal, and marine materials) or forestry materials.

Biochemical Conversion: Fermentation or anaerobic digestion produces fuels and chemicals from organic sources.

Biodiesel: Fuel derives from vegetable oils or animal fats. It is produced when vegetable oil or animal fat chemically reacts with alcohol.

Bioenergy: Renewable energy is produced from organic matter, which may be used directly as fuel or processed into liquids and gases.

Biofuels: Liquid fuels and blending components produced from biomass (plant) feedstocks are used primarily for transportation.

Biogas: A combustible gas derived from decomposing biological waste. Biogas normally consists of 50%–60% methane.

Biomass: Any organic nonfossil material of biological origin constitutes a renewable energy source. It is produced from organic matter available on a renewable or recurring basis, including crops and trees, wood and wood resides, plants (including aquatic plants), grass, animal manure, municipal residues, and other residue materials. Biomass is generally produced sustainably from water and carbon dioxide by photosynthesis. There are three main categories of biomass—primary, secondary, and tertiary.

Biomass Gas (Biogas): A medium Btu gas containing methane and carbon dioxide results from microorganisms' action on organic materials such as a landfill.

Biomaterials: Products derived from organic (as opposed to petroleum-based) products.

Bio-oil: Intermediate fuel derived from fast pyrolysis.

Biopower: The use of biomass feedstock to produce electric power to heat through direct combustion of the feedstock, gasification, and then combustion of the resultant gas or other thermal conversion processes. Fuel, turbines, fuel cells, or other equipment generate power.

Biorefinery: A facility that processes and converts biomass into value-added products. These products can range from biomaterials to fuel, such as ethanol, or important feedstocks for producing chemicals and other materials. Biorefineries can be based on several processing platforms using mechanical, thermal, chemical, and biochemical processes.

Blackbody: It is an ideal substance that absorbs all radiation falling on it, reflecting nothing.

Black Liquor (Pulping Liquor): The alkaline spent liquor was removed from the digesters by chemically pulling wood. After evaporation, the liquor is burned as fuel in a recovery furnace that permits the recovery of certain basic chemicals.

Bone Dry: Having zero-percent moisture content.

Borehole Breakouts: Stress in the rock surrounding the borehole causes the borehole to fail. The breakout is generally located symmetrically in the wellbore, perpendicular to the direction of the most significant horizontal stress on a vertical wellbore.

Brine: A geothermal solution containing appreciable amounts of sodium chloride or other salts.

British Thermal Unit (Btu): This is an essential measure of thermal (heat) energy. A Btu is the energy required to increase the temperature of 1 pound of water by 1°F at normal atmospheric pressure. 1 Btu = 1,055 J.

Bulk Density: Weight per unit of volume is usually specified in pounds per cubic foot.

C

Cap Rocks: Rocks of low permeability overlie a geothermal reservoir.

Capacity Factor: The ratio of the electrical energy produced by a generating unit for the period considered to the electrical energy that could have been made at continuous full-power operation during the same period.

Capacity, Gross: The full-load continuous rating of a generator, prime mover, or other electric equipment under specified conditions designated by the manufacturer. It is usually indicated on a nameplate attached to the equipment.

Capital Cost: The cost of field development, plant construction, and the equipment required for electricity generation.

Carbon dioxide (CO_2): A product of combustion, the most common greenhouse gas.

Carbon monoxide (CO): A colorless, odorless gas produced by incomplete combustion. Carbon monoxide is poisonous when inhaled.

Carbon Sequestration: The absorption and storage of carbon dioxide from the atmosphere by naturally occurring plants.

Carnot Cycle: An ideal heat engine (conceived by Sadi Carnot) is one in which the sequence of operations forming the working cycle consists of isothermal expansion, adiabatic expansion, isothermal compression, and adiabatic compression back to its initial state.

Cascading Heat: A process that uses a stream of geothermal hot water or steam to perform successive tasks requiring low and lower temperatures.

Casing: The pipe is placed in a wellbore as a structural interface between the wellbore and the surrounding formation. It typically extends from the top of the well and is cemented to maintain the diameter of the wellbore and provide stability.

Cast Silicon: Crystalline silicon is obtained by pouring pure molten silicon into a vertical mold and adjusting the temperature gradient along the mold volume during cooling to get slow, vertically advancing crystallization of the silicon. The polycrystalline ingot thus formed is composed of large, relatively parallel, interlocking crystals. The cast ingots are sawn into wafers for further fabrication into photovoltaic cells. Cast-silicon wafers and ribbon-silicon sheets fabricated into cells are usually called polycrystalline photovoltaic cells.

Cathode: The electrode at which reduction (a gain of electrons) occurs. The cathode is the positive terminal for fuel cells and other galvanic cells; the cathode is the negative terminal for electrolytic bells (where electrolysis occurs).

Cation: A positively charged ion.

Cellulose: The main carbohydrate in living plants. Cellulose forms the skeletal structure of the plant cell wall.

Chips: Woody material is cut into short, thin wafers. Chips are used as a raw material for pulping and fiberboard or biomass fuel.

Closed-loop Biomass: Crops grown sustainably to optimize their value for bioenergy and bioproduct uses include annual crops such as maize and wheat as well as perennial crops such as trees, shrubs, and grasses like switchgrass.

Coarse Materials: Wood residues suitable for chipping, such as slabs, edgings, and trimmings.

Co-firing: The practice of introducing biomass into the boilers of coal-fired power plants.

Combined Cycle: An electric generating technology in which electricity is produced from otherwise lost waste heat exiting from one or more gas (combustion) turbines. The exiting heat is routed to a conventional boiler or a heat recovery steam generator for utilization by a steam turbine in the production of electricity. Such designs increase the efficiency of the electric generating unit.

Combined Heat and Power (CHP) Plant: A plant is designed to produce heat and electricity from a single heating source. Note: This term is being used instead of the term "cogenerator" used by EIA in the past. CHP better describes the facilities because some plants did not produce heat and power sequentially and, as a result, did not meet the legal definition of cogeneration specified in the Public Utility Regulatory Policies Act (PURPA).

Commercial Sector: An energy-consuming sector comprises service-providing facilities and equipment for businesses, federal, state, and local governments, as well as other private and public organizations, such as religious, social, or fraternal groups. The commercial sector includes institutional living quarters. It also includes sewage treatment facilities. Common uses of energy associated with this sector include space heating, water heating, air conditioning, lighting, refrigeration, cooking, and running various other equipment. Note: This sector includes generators that produce electricity and/or useful thermal output primarily to support the activities of the above-mentioned commercial establishments.

Commercial Species: Tree species are suitable for industrial wood products.

Compressed Natural Gas (CNG): Mixtures of hydrocarbon gases and vapors consist principally of methane in gaseous form that has been compressed.

Concentrator: A reflective or refractive device that focuses incident insolation onto an area smaller than the reflective or refractive surface, resulting in increased insolation at the point of focus.

Condensate: Water is formed by condensation of steam.

Condenser: Equipment that condenses turbine exhaust steam into condensate.

Conservation Reserve Program (CRP): CRP provides farm owners or operators with an annual per-acre rental payment and half the cost of establishing permanent land cover in exchange for retiring environmentally sensitive cropland from production for 10–15 years. In 1996, Congress reauthorized CRP for an additional round of contracts, limiting enrollment to 36.4 million acres at any time. The 2002 Farm Act increased the enrollment limit to 39 million acres. Producers can offer land for competitive bidding based on an Environmental Benefits Index (EBI) during periodic signups or automatically enroll more limited acreages in processes such as riparian buffers, field windbreaks, and grass strips. CRP is funded through the Commodity Credit Corporation (CCC).

Cooling Tower: A structure in which heat is removed from hot condensate.

Conventional Hydroelectric (Hydropower) Plant: A plant where all the power is produced from natural streamflow as regulated by available storage.

Core: A cylinder of rock recovered from the well by a special coring drill bit.

Crop Failure: It consists mainly of the acreage on which crops failed because of weather, insects, and diseases, but includes some land not harvested due to lack of labor, lower market prices, or other factors. The acreage planted to cover and improve crops not intended for harvest is excluded from crop failure and is considered idle.

Cropland: Total cropland includes five components: cropland harvested, crop failure, cultivated summer fallow, cropland used only for pasture, and idle cropland.

Cropland Harvested: It includes row and closely sown crops; hay and silage crops; tree fruits, small fruits, berries, and tree nuts; vegetables and melons; and miscellaneous other minor crops. In recent years, farmers have double-cropped about 4% of the acreage.

Cropland Pasture: Land used for long-term crop rotation. However, some cropland pastures are marginal for crop uses and may remain in the pasture indefinitely. This category also includes land used for pasture before crops reached maturity and some land used for pasture that could have been cropped without additional improvement.

Cropland Used for Crops: Cropland used for crops includes cropland harvested, crop failure, and cultivated summer fallow.

Crust: Earth's outer layer of rock. Also called the lithosphere.

Cryogenic Liquefaction: The process through which gases such as nitrogen, hydrogen, helium, and natural gas are liquefied under pressure at very low temperatures.

Cull Tree: A live tree, 5.0 inches in diameter at breast height (dbh) or larger than is non-merchantable for saw logs now or prospectively because of rot, roughness, or species.

Cut-in Speed: The speed at which a shaft must turn to generate electricity and send it over a wire.

D

Daylighting: Direct, diffuse, or reflected sunlight provides supplemental lighting for building interiors.

dbh: The diameter is measured at approximately breast height from the ground.

Densification: A mechanical process to compress biomass (usually wood waste) into pellets, briquettes, cubes, or densified logs.

Depletion Factor: Annual percentage of the depletion of the thermal resource.

Digester Gas: Biogas is produced using a digester, an airtight vessel or enclosure in which bacteria decompose biomass in water to produce biogas.

Diode: A solid-state device that acts as a one-way valve for electricity.

Direct Current: An electric current that flows in a constant direction. The magnitude of the current does not vary or has a slight variation.

Direct Methanol Fuel Cell (DMFC): A type of fuel cell in which the fuel is methanol (CH_3OH) in gaseous or liquid form. The methanol is oxidized directly at the anode instead of first being reformed to produce hydrogen. The electrolyte is typically a polymer electrolyte membrane (PEM).

Direct Use: Use geothermal heat without converting it to electricity, such as for space heating and cooling, food preparation, and industrial processes.

Distributed Generation (Distributed Energy Resources): This refers to electricity provided by small, modular power generators (typically ranging in capacity from a few kilowatts to 50 MW) located at or near customer demand.

District Heating: A type of direct use is when a utility system supplies multiple users with hot water or streams from a central plant or well field.

Downwind Turbine: A turbine that does not face into the wind and whose direction is controlled directly by the wind.

Drag Bit: Drilling bit that drills by scraping or shearing the rock with fixed hard surfaces, "cutters."

Drilling: Boring into the Earth to access geothermal resources, usually with oil and gas drilling equipment modified to meet geothermal requirements.

Dry Steam: Very hot steam that doesn't occur with liquid.

Dump Load: A device that allows excess energy to be safely disposed of.

E

E-10: A mixture of 10% ethanol and 90% gasoline based on volume.

E-85: A mixture of 85% ethanol and 15% gasoline based on volume.

Efficiency: The ratio of the sound energy output of a machine or other energy-converting plant to the energy input.

Electric Power Sector: An energy-consuming sector consisting of electricity-only and combined heat and power (CHP) plants whose primary business is to sell electrical energy or electricity and heat to the public—i.e., North American Industry Classification System 22 plants.

Electric Utility: A corporation, person, agency, authority, or other legal entity or instrumentality aligned with distribution facilities for the delivery of electric energy primarily by the public. Investor-owned electric utilities, municipal and State utilities, Federal electric utilities, and rural electric cooperatives are included. A few tariff-based entities that are corporately aligned with companies that own distribution facilities are also included.

Emissions: Anthropogenic releases of gases into the atmosphere. In the context of global climate change, they consist of relatively important greenhouse gases (e.g., the release of carbon dioxide during fuel combustion).

Endothermic: A chemical reaction that absorbs or requires energy (usually in the form of heat).

Energy: The ability to do work.

Energy Crops: Crops are grown specifically for their fuel value. They include food crops such as corn and sugarcane and nonfood crops such as poplar trees and switchgrass. Currently, low-energy crops are under development: short-rotation woody crops, which are fast-growing hardwood trees harvested in five to eight years, and herbaceous energy crops, such as perennial grasses, which are harvested annually after taking two to three years to reach full productivity.

Enhanced Geothermal Systems (EGS): Engineered reservoirs that can extract economic amounts of that from geothermal resources. Rock fracturing, water injection, and water circulation technologies sweep heat from the unproductive areas of existing geothermal fields or new fields lacking sufficient production capacity.

Enthalpy: A substance's thermodynamic property is defined as the sum of its internal energy plus the pressure of the substance times its volume, divided by the mechanical equivalent of heat. The total heat content of air is the sum of the enthalpies of dry air and water vapor per unit weight of dry air, measured in Btu per pound (or calories per kilogram).

Environmental Impact Statement: A document was created from a study of the expected environmental effects of a new development or installation.

Ethanol (also known as Ethyl Alcohol or Grain Alcohol, CH_3-CH_2OH): A clear, colorless, flammable, oxygenated hydrocarbon with a boiling point of 173.5°F in an anhydrous state. However, it readily forms a binary azeotrope with water, with a boiling point of 172.67°F at a composition of 9,557% by weight ethanol. It is used in the United States as a gasoline octane enhancer and oxygenate (maximum 10% concentration). Ethanol can be used in higher concentrations (E-85) in vehicles designed for its use. Ethanol is typically produced chemically from ethylene or biologically from the fermentation of various sugars from carbohydrates found in crops and cellulosic residues from crops or wood. The lower heating value, equal to 76,000 Btu per gallon, is assumed for estimates in this text.

Evacuated Tube: In a solar thermal collector, an absorber tube is contained in an evacuated glass cylinder through which collector fluids flow.

Exothermic: A chemical reaction that gives off heat.

Externality: A cost or benefit is not accounted for in the price of goods and services. Often, "externality" refers to the cost of pollution and other environmental impacts.

F

Fast Pyrolysis: Thermal biomass conversion by rapid heating to 450°C–600°C without oxygen.

Fault: A fracture in rock exhibiting relative movement between the adjoining surfaces.

Feller-Buncher: A self-propelled machine that cuts trees with giant shears near ground level and then stacks the trees into piles to await skidding.

Feedstock: A product used as the basis for the manufacture of another product.

Fenestration: The whole-building design approach will determine what fenestration products should be used—windows, doors, and skylights. You want to select products that accommodate your building's climate, including insulation, daylighting, heating and cooling, and natural ventilation.

Fermentation: Conversion of carbon-containing compounds by microorganisms for producing fuels and chemicals such as alcohol, acids, or energy-rich gases.

Fiber Products: Products derived from fibers of herbaceous wood plant material. Examples include pulp, composition board products, and wood chips for export.

Fine Materials: Wood residues, such as planner shavings and sawdust, are unsuitable for chipping.

Fischer–Tropsch Fuels: Liquid hydrocarbon fuels are produced by combining carbon monoxide and hydrogen. This process converts coal, natural gas, and low-value refinery products into a high-value diesel substitute.

Flash Steam: Steam is produced when the pressure on a geothermal liquid is reduced. It's also called flashing.

Flat Plate Pumped: A medium-temperature solar thermal collector typically consists of a metal frame, glazing, absorbers (usually metal), and insulation. It uses a pump liquid as the heat-transfer medium and is predominantly used in water heating applications.

Flexible-Fuel Vehicle: A vehicle with a single fuel tank designed to run on varying blends of unleaded gasoline with either ethanol or methanol.

Flow Battery: An electrochemical energy storage device that utilizes tanks of rechargeable electrolytes to refresh the energy-producing reaction. Since its capacity is limited only by the size of its electrolyte tanks, it is helpful for large-scale backup systems to supplement other intermittent generation forms.

Forest Land: Land at least 10% stocked by forest trees of any size, including land that formerly had such tree cover and that will be naturally or artificially regenerated. Forest land includes transition zones, such as areas between heavy forested and nonforested lands that are at least 10% stocked with forest trees and forest areas adjacent to urban and built-up lands. Also included are pinyon-juniper and chaparral areas in the West and the afforested regions. The minimum area for classification of forest land is 1 acre. Roadside, streamside, and shelterbelt strips of trees have a crown width of at least 120 ft to qualify as forest land. Unimproved roads, trails, streams, and clearings in the forest areas are classified as forests if they are less than 120 ft wide.

Fracture: Natural or induced break in the rock.

Fracturing Treatments: Fracturing treatments are performed by pumping fluid into the subsurface at pressures above the fracture pressure of the reservoir formation to create a highly conductive flow path between the reservoir and the wellbore.

Fuel Cell: One or more cells can generate an electrical current by converting the chemical energy of fuel directly into electrical energy. Fuel cells differ from conventional electrical cells in that the active materials, such as fuel and oxygen, are not contained within the cell but are supplied from outside.

Fuel Cell Poisoning: A fuel cell's efficiency is lowered due to impurities in the fuel binding to the catalyst.

Fuel Cell Stack: Individual fuel cells are connected in a series. Fuel cells are stacked to increase voltage.

Fuel Treatment Evaluator (FTE): A strategic assessment tool capable of aiding the identification, evaluation, and prioritization of fuel treatment opportunities.

Fuelwood: Wood and wood products, possibly coppices, scrubs, branches, etc., are bought, gathered, and used by direct combustion.

Full Sun: The amount of power density in sunlight is received at the earth's surface at noon on a clear day (about 1,000 W/m^2).

Fumarole: A vent or hole in the Earth's surface, usually in a volcanic region, from which steam, gaseous vapors, or hot gases issue.

G

Gallon: Fuel oil is measured using a volumetric measure equal to four quarts (231 cubic inches). One gallon equals 3.785 L, and one barrel equals 42 gallons.

Gasification: A chemical or heat process converts solid fuel to a gaseous form.

Gasohol: A motor vehicle fuel that is a blend of 90% unleaded gasoline and 10% ethanol (by volume).

Gearbox: Increases the rpm of the low-speed shaft, transferring its energy to the high-speed shaft to provide enough speed to generate electricity.

Generation (Electricity): The process of producing electric energy from other forms of energy; the amount of electric energy produced is expressed in watt-hours (Wh).

Geothermal: Of or relating to the earth's interior heat.

Geothermal Energy: The extraction of the Earth's interior heat from hot water or rocks makes it available to man. At electric power plants, hot water or steam extracted from geothermal reservoirs in the Earth's crust is supplied to steam turbines that drive generators to produce electricity.

Geothermal Gradient: The rate of temperature increase on Earth is a function of depth. Temperature increases an average of 1°F for every 75 ft in descent.

Geothermal Heat Pumps: Devices that take advantage of the relatively constant temperature of the Earth's interior use it as a source and sink of heat for heating and cooling. When cooling, heat is extracted from space and dissipated into the Earth; when heating, heat is extracted from the Earth and pumped into space.

Geothermal Resources: When the earth's natural heat is collected and transported to the surface, it can be used for beneficial purposes.

Geothermal Plant: A plant in which a turbine is driven either by hot water or by natural steam, which derives its energy from heat found in rocks or fluids at various depths beneath the earth's surface. The fluids are extracted by drilling and/or pumping.

Geyser: A spring that shoots jets of hot water and steam into the air.

Glazing: Transparent or translucent material (glass or plastic) used to admit light and/or to reduce heat loss; used for building windows, skylights, or greenhouses or for covering the aperture of a solar collector.

Global Positioning System (GPS): A navigation system using satellite signals to fix the location of a radio receiver on or above the earth's surface.

Grassland Pasture and Range: All open land is used primarily for pasture and grazing, including shrub and brush land types of pasture; grazing land with sagebrush and scattered mesquite; and all team and native grasses, legumes, and other forage used for pasture or grazing. Because of the diversity in vegetative composition, grassland pasture and range are not always clearly distinguishable from other types of pasture and range. At one extreme, permanent grassland may merge with cropland pasture, or grassland may often be found in transitional areas with forested grazing land.

Gravimetry: The use of precisely measured gravitational force to determine mass differences that can be correlated to subsurface geology.

Green Pricing/Marketing: In the case of renewable electricity, green pricing represents a market solution to the various problems associated with regulatory valuation of the non-market benefits of renewables. Green pricing programs allow electricity customers to express their willingness to pay for renewable energy development through direct payments on their monthly utility bills.

Greenhouse Gas: Gases that trap the heat of the sun in the Earth's atmosphere, producing the greenhouse effect. The two major greenhouse gases are water vapor and carbon dioxide. Other greenhouse gases include methane, ozone, chlorofluorocarbons, and nitrous oxide.

Grid: The layout of an electrical distribution system.

Gross Generation: The total amount of electric energy produced by the generating units at a generation station or stations, measured at the generator terminals.

Growing Stock: A classification of timber inventory that includes live trees of commercial species meeting specified standards of quality or vigor. Cull trees are excluded. When associated with volume, it contains only trees 5.0 inches in dbh and larger.

H

HDR: Hot, dry rock. Subsurface geologic formations of abnormally high heat content that contain little or no water.

Hardwoods: Usually broad-leaved and deciduous trees.

Heat Exchanger: A device for transferring thermal energy from one fluid to another.

Heat Flow: Heat moves from the earth to the surface, where it is dissipated by radiation into the atmosphere, surface water, and space.

Heat Pump: A year-round heating and air-conditioning system employing a refrigeration cycle. In a refrigeration cycle, a refrigerant is compressed (as a liquid) and expanded (as a vapor) to absorb and reject heat. The heat pump transfers heat to a space to be heated during the winter period and, by reversing the operation, extracts (absorbs) heat from the same space to be cooled during the summer. The refrigerant within the heat pump in the heating mode absorbs the heat to be supplied to the space to be heated from an outside medium (air, ground, or groundwater) and, in the cooling mode, absorbs heat from the space to be cooled to be rejected to the outside medium.

Heat Pump (Air Source): An air-source heat pump is the most common type of heat pump. The heat pump absorbs heat from the outside air and transfers the heat to the space to be heated in the heating mode. In cooling mode, the heat pump absorbs heat from the space to be cooled and rejects the heat from the outside air. In the heating mode, when the outside air approaches 32°F or less, air-source heat pumps lose efficiency and generally require a backup (resistance) heating system.

Heat Pump (Geothermal): A heat pump in which the refrigeration exchanges heat (in a heat exchanger) with a fluid circulating through an earth connection medium (ground or groundwater). The fluid is contained in a variety of loop (pipe) configurations depending on the temperature of the ground and the ground area available. Loops may be installed horizontally or vertically in the ground or submerged in water.

Heat Pump (efficiency): The efficiency of a heat pump, that is, the electrical energy to operate it, is directly related to the temperatures between which it operates. Geothermal heat pumps are more efficient than conventional heat pumps for air conditioners that use outdoor air since the ground or groundwater is a few feet below the earth's surface and remains relatively constant throughout the year. It is from the atmosphere where the air temperature is much colder, and in summer, waste heat is transferred to the relatively calm ground rather than to the hotter air. Geothermal heat pumps are generally more expensive to install than outside air heat pumps. However, depending on the location, geothermal heat pumps can reduce energy consumption (operating cost) and emissions by more than 20% compared to high-efficiency outside air heat pumps. Geothermal heat pumps also use the waste heat from air-conditioning to provide free hot water heating in the summer.

Heating Value: The maximum amount of energy that is available from burning a substance.

Herbaceous: Non-woody type of vegetation, usually lacking permanent strong stems, such as grasses, cereals, and canola (rapeseed).

High-Speed Shaft: Transmits force from the gearbox to the generator.

High-Temperature Collector: A solar thermal collector is designed to operate at 180°F or higher.

Hub: The center part of the rotor assembly connects the blades to the low-speed shaft.

Hydraulic Stimulation: A stimulation technique was performed using fluid.

Hydrothermal: About hot water.

Hydrothermal Reservoir: An aquifer, or subsurface water, has sufficient heat, permeability, and water to be exploited without stimulation or enhancement.

Idle Cropland: Land in cover, soil improvement crops, and cropland on which no crops were planted. Some croplands are idle each year for various physical and economic reasons. Acreage diverted from corps to soil-conserving uses (if not eligible for and utilized as cropland pasture) under federal farm programs is included in this component. Cropland enrolled in the Federal Conservation Reserve Program (CRP) is included in the idle cropland.

I

Incident Light: Light that shines onto the face of a solar cell or module.

Induced Seismicity: Induced seismicity refers to minor earthquakes and tremors typically caused by human activity altering the stresses and strains on the Earth's crust. Most induced seismicity is of an extremely low magnitude, and in many cases, human activity is merely the trigger for an earthquake that would have occurred naturally.

Industrial Wood: All commercial Roundwood products except Fuelwood.

Injection: The process of returning spent geothermal fluids to the subsurface. Sometimes referred to as reinjection.

Interferometric Synthetic Aperture Radar (InSAR): A remote sensing technique that uses radar satellite images to determine the movement of the earth's surface.

Internal Collector Storage: A solar thermal collector is one in which the storage medium absorbs incident solar radiation.

Irradiance: The direct, diffuse, and reflected solar radiation that strikes the surface.

J

Joule: This is the basic energy unit for the metric system, or in a later, more comprehensive formulation, the International System of Units (SI). It is ultimately defined in terms of the meter, kilogram, and second.

K

Kaplan Turbine: A type of turbine that has two blades whose pitch is adjustable. The turbine may have gates to control the fluid flow angle into the blades.

Kilowatt (kW): One thousand watts of electricity.

Kilowatt-hour (kWh): One thousand watthours.

L

Landfill Gas: Gas is generated by the decomposition of organic material at landfill disposal sites. Landfill gas is approximately 50% methane.

Langley (L): Unit of solar irradiance. One gram calorie per square centimeter. $1\ L = 85.93\ kWh/m^2$

Leeward: Away from the direction of the wind. Opposite of windward.

Life-Cycle Analysis: The analysis focused on a product's environmental impact throughout its life cycle, from resource extraction to post-consumer waste disposal. It is a comprehensive approach to examining a product or package's ecological impacts.

Lignin: Structural constituent of wood and (to a lesser extent) other plant tissues, which encrusts the cell walls and cements the cells together.

Line Shaft Pump: A fluid pump with a pumping mechanism in the wellbore is driven by a shaft connected to a motor on the surface.

Liner: A casing string that does not extend to the top of the wellbore but is anchored or suspended from inside the bottom of the previous casing string.

Liquid Collector: A medium-temperature solar thermal collector, predominantly employed in water heating, uses pumped liquid as the heat-transfer medium.

Lithology: The study and description of rocks in terms of their color, texture, and mineral composition.

Live Cull: A classification that includes live cull trees. When associated with volume, the net volume in live cull trees is 5.0 inches in dbh and larger.

Load: The simultaneous demand of all customers is required at any specified point in an electric power system.

Load Balancing: One of the challenges of wind energy production, which produces energy on a less predictable schedule than other methods, is keeping the amount of electricity produced (the supply) equal to consumption (the demand).

Local Solar Time: A system of astronomical time in which the sun crosses the true north-south meridian at noon and differs from local time according to longitude, time zone, and equation of time.

Logging Residues: The unused portions of growing and non-growing stock trees were cut or killed by logging and left in the woods.

Lost Circulation: Zones in a well that imbibe drilling fluid from the wellbore reduce fluid flow returning to the surface. This loss causes drilled rock particles to build up in the well and can cause problems with cementing the casing.

Low-Speed Shaft: Connects the rotor to the gearbox.

Low-Temperature Collectors: Metallic or nonmetallic solar thermal collectors generally operate at temperatures below 110°F and use pumped liquid or air as the heat transfer medium. They usually contain no glazing or insulation and are often made of plastic or rubber, although some are metal.

M

Magma: Molten rock within the Earth, from which igneous rock is formed by cooling.

Magnetic Survey: Measurements of the earth's magnetic field that are ten mapped and used to determine subsurface geology.

Magneto-telluric: An electromagnetic method of determining structures below the earth's surface using electrical currents and the magnetic field.

Mantle: The Earth's inner layer of molten rock lies beneath its crust and above its liquid iron and nickel core.

Matrix Treatments: Treatment performance is below the reservoir fracture pressure. Matrix treatments are generally designed to restore the reservoir's natural permeability following damage to the near-wellbore area. They typically use hydrochloric or hydrofluoric acids to remove mineral material that reduces flow into the well.

Medium-Temperature Collectors: Solar thermal collectors are designed to operate in the temperature range of 140°F–180°F but can also operate at a temperature as low as 110°F. The collector typically consists of a metal frame, a metal absorption panel with integral flow channels (attached tubing for liquid collectors or integral ducting for air collectors), and glazing and insulation on the sides and back.

Megawatt (WM): One million watts of electricity.

Methane: Methane is a colorless, flammable, odorless hydrocarbon gas (CH_4) that is a major component of natural gas. It is also an essential source of hydrogen in various industrial processes. Methane is a greenhouse gas.

Methanol: Methanol, also known as methyl alcohol or wood alcohol, has the chemical formula CH_3OH. It is usually produced by chemical conversion at high temperatures and pressures. Although it is generally made from natural gas, methanol can also be produced from gasified biomass.

Micro-seismicity: Small movements of the earth cause fracturing and movement of rocks. Such seismic activity does not release sufficient energy for the events to be recognized except with sensitive instrumentation.

Mini-Frac: A minor fracturing was performed before the primary hydraulic fracturing treatment to acquire stress data and test pre-simulation permeability.

MTBE: Methyl Tertiary Butyl Ether is a fuel oxygenate produced by reacting methanol with isobutylene.

MSW (Municipal Solid Waste): Residential solid waste and some non-hazardous commercial, institutional, and industrial wastes.

Multiplier Effect: The multiplier effect is sometimes called the ripple effect because a single expenditure in an economy can have repercussions throughout the entire economy. The multiplier measures how much additional economic activity is generated from an initial expenditure.

N

Nacelle (or Cowling): It contains and protects the gearbox and generator, sometimes large enough for an engineer or technician to stand in while performing maintenance.

Net Metering: An arrangement permits a facility (using a meter that reads inflows and outflows of electricity) to sell any excess power it generates over its load requirement back to the electrical grid to offset consumption.

Net Photovoltaic Cell Shipment: The difference between photovoltaic cell shipments and photovoltaic cell purchases.

Net Photovoltaic Module Shipment: The difference between photovoltaic cell shipments and photovoltaic cell purchases.

Net Summer Capacity: The maximum output, commonly expressed in megawatts (MW), that generating equipment can supply to system load, as demonstrated by a multi-hour test, during summer peak demand. This output reflects a reduction in capacity due to electricity use for station services or auxiliaries.

Nocturnal Cooling: The effect of cooling is the radiation of heat from a building to the night sky.

Nonforested Land: Land that has never supported forests and lands formerly forested where the use of time management is precluded by development for other uses. (Note: Includes areas used for crops, improved pasture, residential areas, city parks, improved roads of any width and adjoining clearings, powerline clearings of any width, and 1–4.5-acre areas of water classified by the Bureau of the Census as land. If intermingled in forest areas, unimproved roads and nonforested strips must be more than 120 ft wide, and clearings, etc., must be more than 1 acre in area to qualify as nonforested land.)

Nonindustrial Private: An ownership class of private lands where the owner does not operate wood-using processing plants.

Nonutility Generation: Electric generation by Nonutility power is produced to supply electric power for industrial, commercial, and military operations or sales to electric utilities.

Nonutility Power Producer: A corporation, person, agency, authority, or other legal entity or instrumentality that owns electric generating capacity and is not an electrical utility. Nonutility power producers include qualifying cogenerators, qualifying small power producers, and other nonutility generators without a designated, franchised service area that do not file forms listed in the Code of Federal Regulations, Title 18, Part 141.

O

Occupied Space: The space within a building or structure that people usually occupy may be conditioned (heated, cooled, and/or ventilated).

Other Biomass: This category of biomass energy includes agricultural byproducts/crops (agricultural byproducts, straw), other biomass gas (digester waste alcohol), other biomass liquids (fish oil liquid acetonitrile, waste, tall oil, waste alcohol), and other biomass solids (medical waste, solid byproducts, sludge waste, and tires).

Other Forest Land: Forest land other than timberland and reserved forest land. It includes available forest land, which cannot produce 20 cubic feet per acre of industrial wood annually under natural conditions because of adverse site conditions such as sterile soils, dry climate, poor drainage, high elevation, steepness, or rockiness.

Other Sources: Sources of Roundwood products that are not growing stock. These include available dead, rough, and rotten trees, trees of non-commercial species, trees less than 5.0 inches dbh, tops, and round wood harvested from nonforest land (for example, fence rows).

P

Paper Pellets: Paper is compressed and bound into uniform-diameter pellets to be burned in a heating stove.

Parabolic Dish: A high-temperature (above 180°F) solar thermal concentrator, generally bow-shaped, with two-axis tracking.

Passive Solar: A system in which solar energy alone is used for the transfer of thermal energy. Pumps, blowers, or thither heat transfer devices that use energy other than solar are not used.

Peak Watt: A manufacturer's unit indicates the amount of power a photovoltaic cell or module will produce at standard test conditions (typically 1,000 W/m^2 and 25°C).

Peaking Plants: Electricity-generating plants that are operated to meet the peak or maximum load on the system. The energy cost from such plants is usually higher than that of baseload plants.

Peat: Peat consists of partially decomposed plant debris. It is considered an early stage in the development of coal. Peat is distinguished from lignite by free cellulose and a high moisture content (+70%). The heat content of air-dried peat (about 50% moisture) is about 9 million Btu per ton. Most U.S. peat is used as a soil conditioner.

Permeability: The ability of a rock to transmit fluid through its pores or fractures when subjected to a difference in pressure is typically measured in darcies or millidarcies.

Photovoltaic (PV) Cell: An electronic device consisting of layers of semiconductor materials fabricated to form a junction (adjacent layers of materials with different electronic characteristics) and electrical contacts capable of converting incident light directly into electricity (direct current).

Photovoltaic (PV) Module: An integrated assembly of interconnected photovoltaic cells is designed to deliver a selected level of working voltage and current at its output terminals. It is packaged to protect against environmental degradation and is suited for incorporation in photovoltaic power systems.

Plate Tectonics: A theory of global-scale dynamics involving the movement of many rigid plates of the Earth's crust. Tectonic activity is evident along the margins of the plates where buckling, grinding, faulting, and vulcanism occur as the plates are propelled by the forces of deep-seated mantle convection current. Geothermal resources are often associated with tectonic activity since they allow groundwater to encounter deep subsurface heat sources.

Pole Timber Trees: Live trees at least 5.0 inches in dbh. But smaller than saw timber trees.

Porosity: The ratio of the aggregate volume of pore spaces in rock or soil to its total volume is usually stated as a percentage.

Process Heating: The direct process end use is when energy is used to raise the temperature of substances involved in the manufacturing process.

Proppant: Sized particles are mixed with fracturing fluid to open fractures after hydraulic stimulation.

Public Utility Regulatory Policies Act of 1978 (PURPA): The National Energy Act, PURPA, contains measures designed to encourage energy conservation, more efficient use of resources, and equitable rates. Among these were suggested retail rate reforms, new incentives for cogenerators to produce electricity, and the use of renewable resources.

Pulpwood: Roundwood, whole tree chips, or wood residues produce wood pulp.

Pumped-Storage Hydroelectric Plant: A plant that usually generates electric energy during peak load periods by using water previously pumped into an elevated storage reservoir during off-peak periods when excess generating capacity is available to do so. When additional generating capacity is needed, water can be released from the reservoir through a conduit to turbine generators located in a power plant at a lower level.

Pyrolysis: The thermal decomposition of biomass at high temperatures (greater than 400°F, or 200°C) in the absence of air. The product of pyrolysis is a mixture of solids (char), liquid (oxygenated oils), and gases (methane, carbon monoxide, and carbon dioxide) with proportions determined by operating temperature, pressure, oxygen content, and other conditions.

Q

Quadrillion Btu (Quad): Equivalent to 10 to the 15th power Btu.

Qualifying Facility (QF): A cogeneration or small power production facility that meets specific ownership, operating, and efficiency criteria established by the Federal Energy Regulatory Commission (FERC) under the Public Utility Regulatory Policies Act of 1978 (PURPA).

R

Rankine Cycle: The thermodynamic cycle is an ideal standard for comparing the performance of heat engines, steam power plants, steam turbines, and heat pump systems that use condensate vapor as the working fluid; efficiency is measured as work done divided by sensible heat supplied.

Recovery Factor: The fraction of total resources that can be extracted for productive uses.

Renewable Energy: Energy is produced using resources that regenerate quickly or are inexhaustible. Wind energy is considered inexhaustible because, while it may blow intermittently, it will never stop.

Renewable Energy Resources: Energy resources that are naturally replenishing but flow-limited. They are virtually inexhaustible in duration but limited in the energy available per unit of time. Renewable energy resources include biomass, hydro, geothermal, solar, wind, ocean thermal, wave action, and tidal action.

Renewable Portfolio Standard (RPS): A mandate requiring renewable energy provides a certain percentage of total energy generation or consumption.

Reservoir: A natural underground container for liquids, such as water or steam (or, in the petroleum contest, oil or gas).

Residues: Bark and woody materials are produced in primary wood-using mills, while roundwood products are converted into other goods. Examples are slabs, edgings, trimmings, sawdust, shavings, veneer cores and clippings, and pulp screenings. Includes bark and wood residues (coarse and fine materials), but excludes logging residues.

Resistivity Survey: Resistivity measures a material's ability to resist or inhibit the flow of an electrical current, measured in ohm-meters. The voltage between two electrodes measures it while an electrical current is generated between two other electrodes. Resistivity surveys can be used to delineate the boundaries of geothermal fields.

Resource Base: All of a given material in the Earth's crust, whether its existence is known or unknown, regardless of cost considerations.

Ribbon Silicon: Single-crystal silicon is derived by fabricating processes that produce sheets or ribbons. These processes include edge-defined film-fed growth, dendritic web growth, and ribbon-to-ribbon growth.

Roller Cone Bit: A drill bit that drills by crushing the rock with studded rotating cones attached to the bit.

Rotten Tree: A live tree of commercial species that does not contain a saw log now or prospectively primarily because of rot (that is when rot accounts for more than 50% of the total cull volume).

Rough Tree: (1) A live tree of commercial species that does not contain a saw log now or prospectively primarily because of roughness (that is, when sound cull, due to such factors as poor form, splits, or cracks, accounts for more than 50% of the total cull volume) or (2) a live tree of non-commercial species.

Roundwood: Wood is cut precisely for use as fuel.

Roundwood Products: Logs and other round timber are generated from harvesting trees for industrial or consumer use.

S

Salinity: A measure of the quantity or concentration of dissolved salts in water.

Saplings: Live trees 1.0 inches through 4.9 inches in dbh.

Secondary Wood Processing Mills: A mill that uses primary wood products to manufacture finished wood products such as cabinets, moldings, and furniture.

Seismic: About the nature of, or caused by, an earthquake or earth vibration, natural or man-made.

Seismicity: The phenomena of earth movements. Also, the frequency, distribution, and intensity of earthquakes.

Seismometer: An electrical device measures the magnitude and direction of seismic events on the surface and within wellbores.

Self-Potential: Self-potential in geothermal systems measures currents induced in the subsurface because of the flow of fluids.

Silicon: A semiconductor material made from silica, purified for photovoltaic applications.

Skylight: A window is located on a structure's roof and provides natural daylight, warmth, and ventilation to interior building spaces.

Slim Hole: Drill holes with a nominal inside diameter of less than 6 inches.

Slotted Liner: Liner that has slots or holes in it to let fluid pass between the wellbore and surrounding rock.

Sludge: A dense, slushy, liquid-to-semi-fluid product that accumulates because of an industrial or technological process designed to purify a substance. Industrial sludges are produced from the processing of energy-related raw materials, chemical products, water, mined ores, sewerage, and other natural and man-made products. Sludges can also form from natural processes, such as the runoff produced by rainfall, which accumulates on the bottom of bogs, streams, lakes, and tidelands.

Solar Energy: The sun's radiant energy can be converted into other forms of energy, such as heat or electricity.

Solar Heat Gain Coefficient (SHGC): Measures how well a product blocks heat caused by sunlight. The SHGC is expressed as a number between 0 and 1. The lower the SHGC, the less solar heat it transmits.

Solar Thermal Collector: A device designed to receive solar radiation and convert it into thermal energy. Typically, a solar thermal collector includes a frame, glazing, an absorber, and the appropriate insulation. The heat collected by the solar thermal collector may be used immediately or stored for later use.

Solar Thermal Collector, Special: An evacuated tube collector or a concentrating (focusing) collector. Special collectors operate at temperatures ranging from low concentration (for pool heating) to several hundred degrees Fahrenheit (high concentration for air conditioning and specialized industrial processes).

Spinning Reserve: A reserve of generation capacity, where generators are kept only, but at idle, in anticipation of an unexpected increase in demand or decrease in supply.

Start-up Speed: The speed at which a rotor begins to rotate.

Stress: The forces act on the rock. In the subsurface, the most significant force or stress is generally vertical, caused by the weight of the overlying rock.

Structural Discontinuity: A discontinuity of the rock fabric can be a fracture, fault, intrusion, or differing adjacent rock type.

Submersible Sump: The pump, including the pumping mechanism and a driving electric motor, is suspended together at depth in the well.

Subsidence: A sinking of an area of the Earth's crust due to fluid withdrawal and pressure decline.

Syngas: A synthesis gas is produced through the gasification of biomass. Syngas are similar to natural gas and can be cleaned and conditioned to form a feedstock for methanol production.

T

TDS: Total dissolved solids. Used to describe the amount of solid materials in water.

Tall Oil: The oily mixture of rosin acids, fatty acids, and other materials is obtained by acid treatment of the alkaline liquors from pine wood's digesting (pulping).

Thermal Drawdown: Decline in formation temperature due to geothermal production.

Thermal Gradient: The temperature increase rate is a function of depth into the earth's crust.

Thermosiphon System: A solar collector system for water heating circulates the collection fluid through a storage loop solely by the temperature and density difference between the hot and cold fluids.

Thin-Film Silicon: A technology in which amorphous or polycrystalline material is used to make photovoltaic (PV) cells.

Tiltmeter: The device can measure minimal changes in its rotation horizontally. The "tilt" measured by an array of tiltmeters placed over a stimulation allows the delineation of inflation and fracturing caused by the stimulation.

Tipping Fee: A fee for the disposal of waste.

Tower: Steel structures support the turbine assembly. Higher towers allow for longer glades and the capture of faster-moving air at higher altitudes.

Tracer: A chemical is injected into the flow stream of a production or injection well to determine fluid path and velocity.

Transformer: Used to step up or step down A-C voltage or A-C current.

Transmission Line: Structures and conductors that carry bulk supplies of electrical energy from power-generating units.

Transmission System (Electric): An interconnected group of electric transmission lines and associated equipment moves or transfers electric energy in bulk between points of supply and points at which it is transformed for delivery over the distribution system lines to consumers or other electrical systems.

Transportation Sector: An energy-consuming sector consists of all vehicles whose primary purpose is transporting people and/or goods from one physical location to another. Included are automobiles, trucks, buses, motorcycles, trains, subways, other rail vehicles, aircraft, ships, barges, and other waterborne vehicles. Vehicles whose primary purpose is not transportation (e.g., construction cranes and bulldozers, farming vehicles, warehouse tractors, and forklifts) are classified in the sector of their primary use.

Turbine: A machine for generating rotary mechanical power from the energy of a stream of fluid (such as water, steam, or hot gas). Turbines convert the kinetic energy of fluids to mechanical energy through the principles of impulse and reaction or a mixture of the two.

U

U-Factor: Measures the rate of heat loss or how well a product prevents heat from escaping. It includes the thermal properties of the frame as well as the glazing. The R-value indicates the insulating value, the inverse of the U-factor. U-factor ratings generally fall between 0.20 and 1.20. The lower the U-factor, the greater a product's resistance to heat flow and the better its insulation value.

Upwind Turbine: A turbine facing the wind requires a wind vane and yaw drive to maintain proper orientation relative to the wind.

Useful Heat: Heat is stored above room temperature (in a solar heating system).

Useful Thermal Output: Thermal energy is available in any industrial or commercial process or heating or cooling application; that is, total thermal energy made available for processes and applications other than electrical generation.

V

Vapor Dominated: A geothermal reservoir system in which subsurface pressures are controlled by vapor rather than by liquid is sometimes referred to as a dry-steam reservoir.

Viewshed: The scenic characteristics of an area, when referred to as a resource.

Visible Light Transmittance: The amount of visible light that passes through the glazing material of a window is expressed as a percentage.

Voltage: The measure of electrical potential difference.

W

Watt (Electric): The electrical unit of power. The rate of energy transfer is equivalent to 1 ampere of electric current flowing under a pressure of 1 V at unity power factor.

Watt (Thermal): A unit of power in the metric system, expressed in terms of energy per second, equals the work done at a rate of 1 J/s.

Watthour (Wh): The electrical energy unit of measure equals 1 watt of power supplied to, or taken from, an electric circuit steadily for 1 hour.

Well Log: Logging includes measurements of the well's diameter and various electrical, mass, and nuclear properties of the rock, which can be correlated with the rock's physical properties. The well log is a chart of the measurements relative to the well's depth.

Wind Energy: Wind motion produces energy that can be converted to mechanical energy for driving pumps, mills, and electric power generators. Wind pushes against sails, vanes, or blades radiating from a central rotating shaft.

Wind Power Plant: A group of wind turbines interconnected to a standard utility system through transformers, distribution lines, and (usually) one substation. Operation, control, and maintenance functions are often centralized through a network of computer monitoring systems, supplemented by visual inspection. This is a term commonly used in the United States. In Europe, it is called a generating station.

Wind Rose: A diagram that indicates the average monthly or annual percentage of time that the wind blows from different directions.

Wind Turbine or Windmill: A device for harnessing the wind's kinetic energy and using it to do work or generate electricity.

Wind Vane: A wind direction measurement device sends data to the yaw drive.

Windward: Into or facing the direction of the wind. Opposite Leeward.

Wood/Wood Waste: The category of biomass energy includes black liquor, wood/wood waste liquids (red liquor, sludge wood, and spent sulfite liquor), and wood/wood waste solids (peat, paper pellets, railroad ties, utility poles, and wood/wood waste).

Wood Energy: Wood and wood products used as fuel, including round wood (cord wood), limb wood, wood chips, bark, sawdust, forest residues, charcoal, pulp waste, and spent pulping liquor

Wood Pellets: Sawdust is compressed into uniform-diameter pellets to be burned in a heating stove.

Y

Yaw: The rotation of a horizontal-axis wind turbine around its tower occurs along the vertical axis.
Yaw Drive: The motor that keeps an Upwind Turbine facing into the wind.

Z

Zonal Isolation: Various methods selectively partition portions of the wellbore for stimulation, testing, flow restriction, or other purposes.
Zone: A zone is an area within the interior space of a building, such as an individual room(s), that can be cooled, heated, or ventilated. It has a thermostat to regulate airflow.

NOTE

1 Definitions retrieved 01/05/2024 @ http://www1.eere.enegy.gov/site_adminstaration/glossary.html from DOE's EERE Glossary of Energy-Related Terms.

Index

Note: **Bold** page numbers refer to tables and *italic* page numbers refer to figures.

Kolibri 360 GmbH
Ferdinand-Jühlke-Straße 7
99095 Erfurt, Deutschland
produktsicherheit@kolibri360.de